GIS in Sustainable Urban Planning and Management

A Global Perspective

T0256327

GIS in Sustainable Urban Planning and Management

A Global Perspective

Edited by
Martin van Maarseveen, Javier Martinez, and
Johannes Flacke

CRC Press
Taylor & Francis Group
Boca Raton London New York

CRC Press is an imprint of the
Taylor & Francis Group, an **informa** business

CRC Press
Taylor & Francis Group
6000 Broken Sound Parkway NW, Suite 300
Boca Raton, FL 33487-2742

First issued in paperback 2022

© 2019 by Taylor & Francis Group, LLC
CRC Press is an imprint of Taylor & Francis Group, an Informa business

No claim to original U.S. Government works

ISBN 13: 978-1-03-247580-6 (pbk)
ISBN 13: 978-1-138-50555-1 (hbk)
ISBN 13: 978-1-315-14663-8 (ebk)

DOI: 10.1201/9781315146638

This book contains information obtained from authentic and highly regarded sources. Reasonable efforts have been made to publish reliable data and information, but the author and publisher cannot assume responsibility for the validity of all materials or the consequences of their use. The authors and publishers have attempted to trace the copyright holders of all material reproduced in this publication and apologize to copyright holders if permission to publish in this form has not been obtained. If any copyright material has not been acknowledged, please write and let us know so we may rectify in any future reprint.

The Open Access version of this book, available at www.taylorfrancis.com, has been made available under a Creative Commons Attribution-Non Commercial-No Derivatives 4.0 license.

Trademark Notice: Product or corporate names may be trademarks or registered trademarks, and are used only for identification and explanation without intent to infringe.

Publisher's Note
The publisher has gone to great lengths to ensure the quality of this reprint but points out that some imperfections in the original copies may be apparent.

**Visit the Taylor & Francis Web site at
http://www.taylorandfrancis.com**

**and the CRC Press Web site at
http://www.crcpress.com**

Contents

Acknowledgments..vii
Editors...ix
List of Contributors..xi

Chapter 1 GIS in Sustainable Urban Planning and Management: A Global Perspective..............1

Martin van Maarseveen, Javier Martinez, and Johannes Flacke

PART I   The 'Sustainable City' and the 'Inclusive City'

Chapter 2 A Planning Support Tool for Scenario Analysis of Low Carbon Transport Plans in a Data-Poor Context 9

Rehana Shrestha, Mark Zuidgeest, and Johannes Flacke

Chapter 3 A Collaborative Planning Framework to Support Sustainable Development: The Case of Municipal Housing in Guatemala City 29

Jose Morales, Johannes Flacke, and Javier Martinez

Chapter 4 An Exploration of Environmental Quality in the Context of Multiple Deprivations: The Case of Kalyan–Dombivli, India.....................................45

Shubham Mishra, Monika Kuffer, Javier Martinez, and Karin Pfeffer

Chapter 5 Relationships between Outdoor Walking Levels and Neighbourhood Built-Environment Attributes: The Case of Older Adults in Birmingham, UK 63

Razieh Zandieh, Johannes Flacke, Javier Martinez, and Martin van Maarseveen

Chapter 6 Knowledge Co-Production and Social Learning on Environmental Health Issues: A Role for Interactive GIS-Based Approaches................................83

Rehana Shrestha, Johannes Flacke, Javier Martinez, and Martin van Maarseveen

Chapter 7 Role of Public Spaces in Promoting Social Interaction in Divided Cities: The Case of Nicosia, Cyprus...103

Marija Kukoleca, Ana Mafalda Madureira, and Javier Martinez

Chapter 8 Spatial Variability of Urban Quality of Life in Kirkos Sub-City (Addis Ababa)....121

Elsa Sereke Tesfazghi, Javier Martinez, and Jeroen Verplanke

Chapter 9 City Morphology and Women's Perception of Travel: A Case Study for Istanbul, Turkey...141

Jowan Khorsheed, Mark Zuidgeest, and Monika Kuffer

Chapter 10 Children's Perception of Their City Centre: A Qualitative GIS Methodological
Investigation in a Dutch City 163

Haifa AlArasi, Javier Martinez, and Sherif Amer

Chapter 11 The Street as a Binding Factor: Measuring the Quality of Streets as Public
Space within a Fragmented City: The Case of Msasani Bonde la Mpunga, Dar
es Salaam, Tanzania 183

*Bolatito Dayo-Babatunde, Javier Martinez, Monika Kuffer, and Alphonce
Gabriel Kyessi*

PART II   The 'Compact-Competitive City' and the 'Resilient City'

Chapter 12 Modelling Urban Growth in the Kathmandu Valley, Nepal 205

Sunita Duwal, Sherif Amer, and Monika Kuffer

Chapter 13 Stakeholder-Based Assessment: Multiple Criteria Analysis for
Designing Cycle Routes for Different Target Populations 225

Amy Butler, Mark Brussel, Martin van Maarseveen and Glen Koorey

Chapter 14 Post-Resettlement Socio-Economic Dynamics: The Case of Ahmedabad, India 245

Rushikesh Kotadiya, Monika Kuffer, Richard Sliuzas, and Sejal Patel

Chapter 15 Planning for Transit Oriented Development (TOD) Using a TOD Index 267

Yamini Jain Singh, Johannes Flacke, Mark Zuidgeest, and Martin van Maarseveen

Chapter 16 Performance Preferences and Policies in Urban Water Supply
in Yogyakarta, Indonesia 283

*Yohannes Kinskij Boedihardja, Mark Brussel, Frans van den Bosch, and
Anna Grigolon*

Chapter 17 Simulating Spatial Patterns of Urban Growth in Kampala, Uganda, and Its
Impact on Flood Risk 295

Eduardo Pérez-Molina, Richard Sliuzas, and Johannes Flacke

Chapter 18 Volunteered Geographic Information (VGI) for the Spatial Planning of Flood
Evacuation Shelters in Jakarta, Indonesia 307

Adya Ninggar Laras Kusumo, Diana Reckien, and Jeroen Verplanke

Chapter 19 Towards Equitable Urban Residential Resettlement in Kigali, Rwanda 325

Alice Nikuze, Richard Sliuzas, and Johannes Flacke

Index 345

Acknowledgments

We want to thank all those individuals and organizations that have supported the development of this book. In particular, we acknowledge our colleagues – as well as MSc and PhD alumni – at the Faculty of Geo-Information Science and Earth Observation (ITC) at the University of Twente, the Netherlands, for their input and support during the conceptualization, writing, review and editing phases of the book. We would especially like to thank Emile Dopheide for his contribution in a very first stage of preparation of the book, Petra Weber for facilitating the communication among authors, Andre da Silva Mano for developing a set of exercises in QGIS that complement several of the case studies in the book and Ian Cressie for the author's editing of the English text. Further thanks go to the PGM department of ITC for funding and to the Open Science Fund of the University of Twente for co-funding the open access publication of the book. We also want to acknowledge collaborating researchers, institutions and host organizations that provided support during the fieldwork, writing and review stages. Of course, we very much want to thank the authors – coming from the Global South and Global North – of each chapter for their dedication and perseverance during the development of their chapters. Through their close links with scientific and professional research projects, all are experts in GIS applications in the field of urban planning and management. Motivated, on the one hand, by the urgency of the new urban agenda set down in Goal 11 of the Sustainable Development Goals of the United Nations and, on the other, by the current potential and accessibility of Geographic Information Systems, collectively the chapters' authors set out to study, design, implement and evaluate alternative methods and solutions in different contexts, thereby meeting the responsibility of the research community for contributing to the process needed to achieve the UN goal. This collaborative effort has produced a book that consolidates practice-based knowledge in a form that will help to train – and, we hope, inspire – practitioners in, and newcomers to, the field, and in so doing will disseminate and amplify the impact of GIS applications in urban planning and management.

Editors

Martin van Maarseveen is Emeritus Professor in the Faculty of Geo-Information Science and Earth Observation (ITC) at the University of Twente, the Netherlands. He holds an MSc (cum laude) in Applied Mathematics and a PhD in Stochastic Systems and Control Theory. He was for almost ten years a senior researcher in transport planning and traffic management at the Netherlands Organization for Applied Scientific Research (TNO) in Delft. Following on from his interest in policy studies and information science, he was one of the founders of the multidisciplinary school of Civil Engineering and Management at the University of Twente and was Chair of Strategic Transport Planning and Sustainable Transport Development. In 2008 he was appointed full Professor in Management of Urban–Regional Dynamics at ITC, where he developed an interest in the application of geo-information science in the field of urban planning and management. He has extensive experience in land-use transport interactions, urban planning and management, transport modeling and assessment and data analysis. He has been involved in scientific projects and capacity-building activities in many countries across Europe, Africa, Asia and Latin America, and has authored more than 250 scientific and professional publications. Since 2003 he has been a guest lecturer at the University of Cape Town, South Africa. He retired from ITC/University of Twente in June 2018.

Javier Martinez is Assistant Professor in the Department of Urban and Regional Planning and Geo-Information Management within the Faculty of Geo-Information Science and Earth Observation (ITC), University of Twente, the Netherlands. He is also coordinator of the Urban Planning and Management specialization of the two-year master's degree in geo-information science and earth observation. He graduated as an architect from the Faculty of Architecture, Planning and Design of Rosario National University (UNR), Argentina, and earned his MSc in Geo-Information for Urban Planning from ITC. He earned his PhD from the Faculty of Geosciences, Utrecht University, for his thesis Monitoring Intra-Urban Inequalities with GIS-Based Indicators: With a Case Study in Rosario, Argentina. Between 1999 and 2001, he worked in Argentina in the Office of Strategic Planning in Rosario (PER) designing its Urban Indicators System. His research, publications and training experience are focused on the application of GIS, mixed methods and indicators for policy-making, urban poverty and quality-of-life and intra-urban inequalities. From 2010 up until November 2014, he was Co-Coordinator of the Network-Association of European Researchers on Urbanization in the South (N-AERUS). Since 2017, he has been a member of the board of directors of the International Society of Quality of Life Studies (ISQOLS).

Johannes Flacke is Assistant Professor for Spatial Planning and Decision Support Systems in the Department of Urban and Regional Planning and Geo-Information Management in the Faculty of Geo-Information Science and Earth Observation (ITC), University of Twente, the Netherlands. He earned a degree in Geography from Ruhr-University Bochum, Germany, and earned his PhD in Geosciences at the Ruhr-University Bochum in 2002 with a dissertation on information systems based on spatial indicators for promoting sustainable land use. Before joining ITC in 2007, he worked as a post-doc researcher in the Department of Urban and Regional Planning, TU Dortmund University, Germany. He has more than 20 years of theoretical and practical experience in urban and regional planning in the Global South and Global North for the development and use of geo-information and planning support systems. Fields of application include land use planning, informal settlements, environmental health-related inequalities, climate-change adaptation, hazard and risk management and poverty alleviation and targeting. His research focuses on the transformation of urban areas into sustainable and resilient places for all. Since 2010, he has been the manager of ITC's Group Decision Room, an interactive lab at the University of Twente for researching stakeholder participation in collaborative planning and decision making.

List of Contributors

Haifa AlArasi
Department of Geography and Planning
University of Toronto
Toronto, Canada

Sherif Amer
University of Twente
Enschede, the Netherlands

Yohannes Kinskij Boedihardja
Program Manager at Indika Foundation
Jakarta, Indonesia

Frans van den Bosch
University of Twente
Enschede, the Netherlands

Mark Brussel
University of Twente
Enschede, the Netherlands

Amy Butler
Storm Water Assistant, City of Missoula
Missoula, Montana, United States

Bolatito Dayo-Babatunde
Federal Capital Territory Administration
Department of Development Control
Abuja, Nigeria

Sunita Duwal
Genesis Consultancy Pvt. Ltd.
Kathmandu, Nepal

Johannes Flacke
University of Twente
Enschede, the Netherlands

Anna Grigolon
University of Twente
Enschede, the Netherlands

Jowan Khorsheed
Architecture Engineering Department, College
of Engineering, University of Duhok
Duhok, Iraq

Glen Koorey
ViaStrada Ltd.
Christchurch, New Zealand

Rushikesh Kotadiya
Indubhai Parekh School of Architecture
Rajkot, India

Monika Kuffer
University of Twente
Enschede, the Netherlands

Marija Kukoleca
Booking.com B.V.
Amsterdam, the Netherlands

Adya Ninggar Laras Kusumo
Department of City Planning, Jakarta Capital
City Government
Jakarta, Indonesia

Alphonce Kyessi
Institute of Human Settlements Studies, Ardhi
University
Dar es Salaam, Tanzania

Martin van Maarseveen
University of Twente
Enschede, the Netherlands

Ana Mafalda Madureira
University of Twente
Enschede, the Netherlands

Javier Martinez
University of Twente
Enschede, the Netherlands

Shubham Mishra
Urban Narratives
New Delhi, India

Jose Morales
SIM-CI Holding B.V.
The Hague, the Netherlands

Alice Nikuze
University of Twente
Enschede, the Netherlands

Sejal Patel
CEPT University
Ahmedabad, India

Eduardo Pérez-Molina
University of Twente
Enschede, the Netherlands

Karin Pfeffer
University of Twente
Enschede, the Netherlands

Diana Reckien
University of Twente
Enschede, the Netherlands

Rehana Shrestha
University of Twente
Enschede, the Netherlands

Yamini Jain Singh
Planit
Enschede, the Netherlands

Richard Sliuzas
University of Twente
Enschede, the Netherlands

Elsa Sereke Tesfazghi
Pro-Hulegeb Socio-Economics and
    Geographic Data Gathering and
    Analysis Centre
Addis Ababa, Ethiopia

Jeroen Verplanke
University of Twente
Enschede, the Netherlands

Razieh Zandieh
University of Manchester
Manchester, UK

Mark Zuidgeest
University of Cape Town
Cape Town, South Africa

1 GIS in Sustainable Urban Planning and Management
A Global Perspective

Martin van Maarseveen, Javier Martinez, and Johannes Flacke

CONTENTS

1.1 Urbanization, Urban Planning and GIS 1
1.2 Aim and Content of the Book 3
References 6

1.1 URBANIZATION, URBAN PLANNING AND GIS

Over the last few decades, the world has seen an increased gathering of its population in urban areas. This trend is far from new. Indeed, its persistence is marked by a remarkable increase in the absolute number of urban dwellers worldwide (UN-HABITAT, 2016). While in 1990 43% (2.3 billion) of the world's population lived in urban areas, by 2015 this had grown to 54% (4 billion). And it continues to increase: aggregated predictions of the United Nations indicate that by 2050 the share of urban population is expected to reach 66% (UNDESA, 2014). Although this figure needs to be interpreted with caution due to inconsistencies in definitions and data availability, it emphasizes the urgent need to pay serious attention to this trend and to consider the positive and negative social, economic and environmental consequences carefully. Urbanization fosters economic growth and is generally associated with greater productivity and a better quality of life for all. Nevertheless, urbanization also often brings with it urban sprawl, environmental degradation, poor living conditions and severe problems of accessibility.

The unceasing migration of rural population to urban regions follows on from the relative attractiveness of cities. Cities provide a wide variety of opportunities and offer a large number of facilities and services – opportunities for generating income, finding a job, pursuing education, accessing health care services, visiting social and cultural events and so forth. Moreover, most urban infrastructures are better developed than their rural counterparts, although it must be noted that some urban dwellers cannot access or benefit from these infrastructures.

Although virtually no region of the world can report a decrease in urbanization, the increase in urban population has not been evenly spread. Nowadays, the highest growth rates can be found in Asia and Africa, in particular in less-developed regions and in middle- and low-income countries. This presents a major challenge to building and sustaining adequate infrastructure and public services for these growing populations. Along with increases in urban population, the land area occupied by cities is increasing at an even higher rate. Projections for the period 2000–2030 indicate that the urban population in developing countries will double, while the area occupied by their cities will triple (Angel et al., 2011).

Urban planning and management is the technical, social and political process concerned with the design, development and maintenance of land use in an urban environment, which includes attention to air and water quality and the infrastructure into and out of urban areas (e.g. transportation, communication and distribution networks). Urban planning encompasses activities such as strategic thinking, research and analysis, public consultation, urban design and policy implementation, whereas urban management mostly consists of tactical and operational issues related to the performance and maintenance of an urban system. The reality of urbanization generates some challenges, among them low-density suburbanization – largely steered by private rather than public interest, and partly facilitated by dependence on car ownership. Urbanization is also energy intensive, contributing dangerously to climate change. Moreover, it generates multiple forms of inequalities, exclusion and deprivation. Often, such urban challenges are exacerbated by urbanization that has been poorly planned and managed. The key objective of adequate urban planning and management is, therefore, to overcome and eradicate the dysfunctions and discontinuities of development. This fits seamlessly with the new Urban Agenda of the United Nations, in particular Goal 11 of the 2030 Agenda for Sustainable Development, which aims to make cities and human settlements inclusive, safe, resilient and sustainable (United Nations, 2017).

Since 1990, a number of different concepts have made their entry into the domain of urban planning and management: sustainability, inclusiveness, compactness, competitiveness and resilience. There is no complete agreement on the definition for what a *sustainable city* should be, nor is there complete agreement upon which components should be included in the paradigm. Generally, developmental experts agree that a sustainable city should meet the needs of the present without sacrificing the ability of future generations to meet their needs. In practice, urban planning in a sustainable city aims to minimise the required inputs of energy, water and food, and the necessary output of waste as well as air and water pollution (Vojnovic, 2014). As such, urban sustainability is a kind of container concept that overlaps with most of the other, later developed concepts. The emergence of slums and social ghettos (gated communities, for example), as well as the very distinct social (and spatial) divisions found within cities, has contributed to the development of the concept of the inclusive city. An *inclusive city* aims to include all citizens in its economic development, political life and access to political power in order to combat exclusion and provide opportunities for the economic and social improvement of the lives of its citizens. The inclusive city is strongly linked with the concepts of justice, equity, democracy and diversity (Fainstein, 2014). The *compact city* is an urban planning concept that promotes high residential density along with mixed land use. This concept is based on an efficient public transport system and an urban layout that encourages walking and cycling (Burton et al., 2016). There is some empirical evidence that compact cities tend to have lower car use. A *competitive city* is one that successfully facilitates its firms and industries to create jobs, raise productivity and, over time, provide income growth for citizens. From an urban planning perspective, it is important to be familiar with factors that help to attract, retain and expand the private sector (World Bank, 2015). A *resilient city* is one that assesses, plans and acts to prepare for and respond to hazards – natural and anthropogenic, sudden and slow in onset, expected and unexpected. Resilient cities are better positioned to protect and enhance people's lives, to secure development gains and to foster a reliable investment climate (UN-HABITAT, 2018). In one way or another, all concepts of the city as described above appear in the elaboration of Sustainable Development Goal 11.

Information, particularly spatial information, plays a fundamental role in the practice of urban planning and management. Most information used by planners is geographic in the sense that it makes use of topographic maps or is linked to a geographic location through a coordinate reference, a street address or an administrative area. A Geographic Information System (GIS) is an information system that integrates, stores, edits, analyses, shares and displays spatial or geographic information. Over the past few decades, these systems have become an increasingly familiar aspect of urban design and planning practice (Maliene et al., 2011). Modern GIS software, hardware and databases are becoming more available to ordinary users, although the complexity of the more powerful applications means accessibility is still a problem. As the range of applications is continually expanding, interest is also constantly growing among planners. The increasing use of GISs in urban planning and design holds the promise of enabling a higher quality of quantitative and qualitative data analysis, thereby improving the evidence base of decision-making as well as the knowledge base of the decision-making process itself – all factors upon which delivering such an aspirational, but highly complex, policy goal as Sustainable Development Goal 11 ultimately depends. Current progress in GIS technology has created favourable conditions for the development of solution-supporting systems at all stages in the planning and design process – although there is still much work to be done if its promised utility is to be realised (Maliene et al., 2011). With respect to planning support systems, this has been dubbed the implementation gap (Geertman, 2017).

1.2 AIM AND CONTENT OF THE BOOK

GIS in Sustainable Urban Planning and Management: A Global Perspective explores and illustrates the capacity that geo-information and GISs have to inform practitioners and other participants in the processes of the planning and management of urban regions. The chapters have been grouped according to the nature and applications of the cases, which correspond with current enquiries about and/or normative visions of the 'Sustainable City' and the 'Inclusive City' (Part I); and the 'Compact–Competitive City' and the 'Resilient City' (Part II).

Part I starts with the sustainability concept. Although this concept has been around for some time, and there is no common understanding of its components, it is still a useful concept for practitioners and students as it raises awareness of the multi-faceted nature of development and promotes understanding of the variety of connotations the word development encompasses. The idea of sustainable development is still very much central in the new development agenda of the United Nations, and in that sense, it is of particular importance for students from both the Global South and Global North. The cases/geo-applications presented in Part I illustrate ways of measuring sustainability and its value in the arena of urban policy. Another key scientific and policy issue is the creation of inclusive cities. The cases presented in this book analyse, evaluate and (aim to) improve the position of disadvantaged groups in society in terms of their overall well-being and prospects and their access to basic services.

Part II acknowledges that urban planning relates to how we shape our cities. Where do we position the various functions of a city to achieve our strategic objectives? The cases presented in this part examine a variety of spatial configurations in cities, the spread of activities and the demands placed on their transport systems. This understanding, in a context of dynamic urban growth, relates to the compact city view. The cases/geo-applications deal with relevant theories and practical applications of land use and transport models that help to analyse and quantify the current situation and future development scenarios. Another relevant discussion in a globalizing world economy is that cities are increasingly challenged to be competitive and offer an appealing living environment, affordable housing, attractive conditions for businesses and a wide range of accessible services. The cases presented in Part II make use of a variety of economic and environmental appraisal techniques.

Part II also focuses on the recognition that urban dwellers are increasingly under threat from natural and anthropogenic hazards. Cities need to develop strategies for making choices that avoid high levels of hazard and target development in ways that limit risks and make them manageable. The cases/geo-applications we present show the relevance of understanding and analysing urban risk to give direction to spatial planning so that cities become more resilient. Resilience in this context is understood to be the ability of a system, community or society exposed to hazards to resist, absorb, accommodate and recover from the effects of a disaster in a timely and efficient manner, including their ability to restore essential basic structures and functions.

The case studies we selected for this book come from the research program 'People, Land and Urban Systems' in the Faculty of Geo-Information Science and Earth Observation (ITC) of the University of Twente, in the Netherlands. Under this program, dynamic interactions in space and time between people, land and urban systems are studied in both a quantitative and qualitative manner using geo-information technology. For specific topics in land administration and land management under this program, readers are referred to the book *Advances in Responsible Land Administration* (Zevenbergen et al., 2016).

TABLE 1.1

Overview of the Book: Chapters, Case Studies, Planning Topics and Cities

		Chapter Title	Keywords	City
2	Part I: the 'Sustainable City' & the 'Inclusive City'	A Planning Support Tool for Scenario Analysis of Low Carbon Transport Plans in a Data-Poor Context	Low-carbon development, low-carbon transport, planning support systems, scenario analysis, sustainable transport	Kathmandu Metropolitan City (KMC), Nepal
3		A Collaborative Planning Framework to Support Sustainable Development: The Case of Municipal Housing in Guatemala City	Collaborative planning framework, sustainable development, housing	Guatemala City, Guatemala
4		An Exploration of Environmental Quality in the Context of Multiple Deprivations: The Case of Kalyan-Dombivli, India	Multiple deprivation, urban environment, environmental quality, remote sensing, spatial indicators	Kalyan-Dombivli, India
5		Relationships between Outdoor Walking Levels and Neighbourhood Built-Environment Attributes: The Case of Older Adults in Birmingham, UK	Outdoor walking levels, neighbourhood built-environment, older adults	Birmingham, UK
6		Knowledge Co-Production and Social Learning on Environmental Health Issues: A Role for Interactive GIS-Based Approaches	Environmental health, interactive planning support, social learning, knowledge co-production	Dortmund and Munich, Germany
7		Role of Public Spaces in Promoting Social Interaction in Divided Cities: The Case of Nicosia, Cyprus	Divided cities, public spaces, social interaction, index	Nicosia, Cyprus
8		Spatial Variability of Urban Quality of Life in Kirkos Sub-City (Addis Ababa)	Quality of life, domains of life, spatial variability, Kirkos sub-city	Addis Ababa, Ethiopia
9		City Morphology and Women's Perception of Travel A Case Study for Istanbul, Turkey	Urban morphology, urban mobility, gender-sensitive perception, safety	Istanbul, Turkey
10		Children's Perception of Their City Centre: A Qualitative GIS Methodological Investigation in a Dutch City	Childhood geographies, qualitative GIS, child-friendly cities, physical environment, social environment	Enschede, the Netherlands
11		The Street as a Binding Factor: Measuring the Quality of Streets as Public Space within a Fragmented City The Case of *Msasani Bonde la Mpunga*, Dar es Salaam, Tanzania	Urban fragments, streets, connectivity, public space, integration	Dar es Salaam, Tanzania

(Continued)

TABLE 1.1 (CONTINUED)
Overview of the Book: Chapters, Case Studies, Planning Topics and Cities

		Chapter Title	Keywords	City
12	Part II: the 'Compact-Competitive City' & the 'Resilient City'	Modelling Urban Growth in the Kathmandu Valley, Nepal	Urban-growth models, land-cover changes, infill, expansion	Kathmandu Valley, Nepal
13		Stakeholder-Based Assessment: Multiple Criteria Analysis for Designing Cycle Routes for Different Target Populations	Multi criteria analysis, target populations' preferences, design of cycle routes	Christchurch, New Zealand
14		Post-Resettlement Socio-Economic Dynamics: The Case of Ahmedabad, India	Involuntary displacement, resettlement, socio-economic dynamics, urban poor	Ahmedabad, India
15		Planning for Transit Oriented Development (TOD) Using a TOD Index	Transit Oriented Development (TOD), sustainable development, mode of travel	City region Arnhem–Nijmegen, the Netherlands
16		Performance Preferences and Policies in Urban Water Supply in Yogyakarta, Indonesia	Choice behaviour, water supply, characteristics and performance of different water sources, consumer behaviour, water provision choice	Yogyakarta, Indonesia
17		Simulating Spatial Patterns of Urban Growth in Kampala, Uganda, and Its Impact on Flood Risk	Land-use planning, flood-risk mitigation, scenario-based planning approach, impact of building densification, pattern of urban growth, flood risk	Kampala, Uganda
18		Volunteered Geographic Information (VGI) for the Spatial Planning of Flood Evacuation Shelters in Jakarta, Indonesia	Volunteered geographic information (VGI), disaster management, spatial planning, flood evacuation shelters	Jakarta, Indonesia
19		Towards Equitable Urban Residential Resettlement in Kigali, Rwanda	Resettlement programmes, natural disasters, impoverishment risks, suitable resettlement sites, multi criteria analysis	Kigali, Rwanda

Integrated and participatory urban planning and management are multidisciplinary and require not only geographic databases but also local knowledge. Therefore, it is important to pay attention to local contexts and ensure public participation to get clear insight into the needs of local communities.

The case studies come from cities all over the world and provide a global perspective of urban development. At least one city from every region of the world has been chosen: from Ahmedabad (India/ Asia) to Guatemala City (Guatemala/the Americas); from Christchurch (New Zealand/Oceania) to Enschede (Netherlands/Europe); from Kampala (Uganda/Africa) to Kathmandu (Nepal/Asia). Table 1.1 presents a brief overview of the topics and cities described.

Each chapter in the book contains two boxes: one describes the geographic context of the case study; the other provides information about the methods applied. The overall aim of this book is to contribute to capacity development in using GIS for urban planning and management topics. With this in mind, in some of the chapters (Chapters 7, 10, 13, 15, 18 and 19) we have included exercises that enable readers to reproduce the analysis with QGIS open-source software. We have called these exercises *Methodological Demonstrations* because the focus is not to replicate results but to understand how the methods used in the chapters can be applied using open-source software tools. The datasets that accompany each of these methodological demonstrations are the same as those used by the chapter's authors in order to emphasize the link between the rationale as it is explained in the chapter and the actual GIS operations that have to be performed. In this effort, we tried to adapt chapters that focus on different types of spatial analysis: qualitative GIS in the case of Enschede; mapping citizens' perceptions in the cases of Christchurch and Nicosia; exploring crowdsourced data in the case of Jakarta; multi criteria analysis for transit accessibility in Arnhem and Nijmegen; and disaster risk assessment in Kigali. The methodological demonstrations and the datasets can be retrieved from https://www.itc.nl/pgm/urbangis.

REFERENCES

Angel, S., Parent, J., Civco, D. L., & Blei, A. M. (2011). *Making Room for a Planet of Cities*. Cambridge, MA: Lincoln Institute of Land Policy.

Burton, E., Jenks, M., & Williams, K. (Eds.). (2016). *The Compact City: A Sustainable Urban Form?* London: Routledge.

Fainstein, S. S. (2014). The just city. *International Journal of Urban Sciences, 18*(1), 1–18. doi:10.1080/1226 5934.2013.834643

Geertman, S. (2017). PSS: Beyond the implementation gap. *Transportation Research Part A: Policy and Practice, 104*, 70–76.

Maliene, V., Grigonis, V., Palevicius, V., & Griffiths, S. (2011). Geographic information system: Old principles with new capabilities. *Urban Design International, 16*(1), 1–6. doi:10.1057/udi.2010.25

UNDESA (United Nations Department of Economic and Social Affairs). (2014). *World Urbanization Prospects – 2014 revision*. New York: United Nations.

UN-HABITAT. (2016). *Urbanization and Development: Emerging Futures – World Cities Report 2016*. Nairobi: UN Habitat.

UN-HABITAT. (2018). *Urban Resilience Hub*. Retrieved from http://urbanresiliencehub.org/what-is-urban-resilience

United Nations. (2017). *Report of the Inter-Agency and Expert Group on Sustainable Development Goal Indicators (E/CN.3/2017/2)*. New York: United Nations, *Annex III*.

Vojnovic, I. (2014). Urban sustainability: Research, politics, policy and practice. *Cities, 41*, S30–SS44. doi: https://doi.org/10.1016/j.cities.2014.06.002

World Bank. (2015). *Competitive Cities for Jobs and Growth*. Washington: The World Bank Group.

Zevenbergen, J., de Vries, W., & Bennett, R. (Eds.). (2016). *Advances in Responsible Land Administration*. CRC Press, Taylor & Francis Group.

Part I

The 'Sustainable City' and the 'Inclusive City'

2 A Planning Support Tool for Scenario Analysis of Low Carbon Transport Plans in a Data-Poor Context

Rehana Shrestha, Mark Zuidgeest, and Johannes Flacke

CONTENTS

2.1 Introduction 9
2.2 Methodology 11
    2.2.1 Study Area 11
    2.2.2 Data Collection and Preparation..12
    2.2.3 Data Preparation for Accessibility Model 13
2.3 A Planning Support Tool for KMC to Assess Various Scenarios of LCT Plans 13
    2.3.1 Accessibility Analysis Model Using a Location-Based Contour Measure.................15
    2.3.2 Carbon Emissions Module Using ASIF Framework 16
    2.3.3 Scenario Definition with Interpretation of LCT Plans for KMC 16
        2.3.3.1 Baseline Scenario in 2020 18
        2.3.3.2 Increase in Patronage of Public Transport..19
        2.3.3.3 Renewal of Trolley Bus System 19
        2.3.3.4 Penetration of Trolley Bus System Together with PT Improvement Plans 20
2.4 Results...20
    2.4.1 Accessibility Under Baseline Scenario in 2020 and Alternative Scenarios for LCT Plans in 2020..20
    2.4.2 Emissions under the Baseline Scenario in 2020 and Alternative Scenarios for LCT Plans in 2020..22
    2.4.3 Sensitivity Analysis 23
2.5 Discussion..24
2.6 Conclusion 25
References..26

2.1 INTRODUCTION

Climate change is a global issue that needs to be addressed by both developed and developing countries. Nevertheless, it is being ignored in many developing countries because of their urgent need to give preference to development activities (Halsnæs and Verhagen, 2007). In addressing this dilemma, the notion of low-carbon development (LCD) has emerged from the discourse on climate change (Tilburg et al., 2011). Definitions of LCD commonly mention reducing CO_2 emissions, intensive use of low-carbon energy sources and ensuring economic growth by reconciling national mitigation priorities with national development needs (Tilburg et al., 2011; Yuan et al., 2011). One key mitigation strategy is to lower their carbon footprint in cities and make them more sustainable.

In low-carbon cities, transport systems are expected to play a key role in tackling issues of climate change because transport is one of the few sectors in which carbon emissions continue to rise. In most developing countries, rapid economic growth, urbanization and motorization account for much of the increase in transport-related emissions. As a result, cities have been called upon to make a transition towards low-carbon transport (LCT) futures. Broadly speaking, LCT is understood to play a role in lowering carbon dioxide emission from transportation while contributing to progress in the context of the broader economic, social and local environmental objectives of sustainable development (Sakamoto et al., 2010).

Transition to low-carbon transport futures demands an integrated approach to transport planning and policy development and implementation. This entails reducing the need for travel, encouraging a shift to cleaner modes of transport, reducing trip lengths and achieving greater fuel efficiencies in transport systems – as captured by the Avoid–Shift–Improve (ASI) strategy (Dalkmann and Brannigan, 2007). In addition, researchers argue for the need to advance transport policy beyond the traditional mobility-centred approach to an accessibility-centred one in order to improve the overall quality of people's lives (Akinyemi and Zuidgeest, 2000, 2002). To this end, improving people's spatial access to opportunities such as jobs and services becomes one of the key performance indicators for monitoring the contribution of the transport sector to low-carbon development.

When dealing with policies and plans related to sustainable transport, it is important to look towards the future, as many interventions require long lead times before becoming effective. As such, it often takes time before impacts become apparent and produce unexpected results (Hickman and Banister, 2007). Owing to the complexity of understanding the future impacts of transport policies and plans, scholars have advocated scenario analysis as an approach for 'thinking the unthinkable', as well as complementing the current range of transport models (Hickman and Banister, 2007; Hickman et al., 2012; Shiftan et al., 2003). The goal of scenario analysis is to provide a coherent, internally consistent and plausible description of a possible future state, to support more effective strategic decision-making and to discover the potentials for achieving breaks against dominant trends. There is also an emerging preference for approaches that make use of computer-supported, GIS-based planning support tools in appraising transport plans and polices (Arampatzis et al., 2004). The strengths of such tools lie in their ability to depict spatial relations, manage spatial information and produce visualizations, which are essential in transport planning, especially when these policies and plans are to be evaluated for their effectiveness in providing spatial accessibility.

Against this backdrop, we contend that a hybrid tool that integrates scenario analysis with a GIS-based planning support tool can be valuable in assessing future impacts and the potential of LCT plans for providing accessibility while lowering carbon dioxide emissions at the urban scale. In this regard, our objective was: (1) to develop a planning support tool that can facilitate spatial accessibility and carbon emissions analyses using a GIS-based platform; and (2) to demonstrate the applicability of the tool in assessing scenarios for LCT plans at city level with respect to two complementary indicators – improving accessibility and lowering CO_2 emissions.

The research for the support tool was undertaken using the case of Kathmandu Metropolitan City (KMC), the capital city of Nepal, which is a low-income developing country. The city is located in the Kathmandu Valley, the largest urban area of Nepal. Rapidly increasing motorization combined with limited transport infrastructure causes traffic congestion and a rising number of road accidents, particularly in the country's urban areas. Furthermore, studies have reported increasing levels of air pollution in the valley, mainly due to the vehicular emissions (Ghimire and Shrestha, 2014; Shrestha and Rajbhandari, 2010). Although CO_2 emissions from the country's transport sector are less significant in a global context, the transport sector is the source of more than half the country's total energy-related CO_2 emissions, and this proportion is reported to continue to increase (Malla, 2014). Moreover, growing dependence on imported fossil fuels has increased the economic vulnerability of the country owing to rising fossil fuel prices. In response, the government of Nepal (GoN) emphasizes long-term national strategies to reduce fossil fuel consumption by diversifying the energy mix through the use of indigenous energy resources such as hydroelectricity (National Environmental and Scientific Services, 2003). With respect to the transport sector, the GoN, through the issue of its National Transport Policy 2001/02, is emphasizing the promotion of an electricity-based transport system throughout the country (Shakya and Shrestha, 2011). Similarly, efforts are being made to promote a sustainable transport system that is accessible, safe and affordable in the Kathmandu Valley (MoPPW and ADB, 2010). With these developments in mind, KMC provides a good case for demonstrating the application of our GIS-based planning support tool for the assessment of various scenarios on LCT plans for the city.

2.2 METHODOLOGY

2.2.1 STUDY AREA

Kathmandu Metropolitan City (KMC) is the largest urban agglomeration in the Kathmandu Valley. KMC is highly urbanised, with around 1 million inhabitants spread over approximately 49.45 km²; the city's population density is nearly 20,289 per km² (CBS, 2018). Urban sprawl is one of the major characteristics of the city. New urban areas are encroaching on outlying, low density areas (Thapa and Murayama, 2012).

Along with increasing urbanization, the city is facing substantial growth in vehicle numbers, especially motorised two-wheelers (2W) (DOTM, 2014). Increasing travel demand, limited road capacity and a shift towards private travel all contribute to problems of traffic congestion in the city and increasing travel times per trip (Malla, 2014). To make matters worse, motorization of the city has always gone hand in hand with the use of poorly maintained vehicles being run on adulterated, low-quality fuels (Ale and Shrestha, 2009).

Currently, public transport (PT) suffers from the stigma of being unreliable, inefficient and uncomfortable. There are several reasons for the poor reputation of the city's PT system (Udas, 2012). Firstly, the PT system is currently owned and operated by private PT entrepreneurs whose focus is maximization of revenue rather than providing efficient services. As a result, passengers increasingly hop on and off vehicles at locations other than designated stops, vehicles are overloaded with large numbers of passengers and traffic regulations are ignored (Udas, 2012). Secondly, both slow-moving, lower occupancy vehicles known locally as *tempos* – typical three-wheeled vehicles in KMC – that are of two types, one running on gas and another on electricity and high-speed, higher occupancy vehicles (large buses, minibuses) use the same routes. This has been identified as one of the reasons for traffic congestion and lower vehicle speeds (MoPPW and ADB, 2010). Thirdly, the routes used by public transport (services originate and end in the city's inner core), poor traffic management, poor driving behaviour and the absence of strict enforcement of vehicle maintenance standards contribute further to the burdens the PT system creates for the city (MoPPW and ADB, 2010). As a consequence, the PT system has remained a mode of transport for captive users, and those who are capable of doing so are shifting to private modes, especially two-wheelers (2W).

BOX 2.1   Case Study Area

Kathmandu Valley is the largest urban area in Nepal and covers an area of approximately 684 km^2. The valley contains Kathmandu Metropolitan City, as well as the towns of Lalitpur, Bhaktapur, Kirtipur and Madhyapur Thimi and other smaller areas.

The case study city, Kathmandu Metropolitan City (KMC) is the largest urban agglomerate in the valley, as well as being the capital city of Nepal and its major administrative, commercial and tourist centre. The city proper has an area of about 49.45 km^2 and a population of around 1 million inhabitants. KMC is one of the densely populated cities in the valley, with nearly 20,289 inhabitants per km^2. Increasing urbanization is a major characteristic of the city, with new areas spilling over onto low-density areas on the city's outskirts. KMC was selected for our study as it is an example of a major city in a developing country that faces substantial growth in the number of vehicles; a high proportion of these are poorly maintained and are run on adulterated, low-quality fuels. Increasing travel demand, limited road capacity and a shift towards private transport, especially motorised two-wheelers, all contribute to problems of traffic congestion. As a result, traveling times per trip in the city are increasing, thereby reducing people's accessibility and adding to the country's total energy-related CO_2 emissions.

2.2.2   DATA COLLECTION AND PREPARATION

Interviews provided the primary data for conceptualizing LCT plans for KMC and for constructing scenarios related to those LCT plans. We interviewed several key informants from government, the private sector and academia. Coincidentally, the Kathmandu Sustainable Urban Transport Project (KSUT) was also in progress at the time of our research as part of the Sustainable Transport Initiative (MoPPW and ADB, 2010). The analyses of our interviews were supplemented with a review of documents from KSUT.

Secondary data were collected to conduct spatial accessibility and carbon emissions analyses of various scenarios chosen for LCT plans. These data were collected from various government departments and consultancies in the form of spatial and non-spatial data. The spatial data collected were the boundaries of the city's administrative wards, its road network, public vehicle routes and stops and points of public interest (hospitals, schools, colleges, temples, etc.). The non-spatial data consisted of population data from the year 2011 at ward level. These data were used to develop an accessibility model. Other data for the accessibility model (e.g. speed according to type of transport mode) and for a carbon emissions model (e.g. vehicle occupancy, modal split, fuel efficiency and carbon content) were collected from KSUT documents and the literature.

For this study, following Amartya Sen's capability approach (Hoffman, 2008; Mazumdar, 2003; Saito, 2003), access to schools and colleges were considered as a proxy for broadening the capabilities of younger generations through access to education. Additionally, the majority of daily trips in the city are related to either work (38.6%) or school/college (34.3%) (Udas, 2012). Thus, only the spatial locations of high schools and colleges were extracted from the point of public interest data.

This gave 151 locations. Likewise, the population aged 15–29 years were considered to be high school and college students and, therefore, users of various modes of transportation. The proportion of the population in this category was calculated for the whole city and then used to compute the number of 15–29-year-olds in each ward.

2.2.3 DATA PREPARATION FOR ACCESSIBILITY MODEL

Our accessibility model uses network analysis to generate serviced areas within specific travel times. As such, the analysis requires comprehensive data on the road network, public vehicle routes, location of stops for different types of public transport, vehicle speeds, waiting times and walking time to stops. The processing of the road network data and public vehicle routes was done in ArcGIS to render usable topology for representing the reality of the city's road network and transport system. The stops for different types of public transport (PT) were, if necessary, digitised manually. The modes of public transport considered were minibuses, microbuses, e-tempos and gas tempos, whereas the two-wheeler (2W) was taken to be representative of the private mode of transport. The network data in the model for PT use was prepared to represent an intermodal network, which is an integrated transportation system comprising two or more modes (Boile, 2000); see Figure 2.1. The total travel time by public transport in such intermodal network is represented by the equation (2.1):

$$\text{Total travel time by public transport} = \text{walking time to stop} + \text{waiting time for vehicle}$$
$$+ \text{travel time in vehicle} + \text{walking time to destination} \tag{2.1}$$

2.3 A PLANNING SUPPORT TOOL FOR KMC TO ASSESS VARIOUS SCENARIOS OF LCT PLANS

We have used a pragmatic approach to conceptualise the planning support tool usable in a data-poor environment. The main purpose of the tool is to assess various scenarios for LCT plans with respect to two indicators: improving spatial accessibility and lowering CO_2 emissions. The tool therefore integrates two models, an accessibility analysis model and a carbon emissions model.

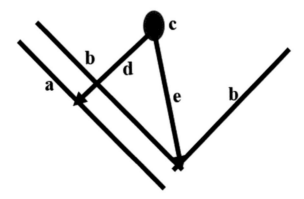

FIGURE 2.1 Intermodal network adapted in the model for representing an integrated transport system consisting of two or more modes: (a) vehicle route coded with an average speed of the mode; (b) road network coded with an average walking speed of a person; (c) a bus stop; (d) connector line in the network data set to connect the bus stop and the vehicle route; (e) connector line in the network data set to connect the bus stop and the (walking) road network.

In order to assess scenarios, relevant scenario themes, scenario uncertainties and spatial and temporal scales with a time horizon first need to be elaborated. The qualitative story lines of these scenarios need to be quantified in order to build a connection between the scenario description and the two models. Therefore, key scenario parameters, decision factors and their effects on various components of the accessibility analysis and carbon emissions models were identified.

The accessibility analysis model allows users to assess the spatial accessibility impact of various scenarios for LCT plans. Given the data-poor context, we considered a location-based measure to be relevant for this study because of its easy conceptualization, operationalization and interpretability with undemanding data (Cerda and El-Geneidy, 2009; Geurs and Van Wee, 2004). The location-based measure in the accessibility model computes the number of persons that can be reached within a given time, which is considered as the number of potential populations provided with opportunities. Potential population is defined as those who are provided services by the transport system within pre-set travel time isochrones and are therefore an indicator of accessibility. The higher the number of people serviced by transport plans, the higher the level of accessibility.

The carbon emissions model calculates CO_2 emissions generated under each scenario of LCT plans. The Activity–Structure–Intensity–Fuel (ASIF) framework (Schipper et al., 2009) is considered relevant for this study as it uses a bottom-up approach to estimate road travel demand and its associated energy use and emissions. Within this framework, travel demand is a function of mode of travel, technology choice, total distance to be travelled, driving style and vehicle occupancy. Similarly, transport energy use is a function of travel demand and fuel efficiency. Thus, emissions from the transport sector are calculated as the product of level of activity (passenger travel), structure (share by mode and vehicle type), fuel intensity (fuel efficiency) and emission factor:

$$G = A \times S \times I \times F \qquad (2.2)$$

Where

G = carbon emissions from transport
A = total transport activity
S = modal structure or share by mode
I = modal energy intensity
F = carbon content of fuels

Figure 2.2 depicts the operational framework of the planning support tool for KMC. The operational framework shows two models, the accessibility analysis and the carbon emissions models and their components, along with key scenario drivers/assumptions that influence various components in the models.

BOX 2.2   Methods Applied in the Chapter

This chapter presents an accessibility analysis combined with an evaluation of LCT scenarios that are aimed at improving spatial accessibility and lowering CO_2 emissions. Each mode of transport is represented by a distinct set of network data that takes into account speed of mode to generate the isochrones of service areas that can be accessed within set travel times. The accessibility analysis is done using network analysis of ArcGIS™ (Version 10), while the various scenarios for LCT plans were evaluated using CommunityViz Scenario 360™ (Version 5.1.2). The dataset used for the analysis comprises spatial data (city ward boundaries, road network, public vehicles routes and stops, points of public interest) and non-spatial secondary data (2011 total population data at ward level and population by age for the whole city) collected from various government departments and consultancies. Primary data for constructing scenarios for LCT plans were collected from interviews with key informants from government, the private sector and academia, and by analysing relevant documents from KSUT. The methodology applied to test the implementation of low-carbon transport policies could also be used for comparing, for example, the costs and benefits of alternative road construction or upgrading plans. The accessibility results, however, greatly depend upon how well the network model represents reality.

2.3.1 ACCESSIBILITY ANALYSIS MODEL USING A LOCATION-BASED CONTOUR MEASURE

The accessibility analysis model was operationalised using ArcGIS Network Analyst. Each mode of transport was represented by a distinct set of network data that took into account mode speed to generate the isochrones of service areas that can be accessed from the location of high schools and colleges within set travel time. Model Builder, a programming tool within ArcToolbox, was used to automate the geoprocessing tasks of Network Analyst in a step-wise and integrated manner by

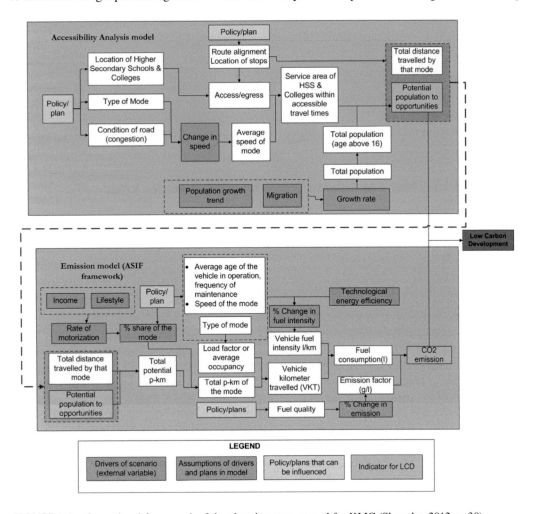

FIGURE 2.2 Operational framework of the planning support tool for KMC (Shrestha, 2012, p. 30).

using a customised tool under Model Builder. The accessibility analysis model under each scenario gives the number of people served within the specified cut-off time and the distance travelled by each mode of travel under consideration in the scenario (see Figure 2.2).

2.3.2 CARBON EMISSIONS MODULE USING ASIF FRAMEWORK

The ASIF framework was operationalised in the open modelling platform of Scenario 360 in Community Viz (Placeways, 2013). The population serviced and the distance travelled by each mode calculated from the accessibility model were input for the activity in the ASIF framework (shown in Figure 2.2). The potential population that are provided access to opportunities by a particular mode are the potential users of the mode. Multiplying the potential users of the mode by the distance travelled to locations gives total transport activity in passenger-kilometres. Depending on the mode share, the passenger-kilometres for each mode is estimated and then, based on average occupancy, assigned to vehicle-kilometres travelled (VKT). Using the fuel-intensity (L/km) of the mode, total fuel consumed is calculated, which is then converted into CO_2 emissions (tonnes) according to emission factors (gCO_2/L); see Figure 2.2.

An interface was created for each type of mode under consideration in each scenario, as shown in Figure 2.3. Through the interface the values of variables in the ASIF framework can be changed in the model: e.g. mode share, average occupancy, fuel intensity, emission factor and speed of the mode under consideration. Additionally, drivers of the scenarios, which in this case is population growth, can also be changed as required through the interface.

Finally, the indicators summarize the potential population provided with accessibility to high schools and colleges within the specified time and the total emissions produced during provision of the service. These results are presented on indicator charts in the CommunityViz platform as shown in Figure 2.4. Thus, the changes can be made iteratively in the assumptions of the model, and the results of that change can subsequently be visualized on the fly through indicator charts. Moreover, alerts in the indicator charts allow the effectiveness of the plans to be analysed against the baseline.

2.3.3 SCENARIO DEFINITION WITH INTERPRETATION OF LCT PLANS FOR KMC

Building on the interviews and the KSUT project, we identified three strategies (avoid-shift-improve) of LCT; shift and improve have been recognised as the most feasible strategies to relieve

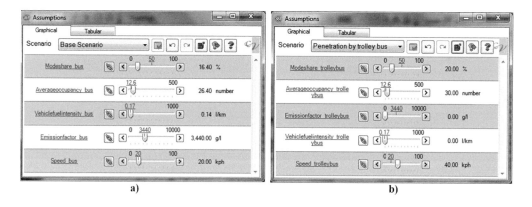

a)   b)

FIGURE 2.3   Interface for each type of mode under consideration, in each scenario. For instance: (a) bus as a mode in the baseline scenario; and (b) trolley bus as a mode in alternative scenario of the penetration of the trolley bus together with the PT improvement.

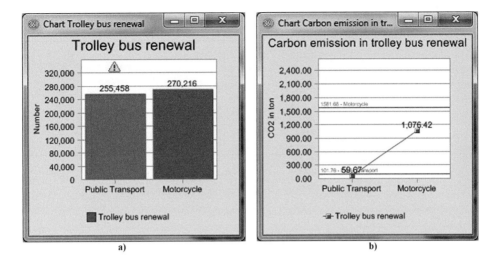

FIGURE 2.4   Indicator charts for each scenario: (a) population serviced in the scenario of trolley bus renewal; and (b) carbon emissions in the scenario of trolley bus renewal.

the city's transport problems. Along with a shift to environment-friendly modes, improvements in the technology and energy efficiency of vehicles is also needed. In line with this, two transport plans were identified as qualifying as LCT plans for KMC: improvement of the public transport (PT) system; and renewal of the trolley bus system.

A combination of policy instruments (planning, regulatory, economic, information and technology) is considered necessary for both the PT system and the trolley bus system in order to improve the patronage of transport system. Assignment of primary routes to PT or trolley buses and secondary routes to other vehicles (e.g. microbuses, minibuses, e-tempos/gas tempos), which in the current situation are being used by all vehicle types, is expected to overcome the mismatch between vehicle type and use of road space, leading to increases in speed and reducing congestion. Likewise, creating dedicated bus-only lanes could further improve the efficiency of PT and trolley bus systems. Nonetheless, such transport systems need to be supported with information and technological instruments to increase patronage among the potential users, for instance by implementing an integrated system of fare collection, proper scheduling of PT, user-friendly terminals, etc. Similarly, regulatory instruments such as restricting use of motorised vehicles in certain parts of the city and congestion charges may encourage people to shift from private vehicles to PT. Finally, financial incentives as an economic instrument, may strengthen patronage of the PT and trolley bus systems. Government tax-reduction policies on the import of public vehicles may attract private investors and may encourage public–private partnerships for the development of sustainable LCT.

We constructed two growth scenarios driven by assumed rates of population growth. Relevant scenario parameters for the city are population growth, migration, income, lifestyle, rate of motorization, and technological energy efficiency. We assumed that uncertainties about future population would produce different results for the potential population to be serviced by LCT plans and, consequently, lead to different levels of emissions at different growth rates. In this study, we considered two growth rates up to year 2020: one of 4.76% (CBS, 2012), and the other of 7.9% (Pradhan, 2004), representing a low-growth scenario and a high-growth scenario, respectively. By combining these two growth rates with the identified LCT plans, eight alternative scenarios for the LCT plans were generated (see Figure 2.5).

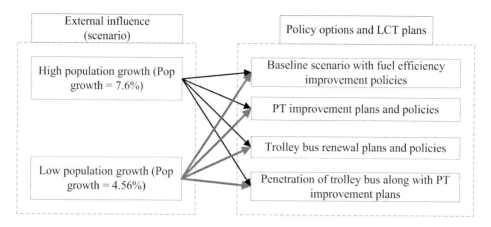

FIGURE 2.5   Construction of alternative scenarios for LCT plans that are based on growth scenarios.

2.3.3.1   Baseline Scenario in 2020

Up to year 2020, GDP was assumed to increase by 5%. As a result, it was estimated that traffic would grow 8.75% per annum from 2010 to 2015 for PT, cars and motorcycles; then from 2016 to 2020 the growth in PT, cars and motorised 2W was estimated at 8.5%, 7.5% and 5.0% per annum respectively (ADB and MoPPW, 2010). This increase in traffic growth shows that by 2020 the number of privately owned cars would increase as compared to motorised 2W. Nonetheless, we assumed that the popularity of 2W among youths would remain and that without strong initiatives to enhance PT patronage and stricter enforcement of laws and policies, the current share of PT and 2W in travel demand would remain the same in 2020. The present mode share for PT – small occupancy vehicles (microbuses and tempos) and high occupancy vehicles (minibuses and buses) – was also assumed to be the same for the baseline scenario in 2020. With regards to fuel efficiency and the emission factor, there has been little variation over the years, as demonstrated by estimates by Dhakal (2006) for 2004 and MoPPW and ADB (2010) for 2010. Nevertheless, taking into account rising incomes and improvements in vehicle technology worldwide, together with the effect of stricter enforcement of some of policies on fuel efficiency, we assumed that there would be a 20% increase in fuel efficiency in the baseline scenario in 2020. Table 2.1 shows the assumptions made for the variables in the model for the baseline scenario in 2020.

TABLE 2.1

Assumptions on the Variables in the Model for the Baseline Scenario in 2020

Type of Mode	Mode Split (%)	Average Speed (kph)	Fuel Efficiency (L/km)	Emission Factor (g/L)	Average Occupancy
Public transport (PT)	52.8	20			
Minibus/bus	31		0.14	3440	26.4
Microbus	47		0.104	2530	12.3
Gas tempo	4.28		0.005	1620	7.6
Electric tempo	5.58		0	0	7.6
Two-wheelers	40.7	20	0.02	3984.8	1.3

Sources: MoPPW and ADB (2010), Bajracharya (2010) and Dhakal (2006)

2.3.3.2 Increase in Patronage of Public Transport

In this alternative scenario, a combination of policy instruments was assumed to have been implemented up until 2020 in order to improve PT patronage. The delineation of primary and secondary routes for large buses and minibuses, respectively, and the assignment of dedicated lanes on primary routes for PT were the main elements considered under this scenario. As a result, average waiting times were expected to decrease to 3 minutes at each stop (ADB and MoPPW, 2010) from a current average of 10 minutes (peak-time assumption). Similarly, according to conservative estimates by KSUT, and with the exception of large vehicles, the current average speed of 20 kph was expected to increase to 25 kph for all vehicle types. Large buses using dedicated bus lanes or primary routes in combination with proper traffic management were expected to reach average speeds of up to 40 kph (Dhakal, 2003). Delineation of routes would require that passengers change vehicles between primary and secondary routes. Such changes would be made convenient by providing user-friendly terminals enhanced with information technology. Additionally, after managing demand and supply, the average occupancy of PT was assumed to decrease by 15%, creating a comfortable environment inside the vehicle. Hence, the mode share of PT is expected to have increased by 20% in 2020 in line with KSUT (MoPPW and ADB, 2010). The increase in PT mode share would be compensated by a simultaneous decrease in the mode share of 2W. Table 2.2 shows the assumptions made for variables in the model for the scenario of PT improvements in 2020.

2.3.3.3 Renewal of Trolley Bus System

Under this alternative scenario, a few primary routes would be dedicated for operating a trolley bus system, in accordance with KSUT proposals. Secondary routes would still be used by minibuses, whereas trolley buses would replace large buses on suitable primary routes. For these routes, therefore, a technologically more efficient trolley bus system with an average speed of 40 kph in dedicated lanes was assumed to be in operation. Strong regulatory and economic instruments – financial incentives and tax breaks – would have to be implemented, in addition to increasing the attractiveness of the trolley bus system. As a result, similarly to the PT improvement scenario, average occupancy would decrease by 15% and mode share would increase by 20% as compared to the baseline scenario for 2020. Furthermore, the trolley bus system would run on electricity so its vehicles would be considered to be 'clean'. A system's CO_2 emissions are generally calculated using a grid emission factor. However, as the trolley bus system would use electricity generated with a hydropower plant it would be considered to run with zero emissions. For this scenario, strong coordination of the electricity authority with other agencies would be required. Table 2.3 shows the assumptions for the variables in the model for the scenario of trolley bus renewal in 2020.

TABLE 2.2
Assumptions Made for Variables in the Model for the Alternative Scenario of PT Improvements in 2020

Type of Mode	Mode Split (%)	Average Speed (kph)	Fuel Efficiency (L/km)	Emission Factor (g/L)	Average Occupancy
Public transport (PT)	63.36	40 (25 kph on secondary route)			
Large bus on primary route	63.36		0.14	3440	30
Minibus on secondary route	63.36		0.104	3440	22.4
Two-wheelers	32.6	25	0.02	3984.8	1.3

Sources: MoPPW and ADB (2010) and Dhakal (2003).

TABLE 2.3

Assumptions Made for Variables in the Model for the Scenario of Trolley Bus Renewal in 2020

Type of Mode	Mode Split (%)	Average Speed (kph)	Fuel Efficiency (L/km)	Emission Factor (g/L)	Average Occupancy
Public transport (PT)	63.36	40 (25 kph on secondary route)			
Trolley bus on selected primary route	63.36		0	0	30
Minibus on secondary route	63.36		0.104	3440	22.4
Two-wheelers	32.6	25	0.02	3984.8	1.3

Sources: MoPPW and ADB (2010), Bajracharya (2010) and Dhakal (2006).

2.3.3.4 Penetration of Trolley Bus System Together with PT Improvement Plans

This alternative scenario comprises the combination of two scenarios: increase in PT patronage and renewal of the trolley bus system. Following a conservative assumption by KSUT, the trolley buses would be able to meet 20% of travel demand in 2020. Therefore, 20% of modal split of large buses on primary routes would be shared by trolley buses that would run with zero emissions. The remaining assumptions on average occupancy, fuel efficiency and emission factors would be the same as for the PT patronage scenario. Table 2.4 shows the assumptions made about the variables in the model for the penetration of the trolley bus system together with the scenario of PT improvement plans in 2020.

2.4 RESULTS

Figure 2.6 shows the serviced areas for 151 schools and colleges using PT and 2W, respectively, for (a, b) baseline scenario; (c, d) PT improvement plans scenario and PT improvement together with penetration of trolley bus scenario; and (e, f) trolley bus renewal plan scenario. Under both high and low population growth baseline scenario and scenarios for LCT plans result into the same serviced areas as in Figure 2.6

2.4.1 Accessibility Under Baseline Scenario in 2020 and Alternative Scenarios for LCT Plans in 2020

In the baseline scenario of 2020, it was found that for the 151 schools and colleges the serviced areas using PT covered most of the city and that these were within 30 minutes of travel time (Figure 2.6a)

TABLE 2.4

Assumptions About Variables in the Model for the Penetration of the Trolley Bus System Together with the Scenario of PT Improvement Plans in 2020

Type of Mode	Modal Split (%)	Average Speed (kph)	Fuel Efficiency (L/km)	Emission Factor (g/L)	Average Occupancy
Public transport (PT)	63.36	40 (25 kph on secondary route)			
Large bus on primary route	43.36		0	0	30
Trolley bus on primary route	20.0				
Minibus on secondary route	63.36		0.104	3440	22.4
Two-wheelers	32.6	25	0.02	3984.8	1.3

Sources: MoPPW and ADB (2010), Bajracharya (2010) and Dhakal (2006).

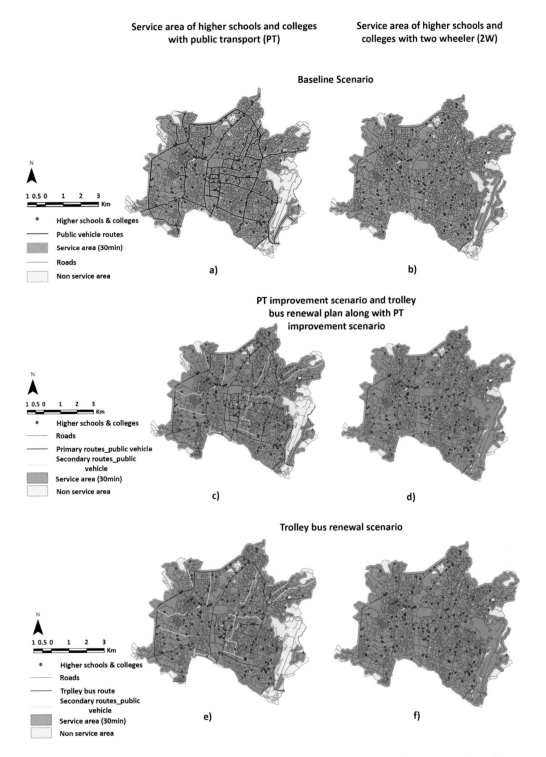

FIGURE 2.6 Serviced areas by public transport (PT) and two-wheelers (2W) for various alternative scenarios in case of both high and low population growth: (a, b) baseline scenario; (c, d) PT improvement scenario and penetration of trolley bus together with PT improvement plan scenario; and (d, e) trolley bus renewal scenario.

of these schools and colleges. In this case, any improvement in the serviced areas for these schools and colleges by using 2W was found to be minimal (Figure 2.6b). Areas in the southeastern part and periphery of the city that are not served by PT would benefit from the use of 2W, however.

The serviced areas for schools and colleges improved to some extent under both alternative scenarios for the LCT plans – the PT improvement plan and penetration of trolley bus system together with the PT improvement plan – as compared to the baseline scenario of PT (Figure 2.6c). These improvements can be observed on the periphery of the city. On the other hand, the trolley bus renewal plan shows some decrease in serviced areas of schools and colleges, mostly in the southeastern part of the city (Figure 2.6e). In all three alternative scenarios, the use of 2W increases the serviced area for the schools and colleges against the baseline scenario of 2W use, particularly in the southeastern part of the city (Figure 2.6d, f). This can be attributed to an increase in the average speed of 2W under all these scenarios as a result of reduced congestion induced by the LCT plans.

These findings fit with the percentage change in the number of people serviced under each alternative scenario, both for low and high population growth against the baseline in 2020 (see Figure 2.7). Alternative scenarios of both the PT improvement plan and penetration of trolley bus system together with PT improvement show an increase in the number of people that are provided with accessibility to the schools and colleges in 2020. In contrast, the alternative scenario of trolley bus renewal leads to a decrease in the number of people against the baseline scenario in 2020. In all scenarios, the percentage change in the number of people serviced against the baseline by using 2W is highest, owing to an increase in serviced areas, mostly in the southeastern part of the city.

2.4.2 EMISSIONS UNDER THE BASELINE SCENARIO IN 2020 AND ALTERNATIVE SCENARIOS FOR LCT PLANS IN 2020

Figure 2.8 shows the percentage change of emissions during high and low population growth scenarios (a) for the baseline scenario in 2020 and alternative scenarios for LCT plans in 2020 against the base year of 2011 and (b) for the alternative scenarios for LCT plans in 2020 against the baseline scenario in 2020. The baseline scenario with improvement only in fuel efficiency clearly shows an undesirable future (Figure 2.8a). All the alternative scenarios for LCT plans under conditions of high population growth show an increase in total emissions against the base year 2011 but are

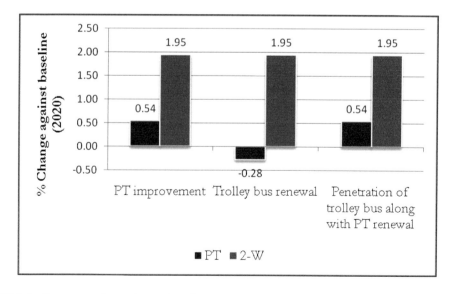

FIGURE 2.7 Percentage change in number of people serviced under alternative scenarios for both low and high population growth.

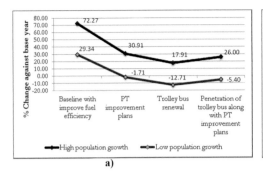 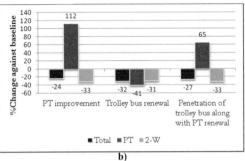

a)   b)

FIGURE 2.8 Percentage change in emissions for both high and low population growth: (a) against the base year 2011 for the baseline scenario 2020 and alternative scenarios for LCT plans 2020; and (b) alternative scenarios for LCT plans 2020 against the baseline scenario 2020.

nonetheless lower than the baseline scenario in 2020. Unlike for high population growth, emissions under the scenario of low population growth decrease against the base year 2011 for all alternative LCT plans. It is noteworthy that the alternative scenario of renewal of the trolley bus system would generate fewer emissions in comparison to the other alternative scenarios for LCT plans alternatives regardless of whether population growth is high or low.

Against the baseline scenario in 2020, under both high and low growth scenarios, all three alternative scenarios for LCT plans show gains in total emission reduction (Figure 2.8b). Of the three alternatives, renewal of the trolley bus system would provide the highest gain in total emission reduction, followed by penetration of the trolley bus system together with PT improvement and, lastly, improvement in PT patronage. Nonetheless, emissions from individual transport modes show that there would be an increase in emissions from PT under the PT improvement plan, which could be halved under the scenario of penetration of the trolley bus together with the PT improvement plan; the trolley bus system would still deliver gains in emission reductions under the trolley bus renewal plan. An increase in the modal share of public transport would be compensated by subsequent decreases in 2W. Therefore, 2W emissions would decrease under all three alternative LCT plans.

2.4.3 Sensitivity Analysis

Sensitivity analysis was performed for two LCT alternatives: PT improvement and penetration of the trolley bus system together with PT improvement. Although the scenario for the trolley bus renewal plan would decrease emissions, it would also result in lower accessibility. Two parameters in the model – average speed and mode share – were varied, as these affect people's accessibility and subsequently gains in emission reduction. Figure 2.9 shows the results of sensitivity analysis for population serviced and gains in emission reduction with respect to change in average speed. Average speed of PT on primary routes in the model, which comprise a large part of public vehicle routes, was lowered from 40 kph to 25 kph. Within this limit of change to a lower average speed, accessibility would decrease to some extent (Figure 2.9a). Nonetheless, alternative scenarios for these LCT plans would still result in greater accessibility against the baseline and gains in total reduction of emissions (Figure 2.9b).

Figure 2.10 shows the sensitivity of gains in reduction of emissions to changes in the mode share. The mode share of PT in the LCT alternatives was increased by only 5% against the baseline scenario subsequent to a shift in 2W. The result shows that the alternative scenario of penetration of the trolley bus together with PT improvement plans would still lead to gains in the total reduction of emissions against the baseline, even when there is only 5% sharing of PT passengers by trolley buses. In the alternative scenario of just PT improvement plans, a shift from the private mode to PT

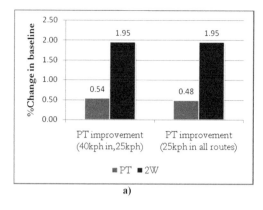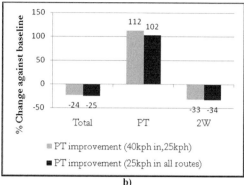

a)   b)

FIGURE 2.9   Sensitivity analysis of (a) the effect on population serviced of change in average speed; and (b) gains in total reduction of emissions.

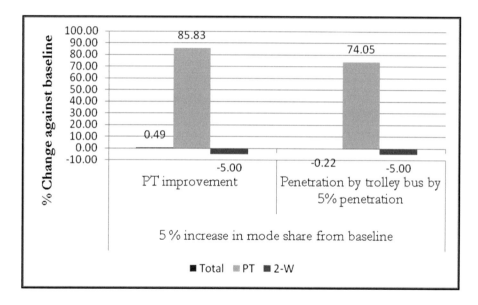

FIGURE 2.10   Sensitivity analysis of gains in total reduction of emissions by changing mode share.

of only 5% would not give gains in total reduction of emissions because the contribution of PT to total emissions would still be high.

2.5   DISCUSSION

Under all LCT plans, using either travel mode – public or private – most areas in the city would be accessible within 30 minutes of travel time. Such a little variation in accessibility can be attributed to the large number of high schools and colleges that are dispersed around the city in such a way that most parts of the city are serviced from at least one of them within the travel time of 30 minutes. Accessibility results are also influenced by the adequacy of the network model to represent reality. Horak and Sedenkova (2005) have shown the importance of lines of transport flow and their characteristics, e.g. the occurrence of one-way and two-way conduits in a network. Similarly, O'Sullivan, Morrison and Shearer (2000) have shown the significant influence of constant average speed along the whole network. We should point out that in the case of the network model prepared for our tool, by using two-way traffic travelling at constant average speed in each road segment and applying a cut-off travel time of 30 minutes, our results may overestimate vehicle speeds in the city centre and underestimate speeds in the other areas. Assumptions of the same waiting time for PT vehicles to arrive at each stop and absence of congestion in the network are other areas of simplification. Nevertheless, to ensure the validity of the results, we compared each scenario with the baseline scenario. Furthermore, we subjected some of the critical model parameters to sensitivity analysis.

It is notable that PT improvement plans would increase emissions from public transport in 2020. More accessibility means more people are able to take part in activities that increase travel demand. The increase in modal share of public transport, the construction of additional primary and secondary routes and the assumptions of the same modal share in both large buses and minibuses resulting from the need to be able to change between routes would all contribute to the increase in emissions against the baseline. Under the same circumstances, the penetration of trolley buses – zero emission vehicles – into primary routes, together with large buses, would halve emissions. On the other hand, in the case of the plan for trolley bus renewal, a trade-off exists between increasing accessibility and gains in emission reduction. The reduction in emissions against the baseline would be at the expense of lower levels of accessibility.

The use of models to analyse different levels of energy consumption and CO_2 emissions in the transport sector is increasing; however, it is often data intensive (Cai et al., 2007; Dhakal, 2003). Nonetheless, a planning support tool such as the one we have developed in this study for a data-poor context is virtually unavailable. There are other advantages. One significant advantage is that the models for accessibility analysis and carbon emissions interact in the same GIS environment with respect to data management, data integration and spatial and non-spatial visualization without any need for data exchange between the two models via computer-system files. Moreover, the tool provides an interface that allows experimentation with assumptions under various scenarios and the dynamic availability of the results in real time.

The tool does have some limitations, however. The accessibility measure built into the tool is based on the locational proximity of opportunities. For now, the tool does not take into account any personal preferences in travel behaviour and ignores the role of individual time-budgeting and space–time constraints in measuring accessibility. In other words, it reflects locational accessibility rather than individual accessibility. The network model still needs to be fine-tuned to consider the effects of linked trips, public transport schedules and transport points, turning restrictions, directionality and other travel impedances of the network links.

2.6   CONCLUSION

In this chapter, we have presented a GIS-based planning support tool that integrates urban accessibility analysis with emission modelling to evaluate various scenarios of LCT plans. We have demonstrated its application in a data-poor context of KMC. The transport plans considered in our study include modal shifts and energy efficiency. To realise low-carbon development for the city, LCT plans should improve accessibility and at the same time minimize CO_2 emissions. Nevertheless, an increase in accessibility means providing more opportunities for people to participate in society, which increases travel demand and hence emissions. Therefore, making low-carbon modes appealing and user-friendly through spatial organization and route reorganization would not be sufficient for minimizing emissions. Behavioural change and a shift from private to public travel modes is necessary to realise substantial gain in emissions reduction. Ultimately, the final choice of travel mode depends on people's preferences, attitudes and behaviour, for which a package of policies that feature 'push and pull' could be more effective.

Our tool could be adapted to incorporate demand-side data such as people's point of origin and destination, people's choice of travel mode, and the value people place on land use and transport components. People's choice of travel mode is influenced by many factors, among them socio-economic profile, cost, travel time, reliability, comfort and convenience of the transport. Techniques – among them stated-preference surveys – need to be designed to understand future travel mode shares within the new system. Another useful extension to the tool could be the integration of space–time accessibility measures as another type of activity-based accessibility measure. A space–time accessibility measure takes into account the number of people travelling to the same destination at a specific time, transportation-benefit measures, and network structure for accessibility measurement. It can provide a useful measure for the analysis of an individual's accessibility to opportunities. By incorporating such a measure, the current limitation of the tool could be lifted to reflect peoples' travel behaviour. Further, the tool in its current state deals with accessibility and low carbon measures in separate models. Future research may enable the development of an integrated model that can facilitate integrated analysis and decision-making.

REFERENCES

ADB, & MoPPW. (2010). Kathmandu sustainable urban transport project: RRP linked documents. Retrieved from http://www.adb.org/Documents/RRPs/?id=44058-01-3

Akinyemi, E. O., & Zuidgeest, M. (2000). Sustainable development and transportation: Past experiences and Future Challenges. *World Transport Policy & Practice*, *6*(1), 31–39. Retrieved from http://worldtransportjournal.com/wp-content/uploads/2015/02/wtpp06.1.pdf

Akinyemi, E. O., & Zuidgeest, M. (2002). Managing transportation infrastructure for sustainable development. *Computer-Aided Civil and Infrastructure Engineering*, *17*(3), 148–161. doi:10.1111/1467-8667.00263

Ale, B. B., & Shrestha, S. O. B. (2009). Introduction of hydrogen vehicles in Kathmandu Valley: A clean and sustainable way of transportation. *Renewable Energy*, *34*(6), 1432–1437. doi:10.1016/j.renene.2008.10.015

Arampatzis, G., Kiranoudis, C. T., Scaloubacas, P., & Assimacopoulos, D. (2004). A GIS-based decision support system for planning urban transportation policies. *European Journal of Operational Research*, *152*(2), 465–475. doi:10.1016/S0377-2217(03)00037-7

Bajracharya, A. R. (2010). Commuter's travel mode and its impact on urban transportation system, a case study. *Urban Development, 17.*

Boile, M. P. (2000). Intermodal transportation network analysis-A GIS application. Paper presented at the 10th Mediterranean Electrotechnical Conference, Information Technology and Electrotechnology for. the. Mediterranean Countries, II. Lemesos, Cyprus. Retrieved from http://ieeexplore.ieee.org/stamp/stamp.jsp?tp=&arnumber=880020

Cai, W., Wang, C., Wang, K., Zhang, Y., & Chen, J. (2007). Scenario analysis on CO_2 emissions reduction potential in China's electricity sector. *Energy Policy*, *35*(12), 6445–6456. doi:10.1016/j.enpol.2007.08.026

CBS. (2012). Final result of population and housing census 2011. Retrieved from http://census.gov.np/

CBS. (2018). National population and housing census 2011. Retrieved from http://cbs.gov.np/sectoral_statistics/population/national_report

Cerda, A., & El-Geneidy, A. (2009). Understanding the relationships between regional accessibility travel behaviour and home values. Paper presented at the Transportation Research Board 89th Annual meeting 2010. Retrieved from http://tram.mcgill.ca/Research/Publications/ACCESS_TRB10.pdf

Dalkmann, H., & Brannigan, C. (2007). Transport and climate change, module 5e, sustainable transport, a sourcebook for policy-makers in developing cities, GTZ global. Retrieved from www.sutp.org

Dhakal, S. (2003). Implications of transportation policies on energy and environment in Kathmandu Valley, Nepal. *Energy Policy*, *31*(14), 1493–1507. doi:10.1016/S0301-4215(02)00205-7

Dhakal, S. (2006). Urban Transport and the environment in Kathmandu Valley, Nepal: Integrating global carbon concerns into local air pollution management. Retrieved from http://www.energycommunity.org/documents/iges_start_final_reprot.pdf

DOTM. (2014). Vehicle registration record. Retrieved from https://www.dotm.gov.np/en/vehicle-registration-record/

Geurs, K. T., & Van Wee, B. (2004). Accessibility evaluation of land-use and transport strategies: Review and research directions. *Journal of Transport Geography*, *12*(2), 127–140. doi:10.1016/j.jtrangeo.2003.10.005

Ghimire, K. P., & Shrestha, S. R. (2014). Estimating vehicular emission in Kathmandu valley, Nepal. *International Journal of Environment*, *3*(4), 133–146. doi:10.3126/ije.v3i4.11742

Halsnæs, K., & Verhagen, J. (2007). Development based climate change adaptation and mitigation – conceptual issues and lessons learned in studies in developing countries. *Mitigation and Adaptation Strategies for Global Change*, *12*(5), 665–684. doi:10.1007/s11027-007-9093-6

Hickman, R., & Banister, D. (2007). Looking over the horizon: Transport and reduced CO_2 emissions in the UK by 2030. *Transport Policy, 14*(5), 377–387. doi:10.1016/j.tranpol.2007.04.005

Hickman, R., Saxena, S., Banister, D., & Ashiru, O. (2012). Examining transport futures with scenario analysis and MCA. *Transportation Research Part A: Policy and Practice, 46*(3), 560–575. doi:10.1016/j.tra.2011.11.006

Hoffman, A. (2008). The Capability Approach and educational policies and strategies: Effective life skills education for sustainable development. In V. Reboud (Ed.), *Amartya Sen: un économiste du développement* (pp. 81–94). Paris: Agence Française de Dévelopement Département de la Recherche.

Horak, J., & Sedenkova, M. (2005). Traffic accessibility evaluation. Evaluation with examples of employers accessibility. Paper presented at the 14th European Colloquium on Theoretical and Quantitative Geography, Tomar, Portugal.

Malla, S. (2014). Assessment of mobility and its impact on energy use and air pollution in Nepal. *Energy, 69* (Supplement C), 485–496. doi:10.1016/j.energy.2014.03.041

Mazumdar, K. (2003). A new approach to human development index. *Review of Social Economy, 61*(4), 535–549. doi:10.1080/0034676032000160895

MoPPW, & ADB. (2010). *Kathmandu Sustainable Urban Transport Project, Final Report.* Kathmandu.

National Environmental and Scientific Services. (2003). *Analysis of HMG Policies and Regulations Affecting Electric Vehicles, Final Report.* Kathmandu.

O'Sullivan, D., Morrison, A., & Shearer, J. (2000). Using desktop GIS for the investigation of accessibility by public transport: An isochrone approach. *International Journal of Geographical Information Science, 14*(1), 85–104. doi:10.1080/136588100240976

Placeways. (2013). CommunityViz. Retrieved from http://placeways.com/communityviz/

Pradhan, P. K. (2004). Population growth, migration and urbanisation. Environmental consequences in Kathmandu Valley, Nepal. In J. D. Unruh, M. S. Krol, & N. Kliot (Eds.), *Environmental Change and Its Implications for Population Migration. Advances in Global Change Research*, Vol. 20 (pp. 177–199). Dordrecht: Springer.

Saito, M. (2003). Amartya Sen's capability approach to education: A critical exploration. *Journal of Philosophy of Education, 37*(1), 17–33. doi:10.1111/1467-9752.3701002

Sakamoto, K., Dalkmann, H., & Palmer, D. (2010). A paradigm shift towards sustainable low-carbon transport – financing the vision ASAP. Retrieved from https://www.itdp.org/publication/a-paradigm-shift-toward-sustainable-transport/

Schipper, L., Fabian, H., & Leather, J. (2009). Transport and carbon dioxide emissions: Forecasts, options analysis, and evaluation. Retrieved from https://think-asia.org/handle/11540/1402

Shakya, S. R., & Shrestha, R. M. (2011). Transport sector electrification in a hydropower resource rich developing country: Energy security, environmental and climate change co-benefits. *Energy for Sustainable Development, 15*(2), 147–159. doi:10.1016/j.esd.2011.04.003

Shiftan, Y., Kaplan, S., & Hakkert, S. (2003). Scenario building as a tool for planning a sustainable transportation system. *Transportation Research Part D: Transport and Environment, 8*(5), 323–342. doi:10.1016/S1361-9209(03)00020-8

Shrestha, R. M., & Rajbhandari, S. (2010). Energy and environmental implications of carbon emission reduction targets: Case of Kathmandu Valley, Nepal. *Energy Policy, 38*(9), 4818–4827. doi:10.1016/j.enpol.2009.11.088

Shrestha, R. (2012). *Toward Low Carbon Development: Urban Accessibility Based Planning Support for Evaluating Low Carbon Transport Plans for Kathmandu Metropolitan City, KMC.* (MSc Thesis). University of Twente.ss

Thapa, R. B., & Murayama, Y. (2012). Scenario based urban growth allocation in Kathmandu Valley, Nepal. *Landscape and Urban Planning, 105*(1–2), 140–148. doi:10.1016/j.landurbplan.2011.12.007

Tilburg, X.v., Würtenberger, L., de Coninck, H., & Bakker, S. (2011). Paving the way for low-carbon development strategies. Retrieved from https://www.ecn.nl/publicaties/PdfFetch.aspx?nr=ECN-E--11-059

Udas, S. (2012). Public transport quality survey. Retrieved from Kathmandu. Nepal: http://www.cen.org.np/uploaded/Public%20Transport%20Survey%20report.pdf

Yuan, H., Zhou, P., & Zhou, D. (2011). What is low-carbon development? A conceptual analysis. *Energy Procedia, 5*, 1706–1712. doi:10.1016/j.egypro.2011.03.290

3 A Collaborative Planning Framework to Support Sustainable Development
The Case of Municipal Housing in Guatemala City

Jose Morales, Johannes Flacke, and Javier Martinez

CONTENTS

3.1   Introduction 29
3.2   Sustainability: Implication for Housing Development and Collaborative Planning 30
3.3   A Framework for Participatory Planning in Analysing Site Suitability for Housing
      Projects 32
3.4   Study Area, Data and Definition of Housing Projects...33
3.5   Methodology for Implementing the Collaborative Planning Process 34
3.6   Results and Discussion 38
3.7   Conclusions...42
References..43

3.1   INTRODUCTION

Giving direction to city development that is in line with the social, environmental and economic dimensions of sustainability is one of the core tasks of spatial planning (Steinebach, 2009). Improving planning and management systems, as well as decision-making processes, is important if one is to address the relevant social and economic dimensions stressed in Agenda 21. It follows that integrated multidisciplinary knowledge, a collaborative/participatory planning environment and the inclusion of data and information at all stages of the planning process are vital (Hall and Pfeiffer, 2000). Crossing professional boundaries to include non-professional, tacit community-based knowledge is important for the success of sustainable development (UNCED, 1992).

Latin American countries possess common roots in their historical, political, cultural and economic transformations that partly explain the path of current urban expansion. Jenkins et al. (2007) describe how the primacy of capital cities, top-down planning structures in housing provision and spatial social segregation in Latin America can be traced back to a strong colonial heritage. In the second half of the last century, high rates of population growth accelerated urban expansion in these cities. This process was, and still is, strongly driven by market-oriented housing provision, accelerating development in peripheral areas and stretching dependency on motorised mobility.

The conditions and dynamics of urban development in Guatemala City (Guatemala) are similar to those of other Latin American countries. In the late 1970s and 1980s, urban expansion took place in the form of sprawl, bringing with it social polarization, loss of valuable natural land, congestion, pollution (Guatemala, 2010) and more. In response, the municipality of Guatemala laid down in the 'Plan Guatemala 2020' (Municipalidad de Guatemala, 2005) its vision of creating an environmentally, socially and economically sustainable city. As a result, policies such as the 'Territorial Ordinance Plan' and the public transportation plan 'Trans-Metro' have been implemented during the last decade. Housing in particular has been identified as a key urban shaper for sustainability. Against this background, the Municipal Enterprise of Housing and Urban Development (MEHUD), a programme and legal entity within the municipality, has been set up to provide the financial tools, management mechanisms and spatial strategies needed for housing development. One of its key tasks is the evaluation from a trans-disciplinary perspective of suitable locations for housing development. MEHUD recognises the need for promoting alliances with other civil actors.

The role of urban planners has been changing to include the building of knowledge bases from a more participatory, stakeholder-based perspective, as well as relying on the inclusion of geo-information systems support for decision-making processes (Jankowski and Nyerges, 2001). In spite of this shift, structured frameworks and methods for collaborative planning were still missing in the planning processes for Guatemala City. Geo-information data technologies such as GIS software applications are, however, available within the municipality, but these are used more or less exclusively by planners for visualizing spatial data. The potential of such technologies for spatial analysis remains virtually untapped.

Our research aim was to investigate how potential sites for municipal housing projects could be evaluated using a collaborative planning framework. In particular, the case of Guatemala City and its MEHUD housing projects were studied. Within this research context, there were three specific research objectives: (1) to identify a process and methodology for supporting the collaborative planning of municipal housing projects; (2) to explore the perspectives for participation of potential stakeholders in housing development and collaborative planning; and (3) to implement the collaborative planning framework developed by identifying and assessing potential sites for municipal housing projects.

3.2 SUSTAINABILITY: IMPLICATION FOR HOUSING DEVELOPMENT AND COLLABORATIVE PLANNING

Camagni (1998) stresses that there is a prevailing ambiguity and vagueness about the implications of the concept of sustainability when applied to urban environments. He defines sustainability as 'a process of synergetic integration and co-evolution among the great subsystems making a city: economic, social, physical and environmental' (Camagni, 1998, page 15). The implications of this concept vary accordingly to the urban scale (e.g. a neighbourhood, a city or a country), as well as the urban phenomena in question. Compact development and housing affordability are concepts commonly found in definitions of sustainable housing (Karuppannan and Sivam, 2009; Sivam and Karuppannan, 2012). In reality, however, such concepts can be rather complex and may be defined differently by the various actors involved. Furthermore, the interdependencies that exist among the various subsystems of a city imply that addressing sustainable housing development should be tackled in a setting in which the perspectives of relevant stakeholders can co-evolve.

A compact city offers positive opportunities for achieving social, economic and environmental sustainability and creating liveability (Camagni, 1998; Shen et al., 2011). Historically, cities with higher populations and development densities have proven themselves to be the wealthiest, most diverse and ecologically sustainable (Hall, 1998; cited by Roberts, 2007). At the same time, such cities are also associated with negative aspects such as congestion, pollution and lack of privacy (Goodchild, 1994). Urban sprawl and low-density residential areas, on the other hand, are negatively associated with lack of sustainability (Zussman et al., 2012).

In the context of sustainable housing development, Goodchild (1994) discussed from a historian's perspective the extremes found in the literature with respect to low and high urban densities. Each extreme is criticised correspondingly based on the negative impacts that cascade from low density and accelerated horizontal expansion, as well as by ignoring the user's preferences in extremely dense areas. In this respect, Marcus and Sarkissian (1986) proposed an intermediate alternative: 'low-rise/high-density clustered housing'. Such an alternative offers the ideal of balanced densification, but

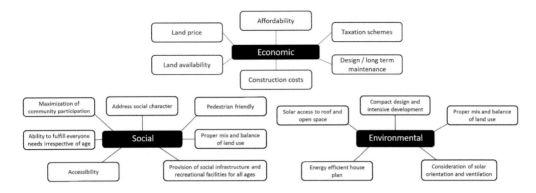

FIGURE 3.1   Issues to be addressed by sustainable housing development (after Sivam and Karuppannan, 2012).

at the same time the users' preference of a countryside lifestyle as described by Goodchild (1991). Figure 3.1 synthetises several of the considerations to be addressed by definitions of sustainable housing development: social, environmental and economic issues. The economic aspect addresses the availability and price of land for economically viable development, and design and legal aspects that might influence the affordability and long-term costs of housing projects. The social aspect addresses the community involvement, universal accessibility and how the housing design solutions could contribute to the urban context. The environmental aspect addresses several design aspects aiming for energy efficient solutions and efficient densification of urban land.

A shift towards collaborative planning involves moving to a more comprehensive and multidimensional approach, i.e. 'helping the planning profession to abandon paternalistic models of planning for the public to planning with the public' (Klosterman et al., 2006, page 81). It assumes that there is more than one designer, each of whom has an idea about what the future design should be (assuming that they have sufficient sense of place and time) (Steinitz, 2012). Dane and van den Brink (2007) compared the different typologies of participation presented in the literature. Levels of participation range from minor and medium involvement (decision-influencers and decision-shapers) to full empowerment in decision-making processes (decision-takers and decision-approvers). Recognition of the different actors in terms of their stake, type of participation and the knowledge they can offer is fundamental in early planning stages (i.e. diagnosis of the current situation, definition of a goal, conceptualization of alternative solutions) for an efficient collaborative planning.

In collaborative approaches, a framework is needed to facilitate participation and proper communication in support of the decision-making process, as well as the selection of proper methods and tools (Groenendijk and Dopheide, 2003; Souza Briggs, 2003). In this sense, a conceptual process facilitates the inclusion of context-specific qualitative and quantitative data, thus allowing the creation of less prescriptive guides for urban design and decision-making (Boyko and Cooper, 2008). Various processes have been conceptualised in the literature (see e.g. Boyko and Cooper, 2008; Malczewski, 1999; Sharifi and Zucca, 2009), but generally speaking this entails the following three phases: intelligence (gathering), design and choice. The 'intelligence' phase concerns the identification of stakeholders and the creation of teams, as well as an analysis of the current situation and formulation of goals. The design phase concerns designing alternatives to reach these goals. The choice phase concerns the processes of evaluation, selection and decision-making.

Data integration and exchange are important elements of collaborative decision-making processes throughout all three phases. This is particularly relevant when moving from a data-poor to a data-rich environment, where tools are needed to filter processes and integrate data and information in support of decision-making processes (Sharifi and Zucca, 2009). Decision Support Systems (DSS) and Planning Support Systems (PSS) are a class of information systems composed of data/information, models and visualization tools that support various tasks, in particular stressing decision-making (choice phase) or planning (design phase) processes, respectively. The combination of both systems creates an Integrated Planning and Decision Support System (IPDSS). Spatial deci-sion-support systems (SDSS) combine the use of geographical information, mathematical and logi-cal formalisms for processing data and human judgments in order to facilitate informed decisions in a collaborative context (Andrienko et al., 2007).

Different methods may serve different purposes when approaching stakeholders and collecting, analysing and communicating information (Groenendijk and Dopheide, 2003; van den Brink et al., 2007). Some methods typically used in qualitative and quantitative surveys range from paper-based, phone or digital questionnaires to interviews, group mapping exercises and focus groups. Other activities may be of a more interactive nature, for example workshops, focus groups and *charrettes*, in which a variety of different group dynamics can be harnessed to achieve different types of infor-mation flow. Lastly, additional to the qualitative–quantitative analytical power embedded in DSS, PSS or IPDSS, geo-visualizations are a common means for reinforcing understanding in multiple stakeholders' situations (Jankowski and Nyerges, 2001).

3.3   A FRAMEWORK FOR PARTICIPATORY PLANNING IN ANALYSING SITE SUITABILITY FOR HOUSING PROJECTS

This section introduces a conceptual–operational framework for participatory planning in Guatemala City, in part addressing the first specific objective of our study. The framework's design relies on the theoretical background discussed in Section 3.2 and its adaptation to the problem of MEUHD housing projects. Figure 3.2 visualises this framework.

The outer rings [1 & 2] correspond to the intelligence phase of the planning process. Ring [1] signifies the identification of and approaches towards relevant stakeholders, as well as their level of involvement in the planning task (i.e. housing development). In the dynamics of housing devel-opment, such stakeholders already interact, but in a 'weak manner' (hollow arrow heads), being guided exclusively by their own knowledge, perspective and interests. Ring [2] implies the estab-lishment of a collaborative environment in which a framework of relevant knowledge is built by stakeholders. For our case, background discussions over the pertaining topic – housing – and crite-ria for evaluating sites were built into this process (i.e. site suitability criteria). Ring [3] encapsulates

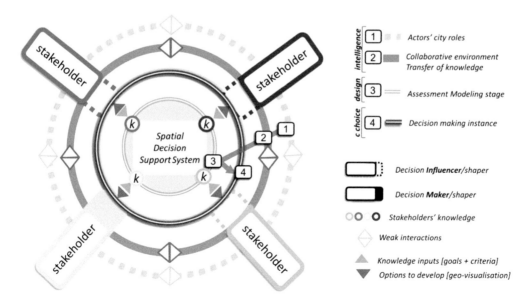

FIGURE 3.2   A conceptual scheme of a collaborative planning framework.

the design phase and incorporates the core of the Planning Support System (PSS). This phase comprises two analytical steps: (1) suitability analysis for the identification of plots for municipal housing; and (2) assessment of the selected plots to rank their suitability for development. While some considerations of sustainable housing are embedded in both steps, Step 2 is designed to stimulate discussion on the intensity of land use (i.e. built up and population density) for such developments. Geo-visualizations are used as a mean to compile knowledge, present results and stimulate communication towards and among stakeholders.

The red and grey arrowheads in the figure suggest a flow of processes, data and information. Stakeholders' collective soft-knowledge provides input for the PSS in both analytical steps of the design phase, the knowledge coming from Rings [1 & 2] to [3]. The output loops back to them in each step: from Ring [3] to Ring [2] for the first analytical step; and from Ring [3] to Ring [4] for the second analytical step. Ring [4] represents the choice phase, where decisions are made, supported using information outputs coming from Ring [3]. Note that the focus of this framework is on the planning process and the production of information to support further decision-making processes that would occur in Ring [4]. In our research, we did not investigate these processes when we applied this framework to the study area. Thus, the outputs of our research are limited to discussions on the outcomes from Ring [3] Step 2. During the research period, stakeholders were not presented with outputs upon which they were expected to make any decisions.

3.4 STUDY AREA, DATA AND DEFINITION OF HOUSING PROJECTS

In line with the visions of MEHUD and the Municipality of Guatemala City (Municipalidad de Guatemala, 2012; URBANISTICA and CIFA, 2010, page 131), a study area was defined within the central postal zones of the city (see Figure 3.3). The study area was delimited by the municipality as part of a larger urban project, 'Corredor Aurora Cañas' (Figure 3.3a). This project is based on a series of public space interventions along a north–south axis of urban revitalization (Figure 3.3b). The shape of the study area polygon (Figure 3.3c) was based on existing natural borders, the Territorial Ordinance Plan, existing land use and the expected influence of the 'Corredor' project. It has a total area of 2,422 ha and includes 2,252 plots with a total plot-area of 1,481 ha.

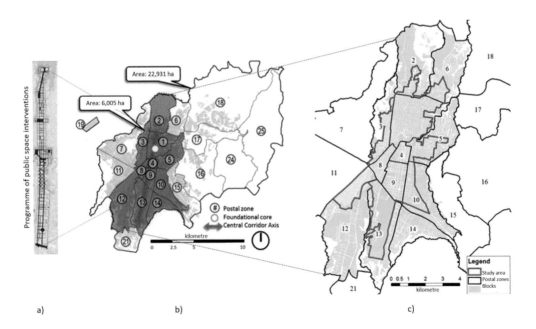

a)   b)   c)

FIGURE 3.3 Guatemala City Central Corridor and delimitation of the study area.

The secondary data used in the research consisted of a dataset of geographical information: i.e. the road network, the municipal transportation system Trans-Metro, the governmental transport system Trans-Urbano, plots, land use data, population census tracts and traffic analysis zones (TAZ). The data was provided by the municipality. Data regarding access to infrastructure (e.g. water, electricity, waste collection) was extracted from municipality website.

MEHUD had already investigated potential alliances and collaborations with civil actors and other institutions. For this research, possibilities for the 'MEHUD–private developers' alliances are explored. The central idea of this is to promote collaboration with private developers through sharing knowledge and participating in investments. The involvement of land owners is not addressed in this research. Even though their role as stakeholders is important for housing development, specific land owners can only be identified after analysing the suitability of land and potential sites for development. The municipal housing projects are envisioned to favour middle- and low-income groups through providing compact urban development and promoting social interaction among households in multi-family settings. Our research focuses on housing for middle-income groups. This is not to suggest that differences in the quality of housing projects should exist for each social group, or that strategies related to social cohesion (where two socio-economic groups live in the same residential area) are not contemplated by MEHUD. Rather, the scope in the research was determined by the fact – according to local knowledge and discussions with real-estate developers – that such types of projects are relatively more viable for development because of the market conditions in the selected study area. In other words, currently the mid-range housing market is where the interests of MEHUD overlap with those of real-estate developers.

BOX 3.1   Case Study Area

Guatemala City is the capital city of Guatemala, a country in Central America. The administrative boundaries of the municipality of Guatemala delimit an area of 996 km². However, the metropolitan area of Guatemala City extends beyond these boundaries.

Approximately 26% of the country's population resides in Guatemala City. The study area for our case was located within the city's inner core and was designated by the municipality to be part of a large urban project called 'Corredor Aurora Cañas'. This project comprises a series of interventions in public space along a north–south axis of urban revitalization. The shape of the area is determined by a combination of natural borders, the Territorial Ordinance Plan and existing land use, and covers a surface area of 2,422 ha; the area includes 2,252 plots with a total surface area of 1,481 ha.

3.5   METHODOLOGY FOR IMPLEMENTING THE COLLABORATIVE PLANNING PROCESS

The methodology is divided into two phases, mirroring those described in Section 3.2: 'intelligence' and 'design'. The first phase concerns the identification of stakeholders and implementation of a collaborative framework (Ring [1 & 2] in Figure 3.2). This task is based on local knowledge provided by municipal planners, market researchers and MEHUD. Consequently, 14 participants were identified as stakeholders, based on their level of involvement, civilian roles and relevant experience (i.e. academics, municipal planners, real-estate developers or target households).

Two methods were used to address the second of our specific objectives (and build up the base for accomplishing the third objective). Stakeholders were to be contacted and a collaborative framework (Ring [2]) established. First, we interviewed stakeholders individually, face-to-face, between 26 September and 12 October 2012 using semi-structured interviews. With their permission, all interviews were recorded, transcribed and analysed using a Thematic Analysis Framework (Bryman, 2012). Second, a workshop was conducted on 12 October with ten stakeholders in a spacious 'living room' setting; see Figure 3.4. The stakeholders were introduced to each other and given an introduction about the study area, its housing projects, the collaborative planning approach and an overview of the methods to be used in further analysis. To stimulate a discussion, participants were asked for their feedback. Slides that included maps and 3D models created from Geographic Information Systems (GIS) were used to support this introduction. The expected output of these processes combined was to collectively establish a set of ranked criteria for analysing the suitability of land for housing projects (to be done in Stage 3). Even though individual stakeholders produced different rankings, influenced by their background (academics, municipal planners, etc.), only results derived from the collective ranking are reported in this chapter.

In the 'design' phase, a first analytical step was undertaken by implementing a Spatial Multi-criteria Evaluation (SMCE). Similar approaches can be found in literature: see, for instance, Al-Shalabi et al. (2006), Mardin (2009) and Zucca et al. (2008). SMCE works well as it allows the combination of related qualitative and quantitative criteria, as well as allowing the influence of each of these criteria (according to their importance – weighting) on the results to be understood (Al-Shalabi et al., 2006; Malczewski, 1999). Following the implementation framework of Malczewski (1999), different spatial analyses were carried out to produce a standardised (positive or negative) value map for each criterion. For example, a value map might be the output of a measurement on a single data layer (e.g. distance to Trans-Metro bus stops) or several layers (e.g. proximity to primary roads and to urban nodes). In the latter case, the layers are weighted so the total value is equal to one. In order to aggregate the results, a rectangular tessellation was applied to the study area with cells of 50 m × 50 m (half a block). A restricted area is indicated in this tessellation by a buffer along the Trans-Metro infrastructure, as agreed with the stakeholders. This means that housing is not considered in that area because it is subject to severe market speculation, therefore making it very difficult to negotiate land acquisition and develop affordable housing.

FIGURE 3.4 Workshop participants.

In the suitability analysis, each value map was weighted according to its importance in the collective criteria ranking. Sensitivity analysis was done by varying the weights of three criteria – i.e. proximity to qualified jobs, proximity to densely populated areas, and distance to municipal interventions – by 20% more and 20% less weight. These three criteria were selected because of divergence observed among stakeholder groups in their criteria rankings. The whole SMCE was fully implemented using the CommunityViz® PSS* extension as it already incorporates a module to perform SMCE in an interactive and dynamic manner. CommunityViz® allows for easy manipulation of criteria weights, which facilitates exploration of alternatives and overall model sensitivity analysis. Figure 3.5 shows a synthesis of this process and the criteria used.

Based on the suitability analysis and considerations extracted from discussions with stakeholders in the 'intelligence' phase (e.g. plot size, current constructions, current and surrounding land use of plots), we proposed 63 suitable sites and extracted them to a separate data layer. These suitability maps were sent to stakeholders as a Google Earth® file for ease of visualization. In addition, a meeting was set up to discuss results and to reach agreement on the proposed sites and the criteria for assessing their suitability for development. This meeting was carried out in a Skype video session between stakeholders and the researcher.

In the second analytical step of the 'design' phase, the focus is on the design and implementation of a site assessment methodology. For our case, we first restricted the next analysis to a homogeneous neighbourhood within the study area. Then, eight sites were selected (sites A to H), which were categorised as G3 or G4, corresponding to their classification in the Territorial Ordinance Plan. A dynamic model in CommunityViz® PSS was created to facilitate exploration of alternative developments (see Figure 3.6). Three groups of input variables were used in the model:

1. *Land use variables*: To express built-up intensity and the ratio between site area and area of open space;
2. *Housing variables*: To express the percentage of the developable area designated to each type of housing unit (flats of 50, 75 or 100 m²), amenities and presence of underground parking. Input values were set according to housing typology and considerations of stakeholder groups.

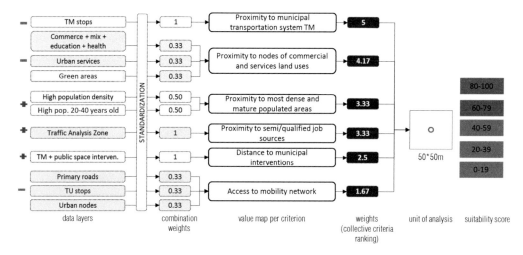

FIGURE 3.5   SMCE implementation in CommunityViz®.

* For more information about the CommunityViz® PSS, visit: http://placeways.com/communityviz/

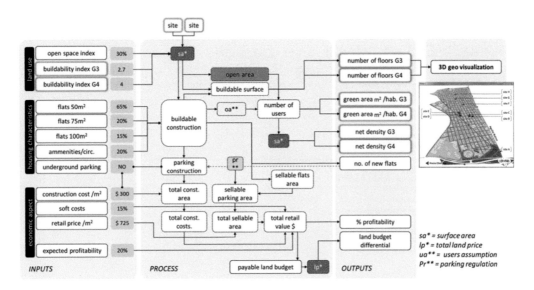

FIGURE 3.6 Dynamic assessment model implemented in CommunityViz® PSS.

3. *Economic variables*: To express the monetary concerns of real estate development, including variables for construction costs, sale prices and expected profitability. The latter variable would express the profitability conditions under which a project would be considered viable and attractive for investors. All values input in this group were adjusted to take into account what was considered affordable by middle-income buyers.

Output variables of the model include height (number of floors), ratio of green area to new inhabitant, net density (persons/m²), number of housing units achieved, profitability realised and land–budget differential. Actual profitability is calculated based on site conditions (i.e. land values) and the development itself (i.e. total retail value, construction costs including the value of land). Land–budget differential is the difference between what would be a payable land value under the assumption that the project would be realised and the actual market price of land. A value equal to zero would mean that the assumed project would generate enough financial resources to fulfil the expected profitability and pay the actual market price of the land. A positive value would imply additional profit or additional budget for construction. In this model, all the input variable values were proposed by the researcher in agreement with the stakeholders. Finally, three scenarios of development were calculated: low-, middle- and high-density housing. The middle-density development scenario was selected for further analysis as the output variables were closer to the idealised density described by planners and target households, as well as being favoured in the literature.

In the final stage of our analysis, we employed a Multi-Criteria Evaluation (MCE) using ILWIS™ to rank the sites and assign a priority for intervention (Figure 3.7). A second set of criteria was developed together with representatives of the stakeholder groups. This time one representative of each group established a criteria rank, resulting in four sets of criteria with different rankings. A preliminary site's rank was calculated from the average score of each site, which was based on the scoring values after applying the MCE four times (once for each group's preference rankings). A sensitivity analysis was done by giving full weight to each group of criteria, to examine how each site performed and to test the robustness of the site's rank.

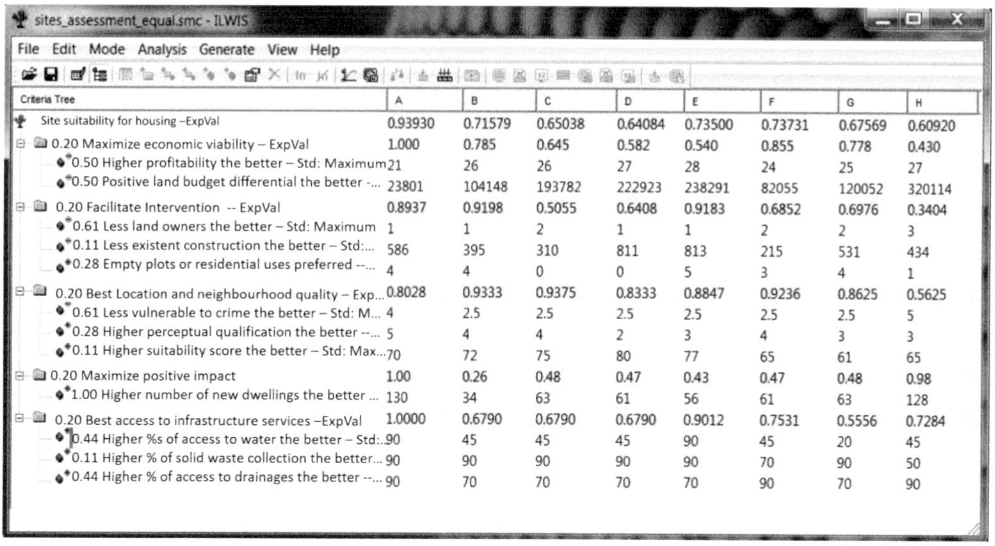

Criteria Tree	A	B	C	D	E	F	G	H
Site suitability for housing –ExpVal	0.93930	0.71579	0.65038	0.64084	0.73500	0.73731	0.67569	0.60920
0.20 Maximize economic viability – ExpVal	1.000	0.785	0.645	0.582	0.540	0.855	0.778	0.430
◆*0.50 Higher profitability the better – Std: Maximum	21	26	26	27	28	24	25	27
◆*0.50 Positive land budget differential the better -...	23801	104148	193782	222923	238291	82055	120052	320114
0.20 Facilitate Intervention -- ExpVal	0.8937	0.9198	0.5055	0.6408	0.9183	0.6852	0.6976	0.3404
◆*0.61 Less land owners the better – Std: Maximum	1	1	2	1	1	2	2	3
◆*0.11 Less existent construction the better – Std:...	586	395	310	811	813	215	531	434
◆*0.28 Empty plots or residential uses preferred --...	4	4	0	0	5	4	4	1
0.20 Best Location and neighbourhood quality – Exp...	0.8028	0.9333	0.9375	0.8333	0.8847	0.9236	0.8625	0.5625
◆*0.61 Less vulnerable to crime the better – Std: M...	4	2.5	2.5	2.5	2.5	2.5	2.5	5
◆*0.28 Higher perceptual qualification the better --...	5	4	4	2	3	4	3	3
◆*0.11 Higher suitability score the better – Std: Max...	70	72	75	80	77	65	61	65
0.20 Maximize positive impact – ExpVal	1.00	0.26	0.48	0.47	0.43	0.47	0.48	0.98
◆*1.00 Higher number of new dwellings the better ...	130	34	63	61	56	61	63	128
0.20 Best access to infrastructure services –ExpVal	1.0000	0.6790	0.6790	0.6790	0.9012	0.7531	0.5556	0.7284
◆*0.44 Higher %s of access to water the better – Std:.	90	45	45	45	90	45	20	45
◆*0.11 Higher % of solid waste collection the better...	90	90	90	90	90	70	90	50
◆*0.44 Higher % of access to drainages the better --...	90	70	70	70	70	90	70	90

FIGURE 3.7   Assessment criteria for sites and MCE implementation in ILWIS.

BOX 3.2   Methods Applied in the Chapter

This chapter demonstrates how the combination of qualitative and quantitative methods can be used to design and implement a collaborative planning framework for identifying and assessing locations for housing development. The GIS methods applied include a Spatial Multi-Criteria Assessment to conduct suitability analysis and a dynamic scenario model to assess a number of suitable sites for housing development. The analyses were implemented using a CommunityViz® extension within the ESRI ArcGIS® application, as well as stand-alone ILWIS. The data used for the analysis comprises secondary geodata (cadastral parcels, road networks, public transport networks, administrative subdivisions) received from the municipality of Guatemala City. Primary qualitative data were collected from structured interviews of stakeholders and a workshop.

3.6 RESULTS AND DISCUSSION

Different aspects of sustainable housing development, as conceptualised in the literature, are pursued separately through the individual interests of stakeholders. Target households are interested in locational characteristics associated with sustainability (e.g. short trips to access to services and jobs in high-quality neighbourhoods). However, they highly value aspects such as housing affordability and proximity to less-polluted environments and open space, conditions that currently are mostly met in peripheral urban areas. Dependency on automobiles is not just due to the need for driving long distances to work, but also because of a cultural belief that the automobile provides status. For reasons similar to those of academics, municipal planners focus on sustainability aspects in an effort to manage accelerated expansion of the city and achieve more balance in housing densities. However, they also recognise that they only have limited knowledge of the market-driven opportunities, barriers and operations related to housing provision. Developers, on the other hand, are state-of-the-art practitioners of project development (to ensure economic value and long-term physical integrity) and have better knowledge about the target households. They were able to point out some opportunities for middle-income, compact housing given a slow, ongoing cultural shift in residential preferences (people being willing to move from horizontal to vertical property formats). There are barriers, however: they point to the speculative real-estate market existing in central areas and excessive bureaucratic procedures – and the uncertainty that generates – for real-estate investment within the city.

There are potential advantages and disadvantages from the perspective of stakeholders. In general, all stakeholders expressed interest in collaborative approaches. They foresee a possible path for overcoming some of the current barriers for housing development, for promoting a positive environment for local economic growth through investment and for knowledge sharing. In contrast, the usefulness of the entire approach might be limited if the interests and behaviour of land owners are incorporated. Planners and real-estate developers agree that public knowledge of the analytical outputs and expressions of interest for specific properties for future investment in housing might trigger land-price speculation. Furthermore, distrust of public institutions owing to political stress and their acknowledged lack of resources militates against collaborative approaches. Stakeholders also acknowledge that differing perspectives and interests might lead to slow, unfertile decision-making processes, resulting in frustration and tensions among participants.

In terms of levels of involvement, MEHUD sees itself as a coordinator of the alliances and responsible for assuring high-quality development, i.e. decision-approvers. Developers, on the other hand, are of the opinion that since they are the investors, they should take full control of the whole process and decision-making. Householders, however, consider that as long-term users they should be active shapers of decisions. Academics are seen by all other stakeholders as a source of valuable research-based knowledge that could be useful for the planning tasks and as potential educators for combating cultural barriers. Academics themselves, and the other stakeholders, agree that their participation should be limited to guidance: they should not take part in the decision-making.

Figure 3.8 shows the weighting schemes for the criteria rankings assigned by each stakeholder group and the agreed overall group-based weighting. Overall, most stakeholder groups tend to place higher value on proximity to public transport systems, nodes of commercial and public services land uses and good access to job opportunities. Municipal planners instead seem to prioritise each criterion evenly. Real-estate developers seem to pay additional attention to the proximity of populated areas and to avoid proximity to municipal intervention sites. Seen from within a collaborative

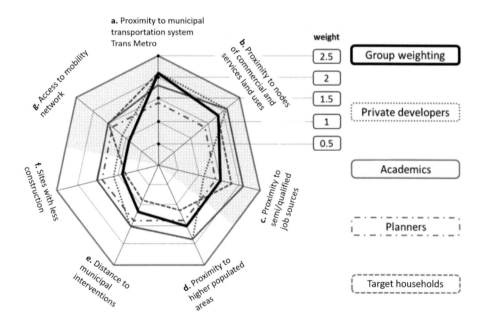

FIGURE 3.8 Weighting schemes.

framework, the figure visualises convergent and divergent interests of stakeholders. Nevertheless, the collaborative approach also allows for the consolidation and agreement of a unified weighting scheme, which then makes it possible to validate the analytical outputs.

Figure 3.9 shows the results of the suitability analysis and the proposed development sites visualised in Google Earth. Additional site information is included in the layers (i.e. site area, built-up area, buildable area based on regulations). The yellow brackets indicate the parcels used for the final assessment. A large extension of suitable area (light- and dark-green shading) can be seen along the central corridor. This is because of the nearby concentration of commerce and public services, and high access to jobs, together with the presence of the municipal transportation system. Proximity to mobility networks seems to contribute mainly along the centre of the north–south corridor. Distance to municipal interventions does not strongly affect suitability. The suitability of the northern area seems to be reduced potentially due to a reduced proximity to densely populated areas and the distribution of primary roads.

In general, the results were as expected. Some interesting, suitable spots were identified in the northeastern, mid-western and southwestern areas. Whilst these were not core areas (i.e. with the highest access to jobs), other criteria such as proximity to densely populated areas and to public transportation systems increased their suitability. In addition, proximity to nodes of commercial and public services and land uses add significantly to suitability. Areas of poor suitability (orange and red shading on Figure 3.9) can be explained by their low weights for most of the criteria, the only exception being distance to municipal interventions.

Table 3.1 shows the criteria established for site assessment, the ranking assigned per stakeholder group and the resulting weighting schemes. The final site ranks shown in Table 3.2 are based on the results of the sensitivity analysis. The sensitivity analysis to some extent conceals the convergences and divergences of the different preferences of each stakeholder group. Site A appears to be the best option and Site H the second-best option, while Sites F, B and G are the last three options. The ranks for all these sites seem consistent with stakeholder group preferences (based on the four sets of weights). Intermediate priority for development was assigned to Sites C, D and E. These three sites were ranked differently in the sensitivity analysis as compared to the preferences of stakeholder groups. An explanation for the difference can be found from observing trade-offs between

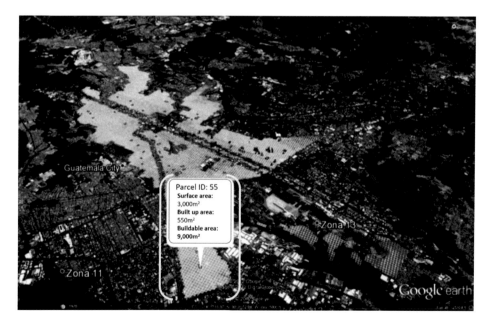

FIGURE 3.9   Land suitability and potential sites for development.

TABLE 3.1

Criteria for Site Assessment, Ranks and Weighting Schemes per Stakeholder Group

AU: Could you please check and confirm whether we have to retain the shades in the "Tables 3.1 & 3.2" as it is or not. If we retain please provide the significance for the same.

Assessment Criteria	EI* Weight	PD* Rank	PD* Weight	TG* Rank	TG* Weight	MP* Rank	MP* Weight	AC* Rank	AC* Weight
Maximize economic viability	☐ 0.2	1	0.2	4	0.09	1	0.46	3	0.16
Higher profitability the better		1	0.5	1	0.5	1	0.5	1	0.5
Greater the positive budget differential the better		1	0.5	1	0.5	1	0.5	1	0.5
Facilitate development	☐ 0.2	1	0.2	5	0.04	5	0.04	4	0.09
Less land owners involved the better		1	0.61	1	0.61	1	0.61	1	0.61
Less existent construction the better		3	0.11	3	0.11	3	0.11	3	0.11
Empty plots or residences preferred		2	0.28	2	0.28	2	0.28	2	0.28
Location and neighbourhood quality	☐ 0.2	1	0.2	1	0.46	2	0.26	2	0.26
Less vulnerable to crime the better		1	0.61	1	0.61	1	0.61	1	0.61
Higher perceptual qualification the better		2	0.28	2	0.28	2	0.28	2	0.28
Higher suitability scores the better		3	0.11	3	0.11	3	0.11	3	0.11
Maximize the positive impact	☐ 0.2	1	0.2	3	0.16	4	0.09	1	0.46
Higher the number of new dwellings the better		1	1	1	1	1	1	1	1
Best access to infrastructural services	☐ 0.2	2	0.2	2	0.26	3	0.16	5	0.04
Higher % of access to water the better		1	0.44	1	0.44	1	0.44	1	0.44
Higher % of services for solid waste collection the better		2	0.11	2	0.11	2	0.11	2	0.11
Higher % of access to drainage the better		1	0.44	1	0.44	1	0.44	1	0.44

* EI = Equal importance *PD = Private developer *TG = Target household *MP = Municipal planner *AC = Academic.

TABLE 3.2
Rank of Sites Based on Sensitivity Analysis

Objective	A	B	C	D	E	F	G	H
Maximize economic viability	0.41	0.63	0.77	0.83	0.87	0.56	0.63	0.98
Facilitate development	0.73	0.75	0.78	0.92	0.64	0.63	0.53	0.51
Location and neighbourhood quality	0.08	0.93	0.94	0.83	0.88	0.92	0.86	0.56
Maximize positive impact	1	0.26	0.48	0.47	0.43	0.47	0.48	0.98
Best access to infrastructural services	1	0.68	0.68	0.68	0.9	0.75	0.56	0.73
Average	0.79	0.65	0.73	0.75	0.74	0.67	0.61	0.75
Rank of sites	1	7	5	3	4	6	8	2

the scores of each site per criterion and the importance that different stakeholder groups assign to each criterion. Still, strong discussion should be stimulated when prioritizing sites for development. For example, Site A scores high on most of the criteria but has a low score for maximizing economic viability (i.e. higher economic risk). Site H, on the other hand, has the highest score for economic viability but the lowest score for access to infrastructure. When considering the preferences of the target households with respect to the quality of the neighbourhood, low access to infrastructure would be relatively less preferred, which might represent a risk in terms of the attractiveness of the project if parallel investments (i.e. neighbourhood improvements) to improve this condition are not considered.

3.7   CONCLUSIONS

In our research, housing and collaborative planning are considered to be two of the drivers of sustainable development. We therefore constructed a conceptual–operational framework for collaborative planning, the core of which was a Planning Support System (PSS) that included both analytical and communication components. We followed this approach to identify and assess sites for municipal housing development using new knowledge gathered as part of the collaborative planning process. That new knowledge gave an insight into the opportunities for and barriers to housing development in Guatemala City through collaborative schemes.

The conceptual–operational framework enabled the visualization of various concerns related to the planning of sustainable housing development. In particular, the framework clearly demonstrated the advantages of providing a comprehensive overview of the challenges to be faced with sustainable development initiatives. To produce that overview, different groups of stakeholders were encouraged to reveal – make more explicit – their perspectives, knowledge and interests. The participation of stakeholders in the collaborative exercises stimulated an attitude of commitment to the planning task: for example, participants actively expressed their doubts, made suggestions of how to carry out the analysis for the suitability map and volunteered the data required.

Some basic challenges and limitations had to be addressed during the research. During the workshop, which provided opportunities for collective interaction among stakeholders, the dominance of certain participants was noticed (i.e. real-estate developers) who were acknowledged to have more information on and experience with the topic (housing development). Similarly, the use of sophisticated terms used by planning experts in conversation was not always understood by the rest, resulting in a loss of interest or difficulties in keeping the participation of the rest of the group active. The appointment of a moderator and establishment of clear planning and meeting objectives is therefore vitally important for maintaining focus on the overall approach.

The experience gained from this research shows that the various methodologies employed, and the use of a planning support system, offer an effective and transparent approach for improving information flows that take place in multi-stakeholder environments during planning processes. Nevertheless, during site assessment, discussions among stakeholders about alternatives for sustainable housing development via a multidimensional approach still need to be stimulated. Due the scope of our research and time limitations, the usability of the CommunityViz® PSS model (designed for site assessment) was not tested with the stakeholders, although its concepts were based on their interests, their professional perspectives and the overall planning objective. We suggest that further research could consider ways of using the model to intuitively explore housing development alternatives that take into account different economic, social and environmental objectives and constraints, thereby facilitating discussion in an interactive environment that encourages participants to exchange perspectives and knowledge.

Future research could include three main points. Considering the analytical principles of the SMCE method for the land suitability analysis and the criteria proposed by stakeholders, the first point would be to broaden the extent of the study area, as results might vary accordingly. This means that other suitable areas, not necessarily restricted to the city's central core, might be identified and explored, perhaps delivering sites that allow some of the economic barriers to sustainable housing development (i.e. high land values in central areas) to be avoided. Second, the inclusion of land owners as an important stakeholder group in collaborative schemes for sustainable housing development in urban areas should be addressed. The identification of potential land owners could be based on the results of the suitability analysis. Third, the models we present in this work (i.e. SMCE for land suitability, the dynamic assessment model via CommunityViz® PSS, and MCE in ILWIS) could be further developed and adapted to address a wider scope of considerations, dependent on the stakeholders, study area selection and availability of data.

REFERENCES

Al-Shalabi, M., Bin Mansor, S., Bin Ahmed, N., & Shiriff, R. (2006). *GIS based multicriteria approaches to housing site suitability assessment*. Paper presented at the Shaping the Change, XXIII FIG Congress, Munich, Germany.

Andrienko, G., Andrienko, N., Jankowski, P., Keim, D., Kraak, M. J., MacEachren, A., & Wrobel, S. (2007). Geovisual analytics for spatial decision support: setting the research agenda. *International Journal of Geographical Information Science, 21*(8), 839–857. doi:10.1080/13658810701349011

Boyko, C., & Cooper, R. (2008). Decision-making processes in urban design. In I. Cooper & M. Smes (Eds.), *Changing professional practice* (Vol. 4, pp. 68–98). London: Routledge.

Bryman, A. (2012). *Social research methods* (Fourth edition ed.). Oxford: Oxford University Press.

Camagni, R. (1998). Sustainable urban development: definition and reasons for research programme. *International Journal of Environmental Pollution, 10*(1), 6–26.

Dane, S., & van den Brink, A. (2007). Perspectives on citizen participation in spatial planning in Europe. In A. van den Brink, R. van Lammeren, R. van de Velde, & S. Dane (Eds.), *Imaging the future: geovisualisation for participatory spatial planning in Europe* (Vol. 3, pp. 33–51). Wageningen: Wageningen Academic Publishers. Mansholt Graduate School of Social Sciences.

Goodchild, B. (1991). Postmodernism and housing: a guide to design theory. *Housing Studies, 6*(2), 131–144. doi:10.1080/02673039108720702

Goodchild, B. (1994). Housing design, urban form and sustainable development: reflections on the future residential landscape. *Town Planning Review, 65*(2), 143.

Groenendijk, E. M. C., & Dopheide, E. J. M. (2003). *Planning and management tools*. Enschede: ITC.

Guatemala, M. d. (2005). Plan Director Guatemala 2020.

Guatemala, M. d. (2010). *Guatemala City New Territorial Ordinance Plan, The Reasons Behind It*.

Hall, P. (1998). *Cities in civilization: culture, innovation and urban order*. London: Weidenfeld & Nicolson.

Hall, P., & Pfeiffer, U. (2000). *Urban future 21: a global agenda for 21st century cities*. London: E & FN SPON.

Jankowski, P., & Nyerges, T. (2001). *Geographic information systems for group decision making: towards a participatory geographic information science* (Vol. *8). London etc.: Taylor and Francis.

Jenkins, P., Smith, H., & Wang, Y. P. (2007). Urban development and housing in Latin America. In *Planning and housing in the rapidly urbanising world* (pp. 235–265). London: Routledge.

Karuppannan, S., & Sivam, A. (2009). *Sustainable development and housing affordability.* Paper presented at the European network for housing research conference.

Klosterman, R., Siebert, L., Kim, J-K., Hoque, M., & Parveen, A. (2006). What if? TM evaluation of growth management strategies for a declining region. *International Journal of Environmental Technology and Management, 6*(1/2), 79–95.

Malczewski, J. (1999). *GIS and multicriteria decision analysis.* New York etc.: Wiley & Sons.

Marcus, C., & Sarkissian, W. (1986). *Housing as if people mattered.* Berkeley, CA: University of California Press.

Mardin, R. (2009). *Collaborative decision making in railway planning, a multi criteria evaluation of JOGLOSEMAR project, central Java – Indonesia.* Enschede: ITC.

Municipalidad de Guatemala. (2012). *Acuerdo COM 3-2012.* Guatemala City: Diario de Centro America.

Roberts, B. (2007). *Changes in urban density: its implications on the sustainable development of Australian cities.* Paper presented at the Proceedings of State of Australian Cities Conference, Adelaide.

Sharifi, M. A., & Zucca, A. (2009). Integrated planning and decision support systems: concepts and application to a site selection problem. In D. Geneletti & A. Abdullah (Eds.), *Spatial decision support for urban and environmental planning: a collection of case studies* (pp. 5–31). Sengalor Adrul Ehsan: Arah.

Shen, L.-Y., Ochoa, J., Shah, M. N., & Zhang, X. (2011). The application of urban sustainability indicators – a comparison between various practices. *Habitat International, 35*(1), 17–29. doi:http://dx.doi.org/10.1016/j.habitatint.2010.03.006

Sivam, A., & Karuppannan, S. (2012). Density, design and sustainable residential development In S. Lehmann & R. Crocker (Eds.), *Designing for zero waste: consumption, technologies and the built environment* (pp. 267–283). New York and London: Earthscan.

Souza Briggs, X. d. (2003). *Strategy tool no. 2. Planning together: how (and how not) to engage stakeholders in charting a course.* Massachusetts: The Community Problem-Solving Project @ MIT.

Steinebach, G., Guhathakurta, S., and Hagen, H. (2009). Planning sustainable living visualizing sustainable planning. In H. Hagen, S. Guhathakurta, & G. Steinebach (Eds.), (pp. 3–36). Berlin: Springer Berlin Heidelberg.

Steinitz, C. (2012). *A framework for geodesign, changing geography by design* (First edition ed.). United States: ESRI.

UNCED. (1992). *Earth summit: convention on climate change: United Nations Conference on Environment and Development UNCED: Rio de Janeiro, Brazil, 3–14 June 1992 = agenda 21.* New York: United Nations.

URBANISTICA, & CIFA. (2010). *La Zona Central, Análisis del Suelo Vacante para Proyectos de Vivienda en la Ciudad de Guatemala.* Guatemala.

van den Brink, A., van Lammeren, R., van de Velde, R., & Dane, S. (Eds.). (2007). *Imaging the future: geo-visualisation for participatory spatial planning in Europe* (Vol. 3). Wageningen: Wageningen Academic Publishers Mansholt Graduate School of Social Sciences.

Zucca, A., Sharifi, A. M., & Fabbri, A. G. (2008). Application of spatial multi-criteria analysis to site selection for a local park: A case study in the Bergamo Province, Italy. *Journal of Environmental Management, 88*(4), 752–769. doi:10.1016/j.jenvman.2007.04.026

Zussman, J., Srinivasan, A., & Dhakal, S. (2012). *Low carbon transport in Asia: strategies for optimizing co-benefits.* London: Earthscan.

4 An Exploration of Environmental Quality in the Context of Multiple Deprivations
The Case of Kalyan–Dombivli, India

Shubham Mishra, Monika Kuffer,
Javier Martinez, and Karin Pfeffer

CONTENTS

4.1 Introduction 45
4.2 Selection of Environmental Aspects for Kalyan–Dombivli (KD)...................47
4.3 Data Sets and Methodology...49
    4.3.1 Urban Thermal Environment...51
    4.3.2 Distribution of Greenery...52
    4.3.3 Building Density..53
    4.3.4 Street Geometry..54
    4.3.5 Data Aggregation..55
4.4 Results and Discussion 55
    4.4.1 Urban Thermal Environment...55
    4.4.2 Distribution of Greenery...56
    4.4.3 Building Density..56
    4.4.4 Street Geometry..56
    4.4.5 Index of Environmental Quality 58
    4.4.6 Sub-Ward Level Analysis 58
4.5 Discussion..59
4.6 Conclusions..60
References...60

4.1 INTRODUCTION

Since Charles Booth's pioneering work of mapping poverty in London in the late 19th century, the understanding of poverty and multiple deprivations has improved enormously (Orford et al., 2002). The debate on measuring and analysing urban poverty in recent times has departed from an earlier traditional and narrow vision (mostly monetary and economically centred) towards a broader range of dimensions (Devas, 2004). Issues such as inadequate access to basic services, social exclusion and lack of entitlements have started receiving considerable scholarly attention. The asset vulnerability framework of Moser (Baud et al., 2009; Moser, 1998), focusing on assets that the poor possess, has built upon this new understanding of urban poverty. These assets exist in the form of physical, human, social, financial and natural capital. Drawing on this understanding and framework, a study by Baud et al. (2009) demonstrated the utility of an index of multiple deprivations to conceptualise the 'multi-dimensional character' of poverty for Delhi, Mumbai and Chennai in India. Despite this informed and integrated view, the natural dimension, which in urban areas translates mainly into environmental quality, often gets neglected in urban studies on multiple deprivations (Baud et al., 2009; Niggebrugge et al., 2005) because of limited environmental data availability.

In some cases, such as the English Indices of Deprivation 2010 (McLennan et al., 2011), the living environment – along with income, employment, health and disability, education and training, barriers to housing and services and crime – is one of the seven domains that make up the overall index. The living environment is further divided into two sub-domains: the indoor environment, which measures the quality of housing, and the outdoor environment, which measures air quality and road accidents. In the past, aspects of the natural environment received little attention. However, with the emergence of the smart city concept (Anguluri and Narayanan, 2017), environmental indicators are also receiving increased attention. One of the issues that persists is that such data are mostly not included in statistical or census databases for countries like India. State-of-the-art remote sensing techniques provide relevant data to address this gap (Taubenböck et al., 2012) as they can generate consistent information on ecological and environmental urban issues (Blaschke et al., 2011).

When analysing urban deprivation and considering the main determinants of good versus poor living conditions of people, the importance of environmental quality is rather obvious. It also affects other livelihood dimensions such as health, which is an important constituent of human capital (Pearce et al., 2011; Rakodi and Lloyd-Jones, 2002). There is a growing body of literature attempting to map local climate zones (LCZ) that relates the urban structure/morphology to urban climatology (Kotharkar and Bagade, 2017; I. D. Stewart and Oke, 2012). Harlan et al.'s (2006) work on heat-related health inequalities in Phoenix (Arizona, USA) has clearly indicated that lower socioeconomic and ethnic minority groups were more likely to be exposed to heat stress because of the warmer neighbourhoods they live in. Absence of vegetation also affects the health and living environment negatively, as the presence of vegetation in or near residential areas has the capacity to reduce air pollution and produce cooling effects. Furthermore, the presence of vegetation contributes positively to the perceived quality of the living environment, while high built-up density normally has an opposite effect, which has for example been addressed by the local climate zone classification system (I. D. Stewart and Oke, 2012). Poor environmental quality thus adds to deprivations faced by the poor and is an indispensable component in the analysis of urban deprivation (Dahiya, 2012).

When measuring environmental quality, the constraints faced are both conceptual and methodological in nature. Poor environmental quality, in this chapter defined as environmental deprivation, is an outcome of the governance of the urban environment. It may, however, be conceptualised differently across space, time and discipline. The selection of indicators therefore requires a careful consideration of the purpose, the context and the geographic location. Methodologically, lack of adequate data is the main constraint, both in terms of temporal and spatial coverage. The census data, which forms the basis for most urban deprivation research (see Baud et al., 2008), hardly provide any information on the urban environment. The few existing environmental data sets, available in countries of the Global South, such as India, are either highly aggregated, inaccessible or represent too few measuring stations to allow any meaningful interpretation of the spatial pattern of environmental deprivation at city scale. Information extracted by remote sensing has a tremendous potential to fill this gap and support local authorities in gaining better insights into the geographical variation of environmental quality within an area, as well as in designing interventions to address that variation.

In response to this conceptual and methodological challenge, we have investigated the opportunities that standard remote sensing tools offer in identifying, mapping and monitoring the environmental quality of an urban setting. The methods were applied to Kalyan–Dombivli (KD) – a Class I Indian city* and one of the fastest growing cities in the Mumbai Metropolitan Region.

---

* Census of India classifies towns and cities as Class I (population > 100,000), Class II (population 50,000–99,999), Class III (population 20,000–49,999), Class IV (10,000–19,999), Class V (population 5,000–9,999) and Class VI (population < 5,000).

Methodologically, our research illustrates the capacity of remotely sensed information to measure environmental capital. Its outcome contributes to the construction of an index of environmental quality for KD as a proxy for natural capital, which has not been included in previous research (Baud et al., 2009) on the Index of Multiple Deprivation for Indian cities. Our research uses selected aspects of the urban environment that have emerged as major concerns in KD and are also recognised as relevant indicators of environmental quality in the literature (Li and Weng, 2007; Nichol and Wong, 2005).

4.2　SELECTION OF ENVIRONMENTAL ASPECTS FOR KALYAN–DOMBIVLI (KD)

KD is located in the Mumbai Metropolitan Region (MMR; see also Figure 4.2) – the largest and one of the fastest growing metropolitan regions in India (Baid, 2008). Among the four municipal corporations that constitute this region, KD (with an area of 137 km^2) has shown the highest growth in terms of population and built-up area since 1991 (Baid, 2008), reaching 1.1 million inhabitants in 2001 (Census of India, 2001) and 1.2 million in 2011 (Census of India, 2011).

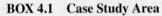

BOX 4.1　Case Study Area

Kalyan–Dombivli (KD), located in the Mumbai Metropolitan Region (a twin city and municipal corporation), is the largest and one of the fastest growing cities within this region (except Mumbai). The city covers an area of approximately 67 km^2 (Figure 4.1) and, according to the 2011 census, has a population of 1,247,327 inhabitants. The city is divided into eight administrative wards, which are relatively large units, so electoral wards have been used in this study. The area was purposely selected because around 44% of its population lives in deprived areas (e.g. slums or areas with slum-like conditions) and several anti-poverty and urban development programmes have been implemented there, e.g. the JNNURM sub-programme on Basic Service Provision for the urban poor and the Rajiv Awas Yojana (RAY) housing programme, which included the mapping of slums. KD has also been shortlisted by the government as a potential smart city. The analysis focuses on Kalyan and does not cover Dombivli.

These growth trends coupled with high real estate prices have led to the proliferation of slums. The city development plan (CDP), prepared as part of the Jawaharlal Nehru National Urban Renewal Mission (JNNURM), estimates that nearly 44% of the city population lives in slums or slum-like conditions. The 75-odd slums, spread over an area of 1.1 km^2, are home to nearly 0.1 million inhabitants (Subhash Patil and Associates, 2007). The 'slum-like housing', (locally known as *chawls*) accommodates an additional 0.4 million people. These areas can best be described as dilapidated, over-crowded, poorly lit and ventilated and lacking in basic public and social services. Their population densities have been increasing because of a constant inflow of people from rural areas. The slums have shown a tendency to grow vertically, by occupying added

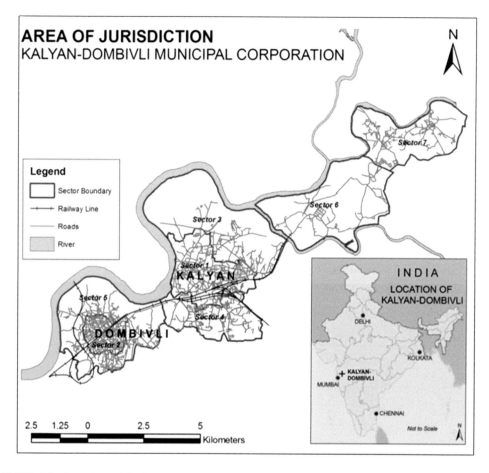

FIGURE 4.1 Location of Kalyan–Dombivli. (Source: GIS databases of Kalyan–Dombivli Municipal Corporation, 2007.)

stories, as well horizontally, by encroaching upon adjoining spaces. Baid (2008) has implicitly pointed towards the negative impacts that this haphazard urban development has on urban environmental quality, specifically increasingly less light and ventilation due to illegal increases in floor area ratios (FAR).

The KD municipal corporation has responded to the growing environmental concerns by carrying out several environmental surveys. These include an annual survey of the state of the environmental situation in the city (KDMC, 2008a) and a tree census (KDMC, 2008b). The urban local body (ULB) also measures air quality (e.g. SO_2, NO, CO) at seven locations within the city, and water quality of water bodies in the city. These efforts exist in isolation, however, and the effects that they may have, individually or collectively, on the habited areas or inhabitants cannot be ascertained. For example, four of the seven air quality sampling locations – a thermal power station, a solid-waste dumping site, a railway station and an *octroi* post – are among the most obvious sources of pollution. Pollutant levels there would give little idea of air quality in residential neighbourhoods. Air quality data from these locations were therefore not used in our study.

Conceptually, the urban environment can be classified into natural and man-made components (Douglas, 1983). They can further be broken down into several sub-components, as illustrated in Figure 4.2. The natural environment consists of wind, water and natural landscape, while the built environment is made up of infrastructure and the built-up landscape. With this concept (Figure 4.2),

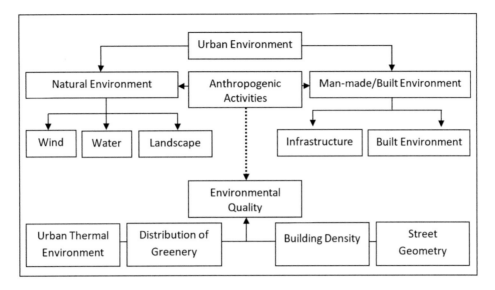

FIGURE 4.2 Main components of urban environmental quality for the case study area. (Adapted from Douglas, 1983.)

urban thermal environment, distribution of greenery, building density and *street geometry* were selected as the main components of the urban environment. Their selection was based on their importance as representatives of urban environmental quality in existing literature (KDMC, 2008a, 2008b; Subhash Patil and Associates, 2007), field observations and expert interviews with local planners, architects and urban designers.

For instance, Yue and Tan (2007) have considered thermal conditions as the most representative indicator of the urban environment. And as vegetation influences humidity, air quality and air temperatures, its presence is also very relevant in an urban setting (Weeks et al., 2007). Building density is a manifestation of human interventions on the natural landscape and therefore is an important aspect of deprivation (Baud et al., 2010). The relevance of building geometry and street orientation in terms of their influence on shade has been demonstrated for Singapore by Nichol (1996), who has also stressed that the high angle created by the sun in low latitude cities creates shorter shadow lengths and comparatively higher ground temperatures.

4.3 DATA SETS AND METHODOLOGY

The research presented in this chapter is aimed at establishing an index of environmental quality to enrich the index of multiple deprivation, as well as represent its spatial heterogeneity within cities. As mentioned in Section 4.2, the selection of indicators is based on the assessment of local environmental conditions and includes *urban thermal environment, distribution of greenery, building density* and *street geometry*. Land surface temperatures (LSTs), Normalised Difference Vegetation Index (NDVI) and roof density were used as indicators to measure the urban thermal environment, distribution of greenery and building density, respectively (see Figure 4.3).

BOX 4.2   Methods Applied in the Chapter

The analysis in this chapter shows how standard earth observation (EO) methods can be used to measure natural capital in support of adding environmental deprivation to the existing index of multiple deprivation. The applied EO methods include an analysis of the urban thermal environment, distribution of greenery, building density and street geometry. The software used for the analysis was Erdas Imagine™ and ArcGIS™, but it can be done in any standard remote sensing software, as well as in QGIS™. The data set consisted of satellite images, namely a thermal infrared image (TIR of Landsat, which is freely available), very-high resolution (VHR) stereo-imageries (Cartosat-1), high resolution (HR) imagery (Resourcesat) and locally available spatial data (e.g. building footprints). The primary data collection included field verification and expert interviews.

Potential field of application of the methods: analysis of environmental deprivation, urban heat islands or environmental discomfort by satellite imagery and standard EO methods.

Potential limitations: the methods rely on the availability of cloud-free imageries (TIR and VHR), which are often not readily available for a given period of time; lack of reliable local reference data.

These were extracted from the data sets listed in Table 4.1. To illustrate the impact of building and street orientation on environmental quality, building orientation and street geometry were investigated using land surface temperature maps (LSTs) and a digital elevation model (DEM) resampled to 3 m resolution; the latter was created from a pair of Cartosat-1 stereo pair images (2.5 m resolution), as indicated in Table 4.1. Being difficult to analyse for the entire city, this indicator was investigated for only a part of the city: it gave an indication of heat stress in different parts of the study area at the time of satellite overpass. The temporal inconsistency of available data sets is a limitation of this research, especially since KD is a rapidly developing city. Yet, as the main aim of the research was to show the potential advantages of adding environmental quality

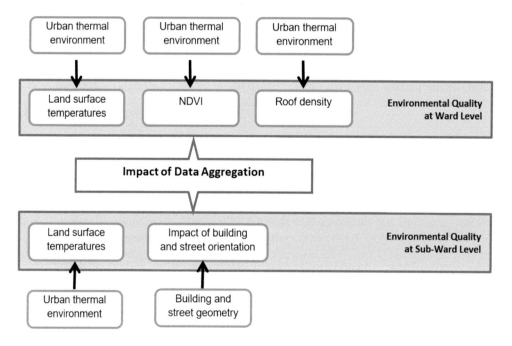

FIGURE 4.3   Environmental quality for two aggregation levels: ward and sub-ward (pixel) levels.

TABLE 4.1
Data Sources

Environmental Aspects	Data Set	Year	Source	Spatial Coverage
Urban thermal environment	Landsat ™ ETM+™	1999–2001	USGS[a]/GLCF[b]	Kalyan–Dombivli
Distribution of greenery	Resourcesat™ IRS P6™	2007	NRSA[+]	Most of Kalyan
Building density	Building footprint data – QuickBird™ image	2007	KDMC[*]	Kalyan–Dombivli
Street geometry (building and street orientation)	Cartosat-1™ stereo pair	2008	NRSA[+]	Kalyan–Dombivli
	Roads/streets	unknown	KDMC[*]	Kalyan–Dombivli

[a]United States Geological Survey; [b]Global Land Cover Facility; [*]Kalyan–Dombivli Municipal Corporation; [+]National Remote Sensing Centre.

to the index of deprivation by using remotely sensed imagery, on this aspect the research produces relevant results.

After computing the various indicators at the pixel level through remote sensing and spatial analysis (described in more detail in Sections 4.3.1–4.3.4), they were aggregated to the electoral ward level and combined using weighted summation (which includes interval standardization, assigning weights using expert knowledge, and aggregation) to form an index of environmental quality for that level. This was done to complement the index of multiple deprivations developed by Baud et al. (2009) for several Indian cities, which in its current form does not include environmental aspects and is based on 2001 census data. Not all data sets covered the entire administrative area of KD; therefore, the final index measures only those wards that are completely covered by all data sets.

4.3.1   URBAN THERMAL ENVIRONMENT

LSTs were extracted from Landsat 7 ETM+ thermal images (resampled to 30-metre spatial resolution) from different seasons (early summer and winter) to account for variability across seasons. Table 4.2 gives details of the Landsat image scenes.

For the extraction of LSTs from the Landsat thermal images, first the digital numbers (DN) were converted into spectral radiance. The radiances obtained were then converted into blackbody temperatures. These were corrected for emissivity, and then surface temperatures (in °C) were extracted following the approach documented by Nichol et al. (2006).

Conversion of DNs into spectral radiance was computed according to:

TABLE 4.2
Landsat Image Scenes

Data Set	Acquisition Date	Source
Landsat ETM+	20.10.99	USGS
	13.04.00	USGS
	23.11.00	USGS
	25.10.01	GLCF

$$L\lambda = \frac{L_{max} - L_{min}}{Q_{calmax} - Q_{calmin}} \times \left(Q_{cal} - Q_{calmin}\right) + L_{min} \tag{4.1}$$

where:

$L\lambda$ = spectral radiance;
Q_{calmin} = minimum DN (1);
Q_{calmax} = maximum DN (255);
Q_{cal} = DN of Band 6;
L_{min} = minimum spectral radiance (1);
L_{max} = Maximum Spectral Radiance (17.04).

Spectral radiance is converted to blackbody temperature by:

$$TB = \frac{K2}{\ln\left(\dfrac{k1}{L\lambda} + 1\right)} \tag{4.2}$$

where:

TB = effective blackbody temperature in °K;
K_1 = calibration constant 1 in $W\,m^{-2}\,sr^{-1}$ (666.09);
K_2 = calibration constant 2 in K (1282.7);
L_λ = spectral radiance in $W\,m^{-2}\,sr^{-1}$.

Surface temperature is corrected for emissivity in the following way:

$$Ts = T \Big/ \left(1 + \frac{\lambda T}{\alpha} \ln \varepsilon\right) \tag{4.3}$$

where:

Ts = land surface temperature in °C;
λ = the wavelength of emitted radiance;
α = hc/K (1.438×10^{-2} m K);
h = Planck's constant (6.26×10^{-34} J s);
c = the velocity of light (2.998×10^{-8} m s^{-1});
K = Bolzmann's constant (1.38×10^{-23} J K^{-1}).

4.3.2 Distribution of Greenery

The second indicator of environmental quality is the distribution of greenery. Remote sensing (Resourcesat IRS P6) is the major data source for mapping the extent of vegetation, as well as in addressing the quality aspects of vegetation or estimating the density of biomass. Our research therefore employed the commonly used Normalised Difference Vegetation Index (NDVI), which is considered a good indicator for the existence and quality of vegetation in an urban environment (see Lo et al., 1997; Rahman and Netzband, 2007). NDVI is the ratio between the red and the near-infrared band. It is expressed mathematically as:

$$NDVI = \frac{NIR - R}{\left(NIR + R\right)} \tag{4.4}$$

where:
NIR = the near-infrared band;
R = the red band.

NDVI ranges from −1 to 1, where the absence of green leaves results in a value below 0, while high positive values indicate high density of green leaves.

4.3.3 BUILDING DENSITY

Building density refers to the physical urban environment. A variety of methods for its computation are to be found in the literature, such as the building coverage ratio (BCR), floor area ratio (FAR) and roof coverage ratio (RCR). Pan et al. (2008) investigated the variation in building densities in Shanghai using high-resolution satellite images. They used BCR and FAR as estimates of density in two and three dimensions, respectively. Lindberg et al. (2003) used building intensity (BI) to assess variations in air and surface temperatures for Goteborg, Sweden. Several studies aiming to identify hotspots of urban deprivation have used RCR as an indicator as slum areas often have a continuous and highly dense roof coverage, which has implications for the living environment (see Kohli et al., 2012; Sliuzas and Kuffer, 2008; G. Stewart and Kuffer, 2007). RCR is computed as the ratio of total area covered by rooftops to the area of any administrative unit, such as wards. RCR can assist in the analysis of several physical aspects, such as ventilation and light. We use RCR as an indicator of building density. The data set used to calculate RCR was building footprints obtained from a 0.6 m spatial resolution QuickBird (QB) image (Figure 4.4). This data set, created by the Kalyan–Dombivli Municipal Corporation (KDMC) as part of its project on mapping the city for property taxation, has been updated by digitization to cover the complete study area. The roof coverage ratio was computed as:

$$\text{Roof coverage ratio} = \frac{\text{Total roof area in a ward}}{\text{Area of the ward}} \tag{4.5}$$

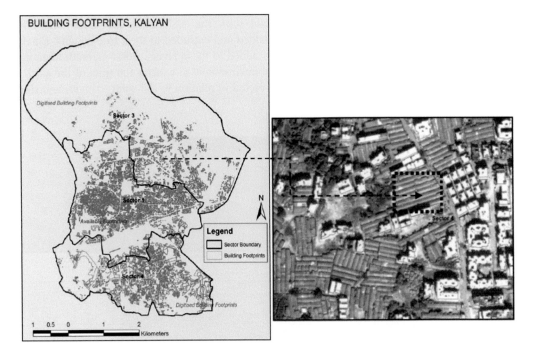

FIGURE 4.4 Building footprints, Kalyan–Dombivli.

4.3.4 STREET GEOMETRY

The fourth indicator of environmental quality is geometry, specifically building and street orientation. Its use assumes that apart from vegetation, building and street orientation also have an impact on the urban thermal environment (Kruger et al., 2011). Nichol and Wong (2005) have demonstrated that both parameters exert significant control over the thermal conditions of Hong Kong, especially in areas devoid of vegetation. In our analysis, the LST maps are overlaid on the 3D model and the road network, which allows temperature patterns to be visualised. Special consideration is given to orientation, in particular to how temperature varies where street orientation runs parallel to the solar azimuth as compared to other areas.

As mentioned earlier, the data set used for this analysis is the digital elevation model (DEM) computed from Cartosat-1 stereo pair images. Creation of the model required an accurate set of Ground Control Points (GCPs). Ideally these would have come from a Differential Global Positioning System (DGPS) survey, but since a DGPS instrument was not available, an alternative method was employed. Specifically, a set of 23 points was obtained that had resulted from a DGPS survey carried out previously by KDMC. Using these, a GPS survey akin to the functioning of DGPS was undertaken. Two hand-held Garmin GPSs were used for this purpose. One of the identified points from the set of 23 DGPS points was taken as the base station, at which the first hand-held unit was set up at the start of the survey. The instrument was set to record x,y,z locations continuously for the entire duration of the survey. With the help of the other GPS unit, x,y,z co-ordinates of certain pre-determined locations were recorded simultaneously, similar to the working of a rover GPS. After completion of the survey, the recorded points from the base, as well as rover stations, were processed. As part of this, the base station readings were converted into UTM and averaged for both days (the duration of this survey) separately. The mean readings were then compared to the actual DGPS reading to ascertain that the deviation was less than 0.5 m.

A similar procedure was carried out for the rover readings. These were corrected using the same error as for the base station. The corrected points were used as the GCPs for generating the DEM (resulting in 3 m resolution) in Leica Photogrammetric Suite™ of ERDAS Imagine™.

The DEM created (Figure 4.5) was overlaid with the LST maps for 2000 and 2001 (Subsection 4.3.1). The building footprints were added and extruded to their true heights. Finally, the road network (from the KDMC data set) was added to these layers. Solar azimuth and angle were set from the metadata to model the thermal environment at exactly the time of the satellite overpass. Thermal conditions were visually interpreted in relation to the solar azimuth and street and building orientation.

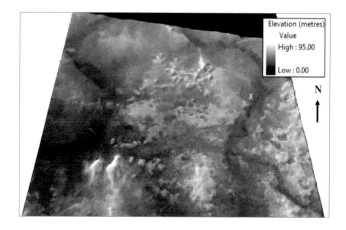

FIGURE 4.5   DEM computed from Cartosat-1 stereo pair images.

4.3.5 Data Aggregation

In India, an electoral ward is the smallest spatial unit at which political decisions are made. A review of the environmental situation at this level can thus highlight dominant issues in a ward and potentially lead to remedial measures. However, data aggregation to such a coarse scale (i.e. electoral wards) averages out the information of fine resolution provided by remote sensing images. Wards, despite being the smallest administrative/political units, are often heterogeneous in their physical characteristics, which in turn can strongly influence the assessment of their environmental quality. We therefore also explore the degree of spatial heterogeneity of selected environmental indicators within and across wards. Thus, an attempt was made to show environmental quality in a disaggregated manner (at the resolution of the imagery) vis-à-vis aggregated environmental condition that the particular ward had been depicting.

4.4 RESULTS AND DISCUSSION

4.4.1 Urban Thermal Environment

Figure 4.6a displays LSTs for October 1999. The temperature range is approximately 3°C. The cooler areas can be observed in the agricultural fields to the north of the city (shown by C_1 in Figure 4.6a)[*] and the river (C_2). Another cool patch can be seen in the central part (C_3). These are some of the planned areas with abundant vegetation. Higher temperatures can be observed towards the west of Central Kalyan[†] (W_1). From here an increasing trend towards the southern part can be seen (W_2). Comparatively little vegetation and a high proportion of built-up area are the reasons for this pattern. Having extracted LSTs from the remaining three images in a similar manner, mean LSTs were calculated at the ward level for all images. All four mean-temperature maps were

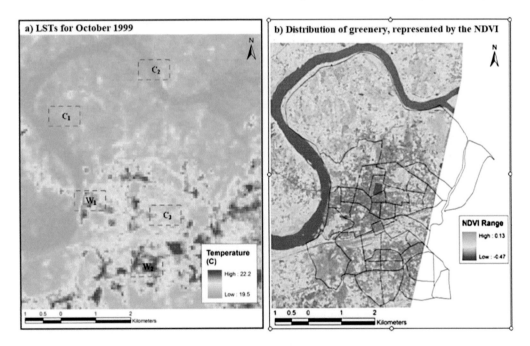

FIGURE 4.6 (a) Example of LSTs for October, 1999; and (b) distribution of greenery, represented by the NDVI, Kalyan.

---

[*] C_i and W_j on the map represent cool and warm areas, respectively.
[†] The analysis focuses on Kalyan and does not cover Dombivli.

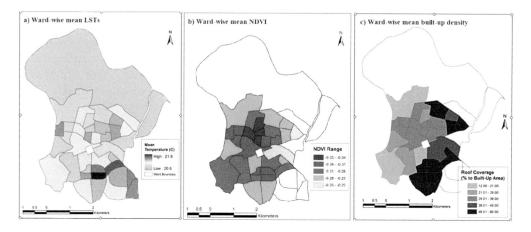

FIGURE 4.7   (a) Ward-wise mean LSTs, October 1999–October 2001; the white area in the middle is a water body; (b) NDVI aggregated at the ward level; the white area in the middle is a water body; and (c) ward-wise mean built-up density.

aggregated into a final temperature map over a period of approximately two years, as shown in Figure 4.7a. Even at the aggregated level of a ward, areas in the centre and south tend to have higher average temperatures than areas on the outskirts of the city.

4.4.2   Distribution of Greenery*

Distribution of greenery is represented by NDVI. Figure 4.6b illustrates that the mean NDVI values are highest in the few patches of vegetation in the city. The lowest values are seen in the water bodies. Similar observations can be made for the central part of the city, which is densely built. Towards the north of the centre, the index shows a slight increase – an area generally being used for agriculture. Like LSTs, the mean values of NDVI were also aggregated at the ward level (Figure 4.7b), which displays a spatial concentration of low values in the central area of the map, a result of very densely built-up areas that leave little or no space for vegetation, even in many planned areas.

4.4.3   Building Density

Building density is an indicator of conditions in the physical environment. We used roof coverage as an indicator for measuring building density. The results are shown in Figure 4.7c. The dark wards represent the areas of the highest building density. Overall, the eastern part displays higher densities than the western part. This is mainly related to the concentration of formal areas of high built-up density and large areas of slums in the east.

4.4.4   Street Geometry

Figure 4.8a shows building footprints and land surface temperatures for April 2000 overlaid on the DEM. Here the purpose was to illustrate the effects of building and street orientation on land surface temperature. To visualise details, a small area with no vegetation – shown in Figure 4.8b – was selected to illustrate the effect of street geometry on LSTs, in general giving high LSTs values.

---

* The scene from the IRS P-6 Resourcesat image, on which this indicator was based, did not cover the eastern part of Kalyan–Dombivli. Therefore, the NDVI values at ward level were aggregated only for wards fully covered.

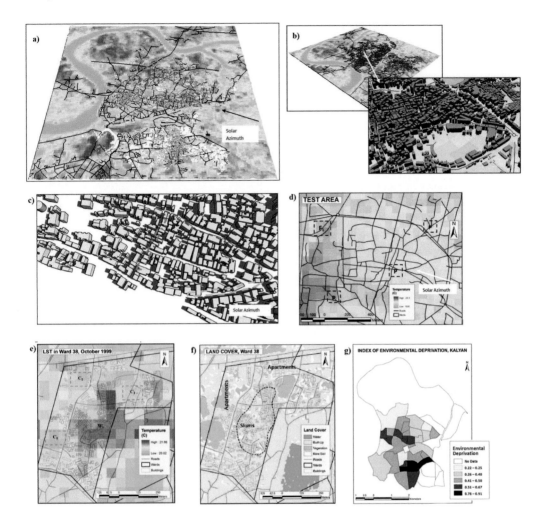

FIGURE 4.8 (a) DEM with LSTs and building footprints; (b) extruded building overlaid with LSTs (the zoomed area displays an area with no vegetation); (c) 3D view of selected area at 09:30 hours, to illustrate the thermal environment – shading; (d) 2D view of the selected area; (e) LSTs of electoral ward 38; (f) land cover of electoral ward 38; and (g) Index of Environmental Quality at electoral-ward level.

The selected area of Figure 4.8b centres on electoral wards 47, 48 and 49 in Kalyan. Located in the central part of the city, the area displays characteristics typical of a downtown Class I Indian city. Land use is generally a mix of residential, retail and small businesses. In terms of physical characteristics, a lot of old buildings with tiled roofs can be found. In recent times, however, a lot of new construction has also been taking place. The number of floor levels usually varies from one to four, although some of the new residential apartments have more floors. Figure 4.8d displays the area in a 2D view to show the variations in thermal conditions as a result of variations in street geometry. Temperature values for streets parallel to the solar azimuth are compared to those perpendicular to it. Several streets parallel to the solar azimuth have high temperatures, but even at this detailed level a clear relation between street geometry and surface temperatures is not visible for all streets. For example, in Figure 4.8d streets marked 'P' are those parallel to the solar azimuth. They display temperature values on the higher side and are roughly in the same temperature zone. However, higher values can also be seen along streets perpendicular to the solar azimuth. The street marked 'F₁' is one such example. High built-up densities dominate this area, where individual roofs are narrowly clustered. The effect of built-up density overrules the impact of the solar azimuth.

Thus, in cities with more irregular street layouts and high built-up density, the influence of built-up density has a stronger influence on the thermal environment (compared to the layout). Urban development in most parts of Kalyan is rather organic, reflected in the narrow and winding streets and buildings of all shapes and sizes, which affects the local thermal environment in varying degrees. For example, depending on the height of the buildings and the building cluster, different stretches of the same street may show different conditions. This is because a stretch may be exposed to the sun while another one could be shaded at the same point in time. See for example Figure 4.8c, which shows a part of the test area at 09:30 hours (13 April 2000), i.e. the time of Landsat overpass. The organic pattern of the streets is clearly visible. Stretches which are roughly parallel to the solar azimuth are not shaded (red broken lines, Figure 4.8c), while streets that are perpendicular are shaded and should be cooler (green broken lines, Figure 4.8c). These streets have buildings on either side, which provide protection. The organic pattern of the streets means that a few of them are neither parallel nor perpendicular to the solar azimuth. Thus, at this time they are partially shaded and partially in the sun (orange broken lines, Figure 4.8c).

From this analysis, one can conclude that in this case the method of assessing the relationship between street geometry and thermal environment does not solely explain temperature variations. Other factors, such as building orientation, building material and, in particular, built density and absence of vegetation, have a large impact on temperature variation. Nonetheless, as some examples show, streets parallel to the solar azimuth do show high temperatures.

4.4.5 Index of Environmental Quality

For constructing the index of environmental quality, the three indicators LST, NDVI and RCR were combined using weighted summation; street orientation was not considered because of the ambiguous results. Lower weights were assigned to both LST and NDVI (0.3 each), because temperatures are quite unpredictable and those taken from the satellite images only indicate thermal conditions at the time of image capture; NDVI, not having been validated on the ground (except for visual comparison at sub-ward level), was assigned a similar weight. The estimation of roof coverage, having been calculated from high resolution data, was considered most accurate and was thus given the highest weight. In other words, a rather pragmatic approach was taken for assigning the weights. A local adoption of the index for planning and decision support would need a more robust weighing that, for instance, makes use of the knowledge of local experts. Figure 4.8g shows the final index of environmental quality at ward level, with the highest levels of deprivation in the southern wards of the city. These areas are dominated by extremely dense clusters of buildings and almost no areas with vegetation cover.

4.4.6 Sub-Ward Level Analysis

The relevance of sub-ward level analysis has been shown by Baud et al. (2010), who explored the spatial heterogeneity of physical deprivation within wards using high resolution satellite images. They found that wards that ranked lowest in the index of multiple deprivation (IMD) still displayed a high diversity in physical deprivation. Building on the work of Baud et al. (2010), we explored the spatial heterogeneity of each indicator within wards.

Diversity in LSTs was quite evident (Figure 4.8e) within the selected ward – ward 38.* Higher temperatures can be seen in the eastern part (W_1). The western edge of the ward from south to north shows comparatively lower temperatures (C_1, C_2 and C_3). This can be explained by the presence of some patches of vegetation in the area. The higher temperatures, on the other hand, can be visually associated with built-up areas (Figure 4.8f). The displayed land cover map is based on a maximum

---

* Ward 38 has a moderate value in terms of environmental deprivation (at ward level) and was selected to explore diversity within the ward.

likelihood classification of the Resourcesat image with an overall classification accuracy of 83%. In terms of physical characteristics, formal residential areas are located along the western and northern edges of the ward. Slums and *chawls* occupy most of the central and eastern parts. A few new multi-storey apartments are being developed in the northern part of the ward.

4.5 DISCUSSION

In this chapter, we argue that environmental quality – as a proxy of natural capital – is an essential dimension when analysing multiple deprivation in urban areas (Baud et al., 2009). To address this issue, we (1) explored the capacity of remotely sensed data for measuring environmental quality through a set of environmental indicators, and (2) investigated the diversity of these indicators within and across electoral wards in the case study area.

With respect to the methodological challenge, the research utilised remote sensing techniques for extracting LSTs from thermal sensors, determining vegetation cover by calculating NDVI and deriving building density from building footprints. Standard techniques were coupled with a more innovative approach using a high-resolution stereo image to extract information on building heights for analysing the urban climate at micro-level (e.g. by highlighting the possible impact of building shading). The combination of a digital elevation model and 3D building footprints, applied to a part of the study area, proved to be useful for the spatial analysis as well as visualization. The visualization, in particular, illustrated the relationship between one environmental quality indicator (LSTs) and street and building geometry, as well as the relationship with built-up density – high density, built-up areas show high LSTs.

While the applied methods have generated satisfying results, there remains some scope for improvement in terms of methods for extracting different environmental indicators. For example, LST is a useful indicator for analysing temperature differences in an urban environment, but our research has also shown that LSTs derived from remote sensing images merely indicate the thermal conditions at the time of satellite overpass. Thus, extensive temporal data representing seasonal and, if possible, diurnal variations could improve the methodology. Other possible improvements in this domain could be mapping complete urban surface and quantifying the heat trapped by different building materials. Unfortunately, the coarse spatial resolution of space-borne thermal sensors limits a detailed analysis of urban–surface temperature variations.

Furthermore, analysis of the distribution of greenery in our research was based on the NDVI. The results support the capacity of NDVI for measuring green areas in the city with the given data. In general, Kalyan has little vegetation cover, which is a threat for the urban climate. However, to extract more detailed information about urban vegetation, techniques such as object-oriented classification (Blaschke et al., 2014; Hamedianfar and Shafri, 2015) for separating different kinds of urban vegetation or the LCZ approach (Bechtel et al., 2015; I. D. Stewart and Oke, 2012) could be applied, depending on the given context and available data.

The research has also highlighted some shortcomings in data collection by government agencies. Lack of coordination between agencies and inconsistency in data collection methods and spatial divisions often leads to outputs that are not shared and do not adequately represent the situation on the ground. The tree census and the environmental status report in Kalyan were a case in point. Despite their richness, their usefulness for planners and policy-makers was limited because of the mentioned shortcomings.

The major conceptual innovation of the research outlined in this chapter is the analysis of within-ward heterogeneity. This has been made possible by the details provided by remote sensing data – as well as the integration of indicators into the index of environmental quality at electoral ward level – thereby adding the missing dimension of natural capital to the livelihood framework (Moser, 1998) in urban areas. The research also illustrates the problems of aggregating electoral wards' environmental data, as environmental deprivation hotspots could only be recognised at a disaggregated level.

Overall, the results of the analysis demonstrate the capacity of the suggested environmental indicators for analysing differences within wards, as well as across wards. Individual environmental indicators and the index of environmental quality were found to be the worst in the southern part of the study area, i.e. the areas occupied mostly by slums and *chawls*. Nevertheless, environmental deprivation also exists in well-off areas, for example in central commercial areas with little vegetation cover.

The long-term implications and hardships caused by the aforementioned environmental aspects of deprivation will be more prominent in slum pockets than in well-off areas. Coupled with income deprivation and/or social deprivation, the inclusion of environmental deprivation puts the urban poor even further down the ladder of livelihood sustenance and vulnerability. At sub-ward level, environmental quality showed diversity that had been concealed due to the aggregation of data. The research in this chapter therefore not only demonstrates a methodology of measuring urban environmental quality using remotely sensed information but also examines its added value in assessing the spatial heterogeneity of environmental deprivation.

4.6 CONCLUSIONS

The research described in this chapter forms an entry point for linking urban environment and multiple deprivation within the livelihood framework. It looks at environmental quality from a relatively broad perspective and considers four indicators. Future research could explore each aspect in detail and analyse its role in relation to other deprivations. For example, urban thermal environment could be further studied in relation to linkages with the health of urban households. Space-borne derived information of the thermal environment could be validated and updated with detailed social surveys in neighbourhoods inhabited by different socio-economic groups. The effect of the thermal environment could then be analysed in terms of its effect on the health of residents of these neighbourhoods.

Policy-makers should be made aware of the paradigm shift of viewing poor households and their living conditions in a multi-dimensional manner, including environmental aspects. More attention should be given to improving both the built and natural environment of slums and *chawls*. The Indian slum surveys and the census of Below Poverty Line population have made an effort to include a diverse mix of indicators. It would be a constructive step to include indicators on environmental quality as well.

REFERENCES

Anguluri, R., & Narayanan, P. (2017). Role of green space in urban planning: Outlook towards smart cities. *Urban Forestry & Urban Greening, 25*, 58–65. doi:10.1016/j.ufug.2017.04.007

Baid, K. (2008). *Spatial Structure of a Subcity.* School of Planning and Architecture, New Delhi.

Baud, I. S. A., Sridharan, N., & Pfeffer, K. (2008). Mapping urban poverty for local governance in an Indian megacity: The case of Delhi. *Urban Studies, 45*(7), 1385–1412. doi:https://doi.org/10.1177/0042098008090679

Baud, I. S. A., Pfeffer, K., Sridharan, N., & Nainan, N. (2009). Matching deprivation mapping to urban governance in three Indian mega-cities. *Habitat International, 33*(4), 365–377. doi:https://doi.org/10.1016/j.habitatint.2008.10.024

Baud, I. S. A., Kuffer, M., Pfeffer, K., Sliuzas, R. V., & Karuppannan, S. (2010). Understanding heterogeneity in metropolitan India: The added value of remote sensing data for analysing sub-standard residential area. *International Journal of Applied Earth Observation and Geoinformation: JAG, 12*(5), 359–374. doi:https://doi.org/10.1016/j.jag.2010.04.008

Bechtel, B., Alexander, P., Böhner, J., Ching, J., Conrad, O., Feddema, J., … Stewart, I. (2015). Mapping local climate zones for a worldwide database of the form and function of cities. *ISPRS International Journal of Geo-Information, 4*(1), 199–219. doi:10.3390/ijgi4010199

Blaschke, T., Hay, G. J., Weng, Q., & Resch, B. (2011). Collective sensing: Integrating geospatial technologies to understand urban systems – an overview. *Remote Sensing, 3*(8), 1743–1776. doi:10.3390/rs3081743

Blaschke, T., Hay, G. J., Kelly, M., Lang, S., Hofmann, P., Addink, E., … Tiede, D. (2014). Geographic object-based image analysis – towards a new paradigm. *ISPRS Journal of Photogrammetry and Remote Sensing, 87*, 180–191. doi:http://dx.doi.org/10.1016/j.isprsjprs.2013.09.014

Census of India. (2001). *Kalyan and Dombivali City Census 2001 Data.* Retrieved from http://www.censusindia.gov.in/DigitalLibrary/Archive_home.aspx New Delhi, India:

Census of India. (2011). *Kalyan and Dombivali City Census 2011 Data.* Retrieved from http://www.censusindia.gov.in/DigitalLibrary/Archive_home.aspx New Delhi, India:

Dahiya, B. (2012). Cities in Asia, 2012: Demographics, economics, poverty, environment and governance. *Cities, 29*(Supplement 2), S44–S61. doi:https://doi.org/10.1016/j.cities.2012.06.013

Devas, N. (2004). *Urban Governance, Voice, and Poverty in the Developing World.* New York, USA: Earthscan LLC.

Douglas, I. (1983). *Urban Environment.* London, UK: Edward Arnold.

Hamedianfar, A., & Shafri, H. Z. M. (2015). Detailed intra-urban mapping through transferable OBIA rule sets using WorldView-2 very-high-resolution satellite images. *International Journal of Remote Sensing, 36*(13), 3380–3396. doi:10.1080/01431161.2015.1060645

Harlan, S. L., Brazel, A. J., Parshad, L., Stefanov, W. L., & Larsen, L. (2006). Neighbourhood microclimates and vulnerability to heat stress. *Social Science & Medicine, 63*(11), 2847–2863. doi:https://doi.org/10.1016/j.socscimed.2006.07.030

KDMC. (2008a). *Environmental Status Report.* Kalyan–Dombivli: Kalyan–Dombivli Municipal Corporation.

KDMC. (2008b). *Tree Census Report.* Kalyan–Dombivli: Kalyan–Dombivli Municipal Corporation.

Kohli, D., Sliuzas, R., Kerle, N., & Stein, A. (2012). An ontology of slums for image-based classification. *Computers, Environment and Urban Systems, 36*(2), 154–163. doi:https://doi.org/10.1016/j.compenvurbsys.2011.11.001

Kotharkar, R., & Bagade, A. (2017). Local Climate Zone classification for Indian cities: A case study of Nagpur. *Urban Climate, 24*, 369–392. doi:https://doi.org/10.1016/j.uclim.2017.03.003

Kruger, E. L., Minella, F. O., & Rasia, F. (2011). Impact of urban geometry on outdoor thermal comfort and air quality from field measurements in Curitiba, Brazil. *Building and Environment, 46*(3), 621–634. doi:10.1016/j.buildenv.2010.09.006

Li, G., & Weng, Q. (2007). Measuring the quality of life in city of Indianapolis by integration of remote sensing and census data. *International Journal of Remote Sensing, 28*(2), 249–267. doi:https://doi.org/10.1080/01431160600735624

Lindberg, F., Eliasson, I., & Holmer, B. (2003). *Urban geometry and temperature variations.* Paper presented at the 5th International Conference on Urban Climate, Łódź, Poland, 1–5 Sept.

Lo, C. P., Quattrochi, D. A., & Luvall, J. C. (1997). Application of high-resolution thermal infrared remote sensing and GIS to assess the urban heat island effect. *International Journal of Remote Sensing, 18*(2), 287–303. doi:https://doi.org/10.1080/014311697219079

McLennan, D., Barnes, H., Noble, M., Davies, J., & Garratt, E. (2011). *The English Indices of Deprivation 2010.* London, UK.

Moser, C. O. N. (1998). The asset vulnerability framework: Reassessing urban poverty reduction strategies. *World Development, 26*(1), 1–19. doi:https://doi.org/10.1016/S0305-750X(97)10015-8

Nichol, J. (1996). High-resolution surface temperature patterns related to urban morphology in a tropical city: A satellite-based study. *Journal of Applied Meteorology, 35*(1), 135–146. doi:https://doi.org/10.1175/1520-0450(1996)035<0135:HRSTPR>2.0.CO;2

Nichol, J., & Wong, M. S. (2005). Modeling urban environmental quality in a tropical city. *Landscape and Urban Planning, 73*(1), 49–58. doi:https://doi.org/10.1016/j.landurbplan.2004.08.004

Nichol, J., Wong, M. S., Fung, C., & Leung, K. K. M. (2006). Assessment of urban environmental quality in a subtropical city using multispectral satellite images. *Environment and Planning B: Planning and Design, 33*(1), 39–58. doi:https://doi.org/10.1068/b31195

Niggebrugge, A., Haynes, R., Jones, A., Lovett, A., & Harvey, I. (2005). The index of multiple deprivation 2000 access domain: a useful indicator for public health? *Social Science & Medicine, 60*(12), 2743–2753. doi:10.1016/j.socscimed.2004.11.026

Orford, S., Dorling, D., Mitchell, R. J., Shaw, M., & Smith, G. D. (2002). Life and death of the people of London: A historical GIS of Charles Booth's inquiry. *Health & Place, 8*(1), 25–35.

Pan, X. J., Zhao, Q. G., Chen, J., Liang, Y., & Sun, B. (2008). Analyzing the variation of building density using high spatial resolution satellite images: The example of Shanghai City. *Sensors, 8*, 2541–2550. doi:10.3390/s8042541

Pearce, J. R., Richardson, E. A., Mitchell, R. J., & Shortt, N. K. (2011). Environment justice and health: A study of multiple environmental deprivation and geographical inequalities in health in New Zealand. *Social Science & Medicine, 73*(3), 410–420. doi:10.1016/j.socscimed.2011.05.039

Rahman, A., & Netzband, M. (2007). *An assessment of urban environmental issues using remote sensing and GIS techniques. An integrated approach: A case study, Delhi, India.* Paper presented at the PRIPODE Workshop on Urban Population, Development and Environment Dynamics in Developing Countries, Nairobi, Kenya. http://www.cicred.org/pripode/CONF/NAIROBI/pdf/Atiqur-RAHMAN_paperNairobi2007.pdf

Rakodi, C., & Lloyd-Jones, T. (2002). *Urban Livelihoods: A People-Centered Approach to Reducing Poverty.* London, UK: Earthscan.

Sliuzas, R. V., & Kuffer, M. (2008). *Analysing the spatial heterogeneity of poverty using remote sensing: typology of poverty areas using selected RS based indicators.* Paper presented at the Remote Sensing: New Challenges of High Resolution, EARSeL 2008, joint workshop, Bochum, Germany. http://intranet.itc.nl/papers/2008/conf/sliuzas_ana.pdf

Stewart, G., & Kuffer, M. (2007). *Where are the poor? A disaggregation approach of mapping urban poverty.* Paper presented at the Proceedings of the 12th International Conference on Urban Planning and Spatial Development in the Information Society and 2nd International Vienna Real Estate Conference: REAL CORP 2007, Vienna, Austria. http://programm.corp.at/cdrom2007/archiv/papers2007/corp2007_STEWART.pdf

Stewart, I. D., & Oke, T. R. (2012). Local Climate Zones for urban temperature studies. *Bulletin of the American Meteorological Society, 93*(12), 1879–1900. doi:10.1175/bams-d-11-00019.1

Subhash Patil & Associates. (2007). *Kalyan Dombivli City Development Plan.* Kalyan–Dombivli: Kalyan–Dombivli Municipal Corporation.

Taubenböck, H., Esch, T., Felbier, A., Wiesner, M., Roth, A., & Dech, S. (2012). Monitoring urbanization in mega cities from space. *Remote Sensing of Environment, 117*(0), 162–176. doi:https://doi.org/10.1016/j.rse.2011.09.015

Weeks, J. R., Hill, A., Stow, D., Getis, A., & Fugate, D. (2007). Can we spot a neighborhood from the air? Defining neighborhood structure in Accra, Ghana. *GeoJournal, 69*(1–2). doi:https://doi.org/10.1007/s10708-007-9098-4

Yue, W., & Tan, W. (2007). The relationship between land surface temperature and NDVI with remote sensing: Application to Shanghai Landsat 7 ETM+ data. *International Journal of Remote Sensing, 28*(15), 3205–3226. doi:https://doi.org/10.1080/01431160500306906

5 Relationships between Outdoor Walking Levels and Neighbourhood Built-Environment Attributes
The Case of Older Adults in Birmingham, UK

Razieh Zandieh, Johannes Flacke, Javier Martinez, and Martin van Maarseveen

CONTENTS

5.1 Introduction 63
5.2 Case Study Area: Birmingham (UK) 65
5.3 Methods 66
    5.3.1 Data on Neighbourhood Built-Environment Attributes 67
    5.3.2 Data Analysis 68
        5.3.2.1 GIS Measures of Macro Built-Environment Attributes 68
        5.3.2.2 Statistical Correlations 70
5.4 Results 72
    5.4.1 Capacity of GIS for Measuring Macro Built-Environment Attributes 72
    5.4.2 Correlations among and between Neighbourhood Built-Environment Attributes 72
    5.4.3 Correlations among GIS-Measured Macro Built-Environment Attributes 74
    5.4.4 Correlations among Perceived Micro Built-Environment Attributes 74
    5.4.5 Correlations between GIS-Measured Macro and Perceived Micro
        Built-Environment Attributes 75
5.5 Discussion 76
5.6 Limitations of Our Study 78
5.7 Conclusion 79
References 79

5.1 INTRODUCTION

Outdoor walking is one of the forms of physical activity that takes place in urban areas. Outdoor walking may be undertaken for different purposes, such as walking as a means of transport and walking for recreation (Zandieh et al., 2017b). Walking has well-known positive impacts on people's health (Lee and Buchner, 2008; Sugiyama and Thompson, 2007), and for this reason people (especially those who are at risk for health problems, such as older adults) are advised to take outdoor walks (Age UK, 2016; Department of Health, 2011).

Previous studies (Handy et al., 2002; Saelens and Handy, 2008) have shown that levels (duration) of outdoor walking are related to neighbourhood built-environment attributes, which can divided into macro built-environment attributes and micro built-environment attributes (Sallis et al., 2011). Macro built-environment attributes, such as residential density, land-use mix and intensity, street connectivity and retail density, shape the overall design and structure of a neighbourhood (Sallis et al., 2011). Micro built-environment attributes, such as safety, pedestrian infrastructure and aesthetics, shape characteristics of walking routes in a neighbourhood (Sallis et al., 2011). Availability of Geographic Information System (GIS) and GIS-based land use and transportation data provides opportunities for developing objective measures of attributes of the macro built-environment (Leslie et al., 2007) and for better reflection about the actual built-environment. GIS-based data on micro built-environment attributes is, however, often unavailable (Leslie et al., 2007), and these built-environment attributes are usually measured subjectively (Sallis et al., 2011).

While significant relationships between outdoor walking levels and GIS-measured macro and perceived micro built-environment attributes have already been investigated (Sugiyama et al., 2015; Van Cauwenberg et al., 2012; Zandieh et al., 2016, 2017a), correlations among (within group) and between (between groups) these built-environment attributes themselves have received less attention in past research on outdoor walking. For example, Zandieh et al. (2016, 2017a) have shown that certain GIS-measured neighbourhood land use (green space and recreation centres) intensity – as macro built-environment attributes – and perceived neighbourhood safety, pedestrian infrastructure (quietness) and aesthetics – as micro built-environment attributes – are positively related to levels of outdoor walking of older adults in Birmingham (UK). They have also shown that certain GIS-measured neighbourhood land use (schools and industries) intensity and street connectivity – as macro built-environment attributes – are negatively related to outdoor walking levels of older adults in Birmingham. Similarly, other studies (Sugiyama et al., 2015; Van Cauwenberg et al., 2012) have found significant relationships between outdoor walking levels and GIS-measured macro and perceived micro built-environment attributes (e.g. street connectivity and safety, respectively) without investigating correlations among and between the neighbourhood built-environment attributes themselves.

It is likely, however, that neighbourhood built-environment attributes also influence each other mutually. For instance, it has been qualitatively found that neighbourhood residential density affects the amount of green space in neighbourhoods (Zandieh et al., 2017a), that high levels of traffic noise reduce perceived neighbourhood quietness (Zandieh et al., 2016) and that the presence of litter and broken windows (low aesthetic level) leads to perceptions of unsafety (Loukaitou-Sideris, 2006). These findings suggest that there could be correlations among macro (e.g. residential density and green space) and among micro (e.g. traffic conditions and quietness) built-environment attributes themselves.

Similarly, correlations may exist between macro and micro built-environment attributes: the presence of (macro) green space offers (micro) beautiful views (Zhou and Parves Rana, 2012), while conversely the presence of industries provides dreary views in neighbourhoods (Zandieh et al., 2016); and the presence of some types of land use in a neighbourhood, such as schools, generates traffic hazards (Zandieh et al., 2017a). Similarly, neighbourhood street connectivity (presence of junctions) affects perceptions of traffic conditions (Gomez et al., 2010) and perceived levels of traffic noise (Quartieri et al., 2010). Such correlations among and between neighbourhood built-environment attributes could influence significant relationships between outdoor walking levels and other neighbourhood built-environment attributes. In other words, some neighbourhood built-environment attributes would be related to outdoor walking levels through their correlations with other neighbourhood built-environment attributes.

Examining correlations among and between neighbourhood built-environment attributes would be helpful in undertaking urban planning interventions for supporting outdoor walking since it makes clear the effects of neighbourhood built-environment attributes on outdoor walking levels. As a consequence, urban planners have a better understanding of how to modify the built environment in ways that increase the positive (or decrease negative) influences of neighbourhood built-environment attributes on levels of outdoor walking.

Frank et al. (2005) have addressed correlations among GIS-measured macro built-environment attributes. However, correlations among perceived micro built-environment attributes have been rarely examined in studies on outdoor walking. It is important to examine correlations among perceived micro built-environment attributes, because people's perceptions of their environment may differ according to their personal details (e.g. age, gender and cultural background). Therefore, correlations among perceived micro built-environment attributes (and also between GIS-measured macro and perceived micro built-environment attributes) may differ for different social groups, including older adults.

The research we describe in this chapter had two aims: (1) to illustrate the capacity of GIS for generating objective measures of macro built-environment attributes; and (2) to use GIS-measured macro built-environment attributes, together with perceived micro built-environment attributes, to examine correlations among and between neighbourhood built-environment attributes in order to better understand how to increase positive (or to decrease negative) influences of neighbourhood built-environment attributes on outdoor walking levels. To achieve our second aim, we focused our research on one specific social group: older adults living in Birmingham (UK). We had three reason for this focus: inactivity has been identified as one of the biggest public health problems in Birmingham (Birmingham Public Health, 2017), inactivity is prevalent among older adults (WHO, 2017) and outdoor walking is an excellent form of physical activity for them (Centers for Disease Control and Prevention, 1999; Cunningham and Michael, 2004).

5.2   CASE STUDY AREA: BIRMINGHAM (UK)

Birmingham is located in the West Midlands of England, UK (Figure 5.1) and lies equidistant from the main ports of England (i.e. Bristol, Liverpool, Manchester and London; William, 2010). It consists of 40 electoral wards (Figure 5.1) and has a population over one million

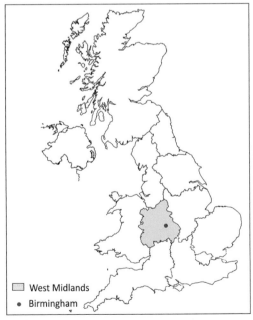
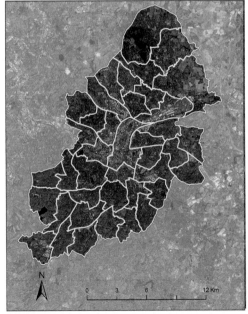

FIGURE 5.1   Left: Geographic locations of the West Midlands and Birmingham in the UK (adapted from (ONS Geography GIS and Mapping Unit, 2011). Right: Birmingham electoral wards (Google Earth™ mapping service (http://earth.google.com) and OS open data Boundary-Line® © Crown copyright/database right 2012).

inhabitants. It is the second-largest city in the UK. Birmingham is a major regional centre for business, retail and leisure activities (Brookes et al., 2012) and is also home to some of Britain's major manufacturing industries: motor vehicles, bicycles, electrical equipment, paints, guns and many other metal products (William, 2010). After Liverpool and Manchester, Birmingham has been identified as the third-most deprived UK Core City (Birmingham City Council, 2010).

Birmingham has a diverse population, with over 42% of its inhabitants being classified as belonging to an ethnic group other than 'white–British' (Birmingham City Council, 2013). Nearly one-third of Birmingham's residents are older adults, almost three-quarters of whom enjoy fairly good to good health, even though this figure is below regional and national averages (Birmingham City Council, 2017).

BOX 5.1   Case Study Area

Birmingham is located in the West Midlands of England (UK). It consists of 40 electoral wards and has a population of over one million inhabitants, making it is the second-largest city in the UK. It is a major regional centre for business, retail and leisure activities (Brookes et al., 2012) and is also home to some of the country's major industries (William, 2010). After Liverpool and Manchester, it has been identified as the third-most deprived UK Core City (Birmingham City Council, 2010). Birmingham has a diverse population, with over 42% of residents having an ethnicity other than 'white–British' (Birmingham City Council, 2013). Nearly one-third of Birmingham's residents are older adults, almost 75% of whom enjoy (fairly) good health, but this figure is below regional and national averages (Birmingham City Council, 2017).

5.3   METHODS

For this study, we made use of secondary data (Zandieh et al., 2016 and 2017a). Macro built-environment attributes were measured using GIS-based land use and transportation data (Zandieh et al., 2017a), while perceived micro built-environment attributes were measured by questionnaire (Zandieh et al., 2016). Detailed information on participant recruitment, data collection and the measurement process has been provided by Zandieh et al. (2016, 2017a). A brief description of data used in the study can be found further on in this section.

High- and low-deprivation areas* were surveyed by questionnaire in Birmingham in 2012, with a convenience sampling approach used to recruit participants from social centres (community centres, University of the Third Age, libraries, etc.). All participants chosen were 65 years or over, able to walk, leading independent daily lives, and mentally healthy; speaking language was not an eligibility criterion. The final sample included 173 participants, whose characteristics are summarised in Table 5.1.

---

* High-deprivation areas are urban areas experiencing high levels of social and economic disadvantage.

TABLE 5.1
Summary of Participants' Characteristics

Participants' Characteristics	Total Sample
No. of participants	173[a]
Age (%):	
75 years and over	48
65–74 years	52
Gender (%):	
Men	43
Women	57
Marital status (%):	
In a relationship	53
Single	47
Ethnicity (%):	
White–British	71
BME groups	29
Educational level (%):	
GCSE and higher	54
Sub-GCSE	35
Perceived health status (%):	
Good	92
Poor	7

Source: Data from Zandieh et al. (2016).

Note: [a] 80 and 93 participants from high- and low-deprivation areas of Birmingham, respectively. BME = black and minority ethnic groups. GCSE = General Certificates of Secondary Education or its equivalent.

5.3.1 DATA ON NEIGHBOURHOOD BUILT-ENVIRONMENT ATTRIBUTES

Data for measuring macro built-environment attributes (i.e. neighbourhood residential density, land-use mix and intensity, street connectivity and retail density) was downloaded from Digimap®/ EDINA (national data centre for UK academics) and the UK Office for National Statistics (for 2001 Census data) (see Table 5.2). This data was used to objectively measure macro built-environment attributes by using GIS (explained in Subsection 5.3.2).

To collect data on perceived micro built-environment attributes (i.e. neighbourhood safety, pedestrian infrastructure and aesthetics), a self-administered questionnaire was used (Zandieh et al., 2016). The questionnaire was based on the Neighbourhood Environment Walkability Scale (NEWS) (Saelens et al., 2003a) and further modified to be applicable for older adults in the UK. The questionnaire contained 29 items (those irrelevant to safety, pedestrian infrastructure and aesthetics in connection with walking were removed from the NEWS questionnaire), and each answer was rated on a six-point Likert scale. Seven subscales were included in the questionnaire: safety; five aspects of pedestrian infrastructure (i.e. traffic condition, pavement condition, presence of amenities (benches and public toilets), quietness and air quality); and aesthetics. A NEWS scoring protocol (Saelens et al., 2003a) was used to score the subscales, which means that higher scores indicate

TABLE 5.2

Data Collected for Measuring Macro Built-Environment Attributes

Macro Built-Environment Attribute	Data (Data Source)
Residential density	Number of household spaces (UK Census, 2001): Number of households in each LSOA[b]
	LSOAs[b] Boundary-Line 2012 (Digimap®/EDINA)[a]: GIS-based data (polygons) representing boundary of LSOAs
Land-use mix and intensity, and retail density	OS Points of Interest (PoI) 2016 (Digimap®/EDINA)[a]: GIS-based data (points) representing locations of non-residential use (retail, catering, etc.)
	OS MasterMap 2016–Topography Layer (Digimap®/EDINA)[a]: The most accurate GIS-based data representing physical features (e.g. roads and buildings) in the UK
Street connectivity	OS MasterMap 2016–Integrated Transport Network (ITN) layer (Digimap®/EDINA)[a]: GIS-based data on road network in the UK
	OS MasterMap 2016–Urban Path Theme (UP) layer (Digimap®/EDINA)[a]: GIS-based data on urban path network appropriate for non-vehicular users (e.g. footpath, subway, cycle path)

Source: Adapted from Zandieh et al., (2017a).

Note: Data on neighbourhood land-use mix and intensity was also used for identifying residential land use and measuring residential density. [a] Digimap®/EDINA is the national data centre for UK academics. [b] LSOA = Lower Super Output Area (a relatively homogenous geographic area with population around 1500 residents (McLennan et al., 2011; Payne, 2012)); OS = Ordnance Survey.

a greater appreciation of micro built-environment attributes – safety, pedestrian infrastructure and aesthetics.

5.3.2   DATA ANALYSIS

5.3.2.1   GIS Measures of Macro Built-Environment Attributes

A Geographic Information System (GIS) – ArcGIS 10.4, ESRI™, Redlands, CA, USA – was used to objectively measure macro built-environment attributes. These neighbourhood built-environment attributes were measured within a home-based neighbourhood – i.e. a 2-km Euclidean buffer area around each participant's home (Figure 5.2a) (Zandieh et al., 2017a). This buffer area is known to be the one most commonly used for taking outdoor walks (Suminski et al., 2014), and each participant's outdoor walking level was measured within this area (Zandieh et al., 2016). The 'buffer' analysis tool in Arc GIS was used to delineate this area for each participant.

Points of Interest (PoI) data (Table 5.2) was used to create a land use map distinguishing residential and non-residential land use. By using ArcGIS, first PoI were overlain with the Topography Layer (Table 5.2). By considering one use for each building (Rodriguez et al., 2009), non-residential buildings were identified and their relevant plots were digitised. After excluding non-residential buildings and their plots, residential buildings were identified and their relevant plots digitised (Figure 5.2b). All identified boundaries were cross-referenced against images from Google Earth. Area of residential land use (calculated by statistical analysis with the assistance of ArcGIS), data on the number of household spaces and LSOAs Boundary-Line 2012 (Table 5.2), and an equation presented by Zandieh et al. (2017a), were used to measure residential density within each participant's home-based neighbourhood:

$$RD = \sum_{i=1}^{n} k_i D_i$$

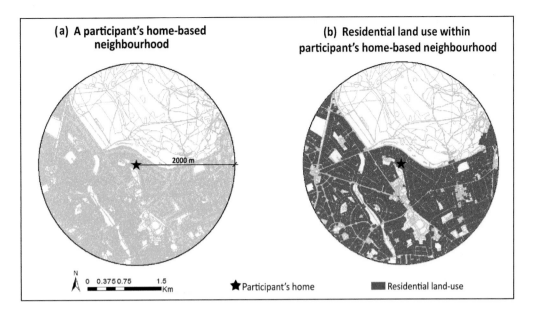

FIGURE 5.2 (a) A participant's home-based neighbourhood, and (b) residential land use within that participant's home-based neighbourhood.

where:

RD = residential density for a home-based neighbourhood

i   = the LSOA

D_i = residential density of the LSOA (total number of household space in the LSOA/total LSOA's residential land use area (hectare))

k_i = the proportion of the LSOA's residential land use area (hectare) located within a home-based neighbourhood against total LSOA's residential land use area (hectare)

n   = the number of LSOAs

(Zandieh et al., 2017a, p.6)

For measuring neighbourhood land-use mix, areas of residential and non-residential land use (including catering (eating/drinking) activities, green spaces, recreation centres, social infrastructure, retail plots and buildings, schools and industries; see Figure 5.3) were used. To generate a land-use mix entropy score, the equation used by Stockton et al. (2016) for calculating land-use mix (or "mixedupness") for London was employed. For measuring land-use diversity, area of non-residential land uses were employed (Zandieh et al., 2017a). Areas of retail plots and buildings within participants' home-based neighbourhoods (Figure 5.3) were measured with the assistance of statistical analysis in ArcGIS and used to calculate neighbourhood retail densities (Zandieh et al., 2017a).

To measure neighbourhood street connectivity, two layers of OS MasterMap (Table 5.2) were employed (Stockton et al., 2016; Zandieh et al., 2017a): an ITN layer (data on the road network of the UK); and a UP layer (data on the urban path network suitable for non-vehicular users; see Figure 5.4a, b). By using ArcGIS, motorways and slip roads were selected and excluded from the ITN layer, as walking on them is prohibited (Department of Transport, 2016; Zandieh et al., 2017a). The 'Network Analyst' extension in ArcGIS was used to combine the two layers to generate a pedestrian-route network data set (Figure 5.4c). The points generated in the set represent junctions (Figure 5.4d). Junctions that connect three or more roads/paths within participants' home-based neighbourhoods were identified and counted. Junction density

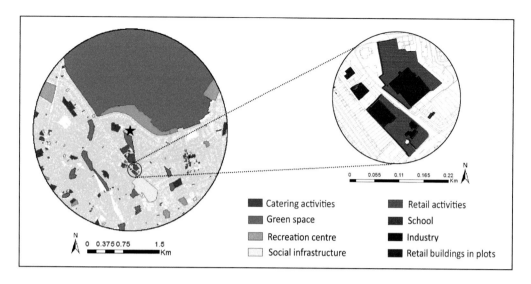

FIGURE 5.3   Non-residential land use within the participant's home-based neighbourhood and a layout of retail plots and buildings.

was then calculated and used as an indicator for neighbourhood street connectivity (Zandieh et al., 2017a).

5.3.2.2   Statistical Correlations

Pearson correlations among and between neighbourhood built-environment attributes were calculated using IBM® SPSS Statistics 24. For this purpose, the logarithmic transformation of data $(x + 1)$ was used in order to improve normality in the data. A p-value < 0.05 was considered statistically significant.

To be able to clearly interpret the degree of correlations among and between neighbourhood built-environment attributes, the strength of Pearson correlations was classified as being either high, fair or weak (see Table 5.3); we used Navarro (2013) as a guide when interpreting correlations. Pearson correlations with p-values ≥ 0.05 were considered to be insignificant correlations. Descriptive statistics were used to compare strength of correlations among and between neighbourhood built-environment attributes.

BOX 5.2   Methods Applied in the Chapter

The methods section of this chapter illustrates the capacity of GIS – ArcGIS 10.4 (ESRI, Redlands, CA) – for generating objective measures of one group of neighbourhood built-environment attributes: macro built-environment attributes (residential density, land-use mix and intensity, street connectivity and retail density). The methods section describes how to combine GIS-based land-use and transportation data by applying simple GIS techniques, such as overlaying data, buffer analysis, digitizing plots, statistical analysis and the 'Network Analyst' (extension) in ArcGIS. Data included the number of household spaces; LSOAs Boundary-Line 2012; OS Points of Interest (PoI) 2016; OS MasterMap 2016–Topography Layer; OS MasterMap 2016–Integrated Transport Network (ITN) layer; and OS MasterMap 2016–Urban Path Theme (UP) layer. Data was collected from the UK Census 2001 and Digimap/EDINA (the national data centre for UK academics).

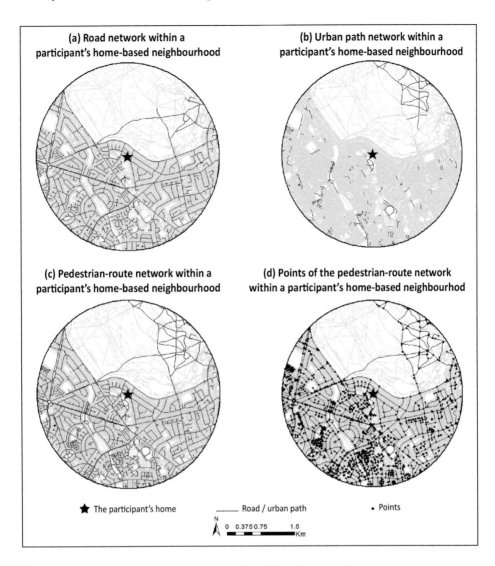

FIGURE 5.4   Generating a pedestrian-route network and identifying junctions.

TABLE 5.3

Classification of Pearson Correlations

Correlation strength	Range correlation coefficient
High	± 0.70 to ± 1.00
Fair	± 0.40 to ± 0.70
Weak	± 0.00 to ± 0.40

The GIS method applied in this chapter has some limitations:

- Data from different years (i.e. 2001, 2012 and 2016) was used.
- Generated data for residential and non-residential land use is coarse because only one use was assigned to each building; some buildings may have more than one use.
- Defined home-based neighbourhood (2-km buffer) for generating GIS-based macro built-environment attributes is not necessarily the same as a respondent's perception of 'neighbourhood' when measuring perceived micro built-environment attributes.

5.4   RESULTS

5.4.1   Capacity of GIS for Measuring Macro Built-Environment Attributes

Subsection 5.3.2 illustrates the capacity of GIS for generating objective measures of macro built-environment attributes. It describes how to use ArcGIS and point data on locations of non-residential uses, from which a land-use map can be generated. It also shows how applying ArcGIS techniques – such as buffer analysis, overlaying data, digitizing, statistical analysis, selecting and extracting data (e.g. non-residential plots and motorways) and Network Analyst – facilitates manipulating and combining data to generate objective measures of macro built-environment attributes.

5.4.2   Correlations among and between Neighbourhood Built-Environment Attributes

Table 5.4 summarises the strength of correlations among and between neighbourhood built-environment attributes. The correlations themselves are presented in Tables 5.5 through 5.7. In total, 153 correlations were examined (including 55 correlations among/within GIS-measured macro built-environment attributes, 21 correlations among/within perceived micro built-environment attributes and 77 correlations between GIS-measured macro and perceived micro built-environment attributes); see Table 5.4.

As Table 5.4 shows, most high correlations were found within GIS-measured macro built-environment attributes. There is no high correlation among perceived micro built-environment attributes. Most correlations among perceived micro built-environment attributes are fair correlations. Almost half of the correlations between GIS-measured macro and perceived micro attributes are weak correlations. Nevertheless, there are some high and fair correlations between GIS-measured

---

TABLE 5.4

Correlation Strength Among/Within and Between GIS-Measured Macro and Perceived Micro Built-Environment Attributes

Correlations	Within Macro	Within Micro	Between Macro and Micro	Total No. of Correlations
Number of correlations (n)	55	21	77	153
High correlation (%)	38	0	7	17
Fair correlation (%)	35	67	27	35
Weak correlation (%)	20	29	48	35
Insignificant correlation (%)	7	4	18	12

Note: Macro=GIS-measured macro built-environment attributes, micro=perceived micro built-environment attributes.

TABLE 5.5

Correlations Among GIS-Measured Macro Built-Environment Attributes

GIS-Measured Macro Built- Environment Attribute	Land-Use Mix	Catering activities	Green Space	Recreation Centres	Social Infrastructure	Retail Activities	Schools	Industries	Street Connectivity	Retail Density
					Land Use Intensity					
Residential density	0.24 *0.002*	0.85 *0.000*	−0.44 *0.000*	−0.81 *0.000*	0.62 *0.000*	0.89 *0.000*	0.86 *0.000*	0.79 *0.000*	0.91 *0.000*	0.61 *0.000*
Land-use mix		0.36 *0.000*	0.19 *0.014*	−0.40 *0.633*	0.57 *0.000*	0.37 *0.000*	0.15 *0.057*	0.07 *0.366*	0.35 *0.000*	0.34 *0.000*
Land-use intensity:			−0.36 *0.000*	−0.65 *0.000*	0.59 *0.000*	0.81 *0.000*	0.82 *0.000*	0.59 *0.000*	0.85 *0.000*	0.82 *0.000*
Catering activities										
Green space				0.58 *0.000*	0.00 *0.996*	−0.47 *0.000*	−0.54 *0.000*	−0.73 *0.000*	−0.58 *0.000*	−0.16 *0.039*
Recreation centres					−0.33 *0.000*	−0.78 *0.000*	−0.79 *0.000*	−0.84 *0.000*	−0.78 *0.000*	−0.43 *0.000*
Social infrastructure						0.54 *0.000*	0.41 *0.000*	0.21 *0.005*	0.54 *0.000*	0.48 *0.000*
Retail activities							0.81 *0.000*	0.83 *0.000*	0.90 *0.000*	0.62 *0.000*
Schools								0.77 *0.000*	0.86 *0.000*	0.67 *0.000*
Industries									0.84 *0.000*	0.36 *0.000*
Street connectivity										0.61 *0.000*

Note: This table shows Pearson correlation values; *p*-values are in *italics*. Dark-grey cells show high correlations (correlation coefficient: ± 0.70 to ±1.00); grey cells show fair correlations (correlation coefficient: ± 0.40 to ± 0.70); light-grey cells show weak correlations (correlation coefficient: ± 0.00 to ± 0.40); and white cells show insignificant correlations.

TABLE 5.6

Correlations Among Perceived Micro Built-Environment Attributes

Perceived Micro Built- Environment Attribute	Pedestrian Infrastructure					
	Traffic Conditions	Pavement Condition	Presence of Amenities	Quietness	Air Quality	Aesthetics
Safety	0.55	0.38	0.17	0.43	0.31	0.47
	0.000	*0.000*	*0.023*	*0.000*	*0.000*	*0.000*
Pedestrian infrastructure:		0.54	0.31	0.41	0.41	0.58
Traffic conditions		*0.000*	*0.000*	*0.000*	*0.000*	*0.000*
Pavement condition			0.21	0.46	0.45	0.52
			0.006	*0.000*	*0.000*	*0.000*
Presence of amenities				0.21	0.10	0.43
				0.006	*0.178*	*0.000*
Quietness					0.40	0.52
					0.000	*0.000*
Air quality						0.41
						0.000

Note: This table shows Pearson correlation values; *p*-values are in *italics*. Grey cells show fair correlations (correlation coefficient: ± 0.40 to ± 0.70); light-grey cells show weak correlations (correlation coefficient: ± 0.00 to ± 0.40); and white cells show insignificant correlations.

macro and perceived micro built-environment attributes. Detailed results on all these correlations are presented in the following subsections.

5.4.3 CORRELATIONS AMONG GIS-MEASURED MACRO BUILT-ENVIRONMENT ATTRIBUTES

Table 5.5 summarises the correlations among GIS-measured macro built-environment attributes. Except for green space and recreation centre intensities, all GIS-measured macro built-environment attributes are positively correlated with each other. GIS-measured green space and recreation centre intensities are positively correlated with each other, but there are negative correlations between these two GIS-measured land-use (i.e. green space and recreation centres) intensities and the other GIS-measured macro built-environment attributes of residential density, intensity of other land uses (e.g. retail activities, schools and industries), street connectivity and retail density. This means that as most GIS-measured macro built-environment attributes – e.g. residential density and the land area dedicated to industries) – increase, the land area dedicated to green space and recreation centres may shrink.

5.4.4 CORRELATIONS AMONG PERCEIVED MICRO BUILT-ENVIRONMENT ATTRIBUTES

Table 5.6 shows correlations among perceived micro built-environment attributes. Except for the correlation between presence of amenities and air quality, there are positive correlations among all perceived micro built-environment attributes. Perceived aesthetics is the only perceived micro built-environment attribute that correlates fairly with all other micro built-environment attributes. After aesthetics, traffic conditions are fairly correlated with other perceived micro built-environment attributes (except for one aspect of pedestrian infrastructure: presence of amenities). Therefore, according to our study, as positive perceptions of traffic conditions improve, with the exception of presence of amenities (one aspect of pedestrian infrastructure), positive perceptions of the other perceived micro built-environment attributes also improve.

5.4.5 CORRELATIONS BETWEEN GIS-MEASURED MACRO AND PERCEIVED MICRO BUILT-ENVIRONMENT ATTRIBUTES

Table 5.7 gives an overview of correlations between GIS-measured macro and perceived micro built-environment attributes. In general, most GIS-measured macro built-environment attributes (i.e. residential density, land-use intensity (but not social infrastructure intensity), street connectivity, and retail density) are correlated with most perceived micro built-environment attributes. GIS-measured land-use mix is the only GIS-measured macro built-environment attribute that is

TABLE 5.7
Correlations Between GIS-Measured Macro and Perceived Micro Built-Environment Attributes

GIS-Measured Macro Built-Environment Attribute	Perceived Micro Built-Environment Attribute						
		Pedestrian Infrastructure					
	Safety	Traffic Conditions	Pavement Condition	Presence of Amenities	Quietness	Air Quality	Aesthetics
Residential density	−0.37	−0.39	−0.40	−0.32	−0.42	−0.29	−0.73
	0.000	0.000	0.000	0.000	0.000	0.000	0.000
Land-use mix	−0.50	0.01	−0.02	0.10	0.01	0.02	−0.07
	0.548	0.856	0.780	0.196	0.927	0.784	0.358
Land-use intensity: Catering activities	−0.39	−0.35	−0.28	−0.31	−0.30	−0.28	−0.62
	0.000	0.000	0.000	0.000	0.000	0.000	0.000
Green space	0.25	0.33	0.31	0.38	0.25	0.12	0.56
	0.001	0.000	0.000	0.000	0.001	0.130	0.000
Recreation centres	0.43	0.47	0.47	0.28	0.48	0.33	0.75
	0.000	0.000	0.000	0.000	0.000	0.000	0.000
Social infrastructure	−0.09	−0.11	−0.09	0.09	−0.19	−0.02	−0.31
	0.245	0.165	0.218	0.256	0.012	0.811	0.000
Retail activities	−0.41	−0.37	−0.40	−0.23	−0.39	−0.28	−0.67
	0.000	0.000	0.000	0.003	0.000	0.000	0.000
Schools	−0.44	−0.44	−0.36	−0.37	−0.40	−0.27	−0.71
	0.000	0.000	0.000	0.000	0.000	0.000	0.000
Industries	−0.43	−0.45	−0.49	−0.37	−0.41	−0.31	−0.75
	0.000	0.000	0.000	0.000	0.000	0.000	0.000
Street connectivity	−0.44	−0.41	−0.38	−0.34	−0.36	−0.25	−0.73
	0.000	0.000	0.000	0.000	0.000	0.001	0.000
Retail density	−0.26	−0.20	−0.15	−0.17	−0.16	−0.21	−0.40
	0.000	0.008	0.052	0.024	0.033	0.005	0.000

Note: This table shows Pearson correlation values; *p*-values are in *italics*. Dark-grey cells show high correlations (correlation coefficient: ± 0.70 to ± 1.00); grey cells show fair correlations (correlation coefficient: ± 0.40 to ± 0.70); light-grey cells show weak correlations (correlation coefficient: ± 0.00 to ± 0.40); and white cells show no correlation. The thick black boundary contains positive significant correlations. All significant correlations outside the thick black boundary are negative correlations.

insignificantly correlated with all perceived micro built-environment attributes. Interestingly, the correlation between GIS-measured green space intensity and perceived air quality is insignificant.

As Table 5.7 shows, excluding GIS-measured green space and recreation centre intensities (thick black boundary), all significant correlations between GIS-measured macro and perceived micro built-environment attributes are negative. Therefore, as GIS-measured macro built-environment attributes (excepting green space and recreation centre intensities) increase, perception of micro built-environment attributes, for example quietness and aesthetics, may decline.

5.5 DISCUSSION

It is important for urban planners to identify relationships between outdoor walking and neighbourhood built-environment attributes, both macro and micro. In doing so, GIS provides urban planners with an opportunity to study these relationships by objectively measuring neighbourhood built-environment attributes. This method is mostly used to measure macro built-environment attributes, since GIS-based land use and transportation data is often available. This chapter shows how applying simple GIS techniques – overlaying data, buffer analysis, digitizing plots, statistical analysis and network analyst in ArcGIS – can help urban planners to integrate various data and produce measures of macro built-environment attributes. Similar methods (applying GIS) for measuring macro built-environment attributes have already been used in studies on outdoor walking (or physical activities) of adults in a number of countries, including the UK (Stockton et al., 2016), the USA (Frank et al., 2005), Australia (Gebel et al., 2009) and New Zealand (Hinckson et al., 2017)).

This chapter also used the generated GIS-measured data to study correlations among and between neighbourhood built-environment attributes in the hope of helping urban planners to better understand how to increase the positive (or decrease the negative) influences of neighbourhood built-environment on outdoor walking.

The results of our study demonstrate strong correlations within GIS-measured macro built-environment attributes that are consistent with previous research (Frank, 2000; Frank et al., 2005). Comparing perceived micro built-environment to GIS-measured macro built-environment attributes, in our study, the correlations among perceived micro built-environment attributes are less strong. Moreover, GIS-measured macro built-environment attributes (except for neighbourhood land-use mix and – largely – social infrastructure intensity) are correlated with perceived micro built-environment attributes.

These correlations may influence relationships between neighbourhood built-environment attributes and the outdoor walking levels of survey participants. Findings of previous studies on significant relationships between outdoor walking levels of older adults and neighbourhood built-environment attributes in Birmingham, UK (Zandieh et al., 2016, 2017a) are presented in Table 5.8. Figure 5.5 uses these findings to show in visual form significant relationships between neighbourhood built-environment attributes and older adults' outdoor walking levels, as well as correlations among and between neighbourhood built-environment attributes.

Figure 5.5 includes all high, fair and weak correlations between built neighbourhood attributes. In Boxes A, B and C, each built-environment attribute is positively correlated with at least one built-environment attribute within the box (there are no negative correlations within these boxes). The dark boxes show older adults' outdoor walking levels and neighbourhood built-environment attributes that are significantly related to older adults' outdoor walking levels. Arrows show significant relationships and lines show correlations: '+' shows positive relationships/correlations and '–'shows negative relationships/correlations. GIS-measured land-use mix is related to outdoor walking levels only in areas where levels of deprivation are low (represented by a white box and a broken line/arrow).

Our study shows that neighbourhood built-environment attributes that are significantly related to outdoor walking levels of older adults – i.e. GIS-measured land-use intensity (green space, recreation centres, schools and industries) and street connectivity, and perceived safety, pedestrian infrastructure (quietness) and aesthetics – are correlated with each other (Table 5.7 and Figure 5.5) as

TABLE 5.8

Summary of Previous Studies in Birmingham (UK) on Relationships Between Outdoor Walking Levels of Older Adults and Neighbourhood Built-Environment Attributes (Data from Zandieh et al. 2016, 2017a)

Neighbourhood Built-Environment Attributes	Positive Relationship With Walking[a]	Negative Relationship With Walking[b]	No Relationship With Walking[c]
GIS-Measured Macro Built-Environment Attributes:			
Residential density			*
Land-use mix			*[d]
Land-use intensity			
Catering activities			*
Green space	*		
Recreation centres	*		
Social infrastructure			*
Retail activities			*
Schools		*	
Industries		*	
Street connectivity		*	
Retail density			*
Perceived Micro Built-Environment Attributes:			
Safety	*		
Pedestrian infrastructure			
Traffic condition			*
Pavement condition			*
Presence of amenities			*
Quietness	*		
Air quality			*
Aesthetics	*		

Notes: [a] Positive significant relationship between older adults' outdoor walking levels and neighbourhood built-environment attribute, independent of area deprivation.

[b] Negative significant relationship between older adults' outdoor walking levels and neighbourhood built-environment attribute, independent of area deprivation.

[c] No significant relationship between older adults' outdoor walking levels and neighbourhood built-environment attribute.

[d] GIS-measured land-use mix is positively related to older adults' outdoor walking levels only in areas where deprivation levels are low.

well as with other neighbourhood built-environment attributes (Figure 5.5, Boxes B and C). These findings support previous studies that discuss relations between land use and safety (Stucky and Ottensmann, 2009), land use and noise (Sanz et al., 1993; Wolch et al., 2014), land use and aesthetics (Zhou and Parves Rana, 2012) and perceived safety and aesthetics (Loukaitou-Sideris, 2006). Negative correlations between GIS-measured street connectivity and perceived aesthetics may be explained by findings of previous studies: high levels of street connectivity may provide short, direct routes (Saelens et al., 2003b), but direct routes may offer walkers fewer 'visual mysteries' and have less visual appeal than curved streets (Nasar and Cubukcu, 2011).

Correlations among and between neighbourhood built-environment attributes may reinforce (or attenuate) the influence of other neighbourhood built-environment attributes on outdoor walking levels. For example, increasing positive aesthetic perceptions (by, for example, planting trees in streets) may attenuate the negative impacts of GIS-measured street connectivity on outdoor walking

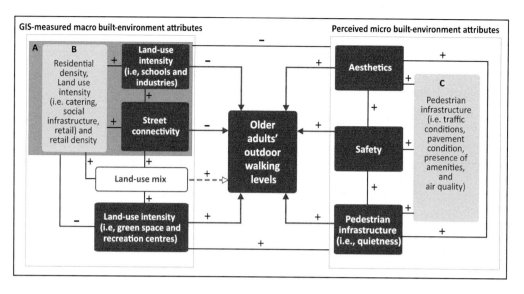

FIGURE 5.5 Significant relationships between neighbourhood built-environment attributes and older adults' outdoor walking levels and correlations among and between neighbourhood built-environment attributes.

levels – owing to the negative correlation between perceived aesthetics and GIS-measured street connectivity. Similarly, improving positive perceptions of traffic conditions may attenuate negative relationship between GIS-measured land use (schools) intensity and older adults' outdoor walking levels – owing to negative correlations between perceived traffic conditions and GIS-measured school-use intensity. Therefore, most GIS-measured macro and perceived micro built-environment attributes are – significantly or indirectly (through another built environment attribute) – related to outdoor walking levels of older adults and have a synergy in creating a neighbourhood built-environment that supports outdoor walking among older adults.

These findings may help urban planners and policy-makers to prioritise their interventions. With the findings of our study in mind, a focus on modifying built-environment attributes significantly related to outdoor walking levels of older adults irrespective of their correlations with other neighbourhood built-environment attributes may not be the most effective strategy for supporting outdoor walking. Moreover, compared to macro built-environment attributes, generally micro built-environment attributes can be changed more rapidly and at less cost (Sallis et al., 2011). The findings of our study on correlations between GIS-measured macro built-environment attributes and perceived micro built-environment attributes may help urban planners (and other professionals) to reduce costs and speed up the implementations of actions for encouraging outdoor walking among older adults.

5.6 LIMITATIONS OF OUR STUDY

This study has some limitations, related to: using data from different years (i.e. 2001, 2012 and 2016); generating coarse data for residential and non-residential land use by assigning only one use to each building; defining a home-based neighbourhood for generating GIS-based macro built-environment attributes that is not necessarily the same as the perceived neighbourhood (of respondents) used for measuring perceived micro built-environment attributes; not showing causality between neighbourhood built-environment attributes; and focusing on one specific social group (in our case, older adults) in one city.

Moreover, this study examined correlations among perceived micro built-environment attributes. However, it is still unknown whether these correlations really exist among actual or GIS-measured micro built-environment attributes. Future studies may have access to GIS-based data (for example, on air quality and noise) and include them in their research to measure GIS-measured micro built-environment attributes. In view of the unavailability of GIS-based data for some aspects of micro built-environment attributes (e.g. aesthetics), further efforts and studies are needed to produce that GIS-based data and to understand how to generate GIS-measured micro built-environment attributes.

Despite these limitations, this chapter provides an insight into the relationship between outdoor walking levels and neighbourhood built-environment attributes. In doing so, it also illustrates a GIS method that can be applied in studies on physical activities in other cities for which GIS-based data are available.

5.7 CONCLUSION

This chapter illustrates the capacity of GIS to contribute to urban planning studies on outdoor walking. GIS-based land use and transportation data were used to generate GIS-measures of one particular type of neighbourhood built-environment attributes: macro built-environment attributes. Subsequently, the GIS-measured macro built-environment attributes generated were used to examine correlations among those same neighbourhood macro built-environment attributes and between the macro and (perceived) micro built-environment attributes. Strong correlations were found among the GIS-measured macro built-environment attributes and also between GIS-measured macro and perceived micro built-environment attributes. Our study shows that most GIS-measured macro built-environment attributes are negatively correlated with perceived micro built-environment attributes. These correlations are likely to influence relationships between outdoor walking levels of older adults and neighbourhood built-environment attributes. The method we have illustrated and the results from the case study could well support the efforts of urban planners and policy-makers in prioritizing interventions, reducing costs and accelerating the implementation of actions to encourage outdoor physical activities, including walking levels of older adults.

REFERENCES

Age UK. (2016). *Staying steady: Keep active and reduce your risk of falling.* Retrieved from http://www.ageuk. org.uk/Documents/EN-GB/Information-guides/AgeUKIG14_staying_steady_inf.pdf?epslanguage= en-GB?dtrk=true

Birmingham City Council. (2010). *Index of deprivation 2010 – an analysis of Birmingham local statistics.* Retrieved from http://www.birmingham.gov.uk/cs/Satellite?blobcol=urldata&blobheader=application %2Fpdf&blobheadername1=Content-Disposition&blobkey=id&blobtable=MungoBlobs&blobwhere= 1223494646963&ssbinary=true&blobheadervalue1=attachment%3B+filename%3D311889Index_of_ Deprivation_2010.pdf

Birmingham City Council. (2013). *Census 2011: What does this mean for Birmingham?* Retrieved from https://www.birmingham.gov.uk/downloads/file/4562/census_2011_corporate_briefingpdf

Birmingham City Council. (2017). *Population 2001.* Retrieved from https://www.birmingham.gov.uk/direc tory_record/75388/population_2001

Birmingham Public Health. (2017). *Inactivity is a bigger challenge than obesity.* Retrieved from http://bir minghampublichealth.co.uk/news.php?id=78&mid=39#

Brookes, N., Kendall, J., & Mitton, L. (2012). City report: Birmingham. *WILCO Publication, 17,* 1–30.

Centers for Disease Control and Prevention. (1999). *Physical activity and health: A report of the surgeon general.* Retrieved from http://www.cdc.gov/nccdphp/sgr/olderad.htm

Cunningham, G. O., & Michael, Y. L. (2004). Concepts guiding the study of the impact of the built environment on physical activity for older adults: A review of literature. *American Journal of Health Promotion, 18*(6), 435–443. doi:10.4278/0890-1171-18.6.435

Department of Health. (2011). *Start active, stay active: A report on physical activity from the four home countries' Chief Medical Officers.* Retrieved from https://www.gov.uk/government/uploads/system/uploads/ attachment_data/file/216370/dh_128210.pdf

Department of Transport. (2016). *Rules for pedestrians (1 to 35).* Retrieved from https://www.gov.uk/guidance/ the-highway-code/rules-for-pedestrians-1-to-35

Frank, L. D. (2000). Land use and transportation interaction implications on public health and quality of life. *Journal of Planning Education and Research, 20*(1), 6–22. doi:10.1177/073945600128992564

Frank, L. D., Schmid, T. L., Sallis, J. F., Chapman, J., & Saelens, B. E. (2005). Linking objectively measured physical activity with objectively measured urban form- Findings from SMARTRAQ. *American Journal of Preventive Medicine, 28*(2S2), 117–125. doi:10.1016/j.amepre.2004.11.001

Gebel, K., Bauman, A., & Owen, N. (2009). Correlates of non-concordance between perceived and objective measures of walkability. *Annals of Behavioral Medicine, 37*(2), 228–238. doi:10.1007/s12160-009-9098-3.

Gomez, L. F., Parra, D. C., Buchner, D., Brownson, R. C., Sarmiento, O. L., Pinzon, J. D., … Lobelo, F. (2010). Built environment attributes and walking patterns among the elderly population in Bogota. *American Journal of Preventive Medicine, 38*(6), 592–599. doi:10.1016/j.amepre.2010.02.005

Handy, S. L., Boarnet, M. G., Ewing, R., & Killingsworth, R. E. (2002). How the built environment affects physical activity: Views from urban planning. *American Journal of Preventive Medicine, 23*(2S), 64–73. doi:10.1016/S0749-3797(02)00475-0

Hinckson, E., Cerin, E., Mavoa, S., Smith, M., Badland, H., Witten, K., … Schofield, G. (2017). What are the associations between neighbourhood walkability and sedentary time in New Zealand adults? The URBAN cross-sectional study. *BMJ Open, 7*(10), e016128. doi:10.1136/bmjopen-2017-016128

Lee, I.-M., & Buchner, D. M. (2008). The importance of walking to public health. *Medicine and Science in Sports and Exercise, 40*(7 Suppl), S512–518. doi:10.1249/MSS.0b013e31817c65d0

Leslie, E., Coffee, N., Frank, L., Owen, N., Bauman, A., & Hugo, G. (2007). Walkability of local communities: Using geographic information systems to objectively assess relevant environmental attributes. *Health & Place, 13*, 111–122. doi:10.1016/j.healthplace.2005.11.001

Loukaitou-Sideris, A. (2006). Is it safe to walk? Neighborhood safety and security considerations and their effects on walking. *Journal of Planning Literature, 20*(3), 219–232. doi:10.1177/0885412205282770

McLennan, D., Barnes, H., Noble, M., Davies, J., Garratt, E., & Dibben, C. (2011). *The English indices of deprivation 2010*. Retrieved from https://www.gov.uk/government/uploads/system/uploads/attachment_data/file/6320/1870718.pdf

Nasar, J. L., & Cubukcu, E. (2011). Evaluative appraisals of environmental mystery and surprise. *Environment and Behavior, 43*(3), 387–414. doi:10.1177/0013916510364500

Navarro, D. (2013). *Learning statistics with R: A tutorial for psychology students and other beginners: Version 0.5*. Retrieved from http://www.fon.hum.uva.nl/paul/lot2015/Navarro2014.pdf

ONS Geography GIS and Mapping Unit. (2011). *United Kingdom regions*. Retrieved from http://geoportal.statistics.gov.uk/datasets?q=Maps&sort_by=updated_at&sort_order=dsc

Payne, R. A. (2012). UK indices of multiple deprivation – a way to make comparisons across constituent countries easier. *Health Statistics Quarterly, 53*, 1–16. Retrieved from PubMed database.

Quartieri, J., Mastorakis, N. E., Guarnaccia, C., Troisi, A., D'Ambrosio, S., & Iannone, G. (2010). Traffic noise impact in road intersections. *International Journal of Energy and Environment, 4*(1), 1–8. Retrieved from https://www.researchgate.net/publication/229067303_Traffic_Noise_Impact_in_Road_Intersections

Rodriguez, D. A., Evenson, K. R., Roux, A. V. D., & Brines, S. J. (2009). Land use, residential density, and walking: The multi-ethnic study of Atherosclerosis. *American Journal of Preventive Medicine, 37*(5), 397–404. doi:10.1016/j.amepre.2009.07.008

Saelens, B. E., & Handy, S. L. (2008). Built environment correlates of walking: A review. *Medicine and Science in Sports and Exercise, 40*(7 Suppl), S550–S566. doi:10.1249/Mss.0b013e31817e67a4

Saelens, B. E., Sallis, J. F., Black, J. B., & Chen, D. (2003a). Neighborhood-based differences in physical activity: An environment scale evaluation. *American Journal of Public Health, 93*(9), 1552–1558. doi:10.2105/AJPH.93.9.1552

Saelens, B. E., Sallis, J. F., & Frank, L. D. (2003b). Environmental correlates of walking and cycling: Findings from the transportation, urban design, and planning literatures. *Annals of Behavioral Medicine, 25*(2), 80–91. doi:10.1207/S15324796ABM2502_03

Sallis, J. F., Slymen, D. J., Conway, T. L., Frank, L. D., Saelens, B. E., Cain, K., & Chapman, J. E. (2011). Income disparities in perceived neighborhood built and social environment attributes. *Health & Place, 17*(6), 1274–1283. doi:10.1016/j.healthplace.2011.02.006

Sanz, S. A., García, A. M., & García, A. (1993). Road traffic noise around schools: A risk for pupil's performance? *International Archives of Occupational and Environmental Health, 65*(3), 205–207. doi:10.1007/BF00381157

Stockton, J. C., Duke-Williams, O., Stamatakis, E., Mindell, J. S., Brunner, E. J., & Shelton, N. J. (2016). Development of a novel walkability index for London, United Kingdom: Cross-sectional application to the Whitehall II Study. *Bmc Public Health, 16*(1), 416. doi:10.1186/s12889-016-3012-2

Stucky, T. D., & Ottensmann, J. R. (2009). Land use and violent crime. *Criminology, 47*(4), 1223–1264. doi:10.1111/j.1745-9125.2009.00174.x

Sugiyama, T., & Thompson, C. W. (2007). Outdoor environments, activity and the well-being of older people: Conceptualising environmental support. *Environment and Planning A, 39*, 1943–1960. doi:10.1068/a38226

Sugiyama, T., Howard, N. J., Paquet, C., Coffee, N. T., Taylor, A. W., & Daniel, M. (2015). Do relationships between environmental attributes and recreational walking vary according to area-level socioeconomic status? *Journal of Urban Health: Bulletin of the New York Academy of Medicine, 92*(2), 253–264. doi:10.1007/s11524-014-9932-1

Suminski, R. R., Wasserman, J. A., Mayfield, C. A., Kheyfets, A., & Norman, J. (2015). Walking during leisure-time in relation to perceived neighborhoods. *Environment and Behavior, 47(7),* 1–15. doi:10.1177/0013916513520605

Van Cauwenberg, J., Clarys, P., De Bourdeaudhuij, I., Van Holle, V., Verte, D., De Witte, N., … Deforche, B. (2012). Physical environmental factors related to walking and cycling in older adults: The Belgian aging studies. *Bmc Public Health, 12*, 1–13. doi:10.1186/1471-2458-12-142

WHO. (2017). *Active ageing: Physical activity promotion in elderly.* Retrieved from http://www.euro.who.int/en/health-topics/disease-prevention/physical-activity/activities/hepa-europe/hepa-europe-projects-and-working-groups/active-ageing-physical-activity-promotion-in-elderly

William, D. (2010). *UK cities: A look at life and major cities in England, Scotland, Wales and Northern Ireland.* New Africa Press.

Wolch, J. R., Byrne, J., & Newell, J. P. (2014). Urban green space, public health, and environmental justice: The challenge of making cities 'just green enough'. *Landscape and Urban Planning, 125*, 234–244. doi:10.1016/j.landurbplan.2014.01.017

Zandieh, R., Martinez, J., Flacke, J., Jones, P., & van Maarseveen, M. (2016). Older adults' outdoor walking: Inequalities in neighbourhood safety, pedestrian infrastructure and aesthetics. *International Journal of Environmental Research and Public Health, 13*(12), 1179. doi:10.3390/ijerph13121179

Zandieh, R., Flacke, J., Martinez, J., Jones, P., & Van Maarseveen, M. (2017a). Do inequalities in neighbourhood walkability drive disparities in older adults' outdoor walking? *International Journal of Environmental Research and Public Health, 14*(7), 740. doi:10.3390/ijerph14070740

Zandieh, R., Martinez, J., Flacke, J., & van Maarseveen, M. (2017b). The associations between area deprivation and objectively measured older adults' outdoor walking levels. *SAGE Open 7*(4), 1–13. doi:10.1177/2158244017740172.

Zhou, X., & Parves Rana, M. (2012). Social benefits of urban green space: A conceptual framework of valuation and accessibility measurements. *Management of Environmental Quality: An International Journal, 23*(2), 173–189. doi:10.1108/14777831211204921

6 Knowledge Co-Production and Social Learning on Environmental Health Issues

A Role for Interactive GIS-Based Approaches

Rehana Shrestha, Johannes Flacke, Javier Martinez, and Martin van Maarseveen

CONTENTS

6.1   Introduction 83
6.2   Related Works 85
6.3   Context 85
6.4   Methodology 86
6.5   Interactive Spatial Understanding Support System Approach (ISUSS) 88
    6.5.1   Conceptual Framework of the ISUSS Approach 88
    6.5.2   The ISUSS Approach for Dortmund 89
    6.5.3   Findings from the ISUSS Workshop in Dortmund 90
6.6   Interactive Cumulative Based Assessment Approach (Interactive-CuBA) 92
    6.6.1   Conceptual Framework of the Interactive-CuBA 92
    6.6.2   Interactive-CuBA Approach for Dortmund and Munich 92
    6.6.3   Findings from the Interactive-CuBA Workshops in Dortmund and Munich 94
6.7   Supporting Knowledge Co-Production and Social Learning by Means of Interactive GIS-Based Approaches 96
6.8   Conclusions and Outlook 98
References 98

6.1   INTRODUCTION

Environmental health issues are multi-dimensional and multi-factorial in nature, touching on those aspects of human health that are determined or influenced by environmental factors (Marmot et al., 2008; Schulz and Northridge, 2004). These aspects include both the physical environment (e.g. density of built structures, presence of green and open space, increasing air and noise pollution) and the social environment (e.g. the socio-economic status (SES) of the population). Health problems associated with environmental factors usually have complex socio-spatial characteristics. For instance, the unequal distribution of environmental burdens such as air and noise pollution, as well as environmental resources such as parks and forests, have been found to evoke intra-urban spatial inequalities affecting certain SES groups more than others (Diez Roux and Mair, 2010; Pearce et al., 2010; Schulz et al., 2005). To add to this complexity, it is now recognised that people are exposed not just to single environmental factors. Instead, people's health is influenced by the cumulative burdens of environmental hazards affecting certain socially vulnerable groups more than others (Huang and London, 2012; Kühling, 2012; Morello-Frosch et al., 2011; Sadd et al., 2011; Sexton, 2012).

Conventional tools and analytical methods based on geographic information systems (GIS) can provide useful support for understanding the spatial aspects of environmental health issues. For instance, GIS-based proximity analysis of hazardous areas (Chakraborty et al., 2011), accessibility analysis to green parks (Zhang et al., 2011) and pollution- and noise-exposure analysis (Jerrett et al., 2001; Seto et al., 2007) can provide evidence of the unequal distribution of environmental burdens and resources across various social groups. Using visualization and mapping methods in GIS, cumulative burden assessment provides screening tools to identify 'hotspots' of place and population that require additional study, investments or other precautionary actions (Huang and London, 2012; Lakes et al., 2014). However, these GIS-based tools and methods fall short when the issues being considered are 'wicked' in nature (Huang and London, 2016; Pretorius, 2017); environmental health issues are increasingly being recognised as such 'wicked problems' (Kreuter et al., 2004).

The term 'wicked problem' is generally used to characterise problems that are multi-factorial, dynamic and resistant to solution (Kreuter et al., 2004). Rittel and Webber (1973) offer several characteristics of a 'wicked problem' that are also applicable to environmental health issues. For instance, the complex interdependencies and intertwining of causes and effects inherent in environmental issues make the definition of problems difficult. Moreover, it is often difficult to reach an immediate agreement about what the problem itself is, let alone find a solution, as various stakeholders may have different perspectives (Kreuter et al., 2004). Involvement of several sectors and domains in a transdisciplinary process is often recognised as important, albeit challenging, in environmental health. In practice, activities on health promotion and prevention are usually hampered by 'siloed' problem-solving attempts from various organizations and professions (Abernethy, 2016).

Addressing such 'wicked' characteristics of environmental health demands new forms of collaboration – bringing researchers and practitioners together – that can constructively integrate different ways of 'knowing' from various domains and sectors. Moreover, researchers argue for deliberation among various stakeholders, so that rather than trying to 'solve' the 'problem' through standard science-based approaches, these problems are 'tackled, managed and dealt with' by sharing knowledge and engaging in a social learning process (Head and Xiang, 2016).

Social learning is described as learning in and with social groups through the interaction (Siebenhüner, 2005) that occurs when participants share their experiences, ideas and environment with others during some group activity (Armitage et al., 2008). When participants engage in deliberation within a collaborative dialogue, they learn as individuals and as a group about the problem, their goals, the perspectives of other participants and the context of the matter at hand (Innes and Booher, 2016). As a result, they engage in cooperative endeavours of knowledge co-production in which knowledge is produced through interaction and learning among people with different perspectives (Fazey et al., 2013). Nonetheless, researchers also argue that social learning does not emerge with every group interaction (Tippett et al., 2005). It needs to be nurtured among the participants. In this regard, stakeholders need to be equipped to engage with the environmental health issues on equal footing, thereby examining such issues using a 'panoramic social lens rather than a scientific microscope, and working on it in an open and heuristic process of collective learning, exploration and experimentation' (Xiang, 2013).

Various studies on social learning have discussed certain process components and process attributes that are necessary in a participatory activity to foster social learning among the participants and co-production of knowledge (Dana and Nelson, 2012; Johnson et al., 2012; Schusler et al., 2003). In general, the process components include facilitation, a democratic structure and a diverse set of participants. The inclusion of these process components is assumed to generate the necessary process attributes: open communication, constructive conflict, inclusion of diverse knowledge types, unrestrained thinking and extended engagement. An elaborate discussion of these process components and attributes can be found in Shrestha (2018). The hypothesis is that if these process components and attributes are present in a participatory activity, this can lead to social learning and knowledge co-production as an outcome of that activity.

This chapter describes two approaches we have developed for engaging stakeholders in the process of knowledge co-production and social learning related to environmental health issues. One of the approaches is called the Interactive Spatial Understanding Support System (ISUSS) that aims at supporting stakeholder dialogue during a problem understanding phase in a planning process. The other approach is called the Interactive Cumulative Burdens Assessment (Interactive-CuBA) that aims at facilitating stakeholder dialogue in assessing cumulative burdens due to exposure to number of environmental factors in the neighbourhood. Both approaches integrate interactive GIS-based support systems that are implemented in MapTable. These approaches have been tested with stakeholders in two German cities, Dortmund and Munich. Building upon the process components and attributes for engendering social learning in a participatory activity, and the lessons learned during the development and implementation of the two approaches in Dortmund and Munich presented in this chapter, we present a conceptual framework for knowledge co-production and social learning by means of an interactive GIS-based approach.

6.2 RELATED WORKS

Progress made developing computer-related hardware and software is contributing enormously to shaping the way technology can be used to engage stakeholders in collaborative activities. In particular, Planning Support Systems (PSS) and Spatial Decision Support Systems (SDSS) have been developed, applied and tested in various fields (Geertman and Stillwell, 2003, 2009). These tools are defined as a subset of geo-information technologies, many of them offering a visually attractive platform that structures the mutual exchange of knowledge among many actors (Te Brömmelstroet, 2017).

Researchers are exploring issues related to these tools that focus explicitly on improving the communicative, informative and analytical aspects of stakeholder dialogue in planning processes. Pelzer et al. (2014) have explored perceived added values of such interactive PSS among planning practitioners as being useful in initiating and strengthening interaction and collaboration. Arciniegas et al. (2013) explored the effectiveness of collaborative map-based SDSS on three criteria: usefulness of the tool, clarity of the information provided by the tool and impact on decisions. The bottlenecks of using PSS have been explored extensively by Vonk et al. (2005).

Recently, researchers have been paying attention to exploring, developing and testing such tools for structuring the exchange of different types of knowledge, encouraging social learning and supporting shared understanding. Kaiser et al. (2017) developed a platform to manage knowledge co-production with respect to PSS in the context of complex decisions related to land use. Goodspeed (2013) has explored the extent to which the PSS can support social learning in planning workshops, while Pelzer and Geertman (2014) explored the interdisciplinary learning effects of PSS among various disciplines. Drawing on these works, the two approaches outlined in this chapter focus on engaging stakeholders in the processes of knowledge co-production and social learning to develop shared understanding and insights with respect to environmental health issues.

6.3 CONTEXT

BOX 6.1   Case Study Area

Two German cities, **Dortmund and Munich**, were chosen as case study cities. Both are among the cities with the ten largest populations in Germany; Dortmund is the eighth-largest city and Munich the third.

Dortmund is located in the western part of Germany, where it is situated in the highly urbanised region of the Ruhr, in the federal state of North-Rhine Westphalia. The city covers an area of 280 km^2 and has a population of around 600,000 inhabitants. Munich, the capital city of the federal state of Bavaria, is located in southern Germany and covers an area of around 311 km^2. It has a population of around 1.5 million inhabitants. Dortmund has a long history of industrialization and subsequent deindustrialization. Consequently, the city has gone through a long-lasting period of economic transformation accompanied by high unemployment rates. Munich, on the other hand, is a major centre of industrial production, services and education. It has low unemployment rates. The cities also differ significantly in terms of spatial and socio-demographic characteristics. Dortmund shows strong social and ethnic segregation and is characterised by a spatial structure comprising better-off neighbourhoods in the south and disadvantaged neighbourhoods in the north. Munich has a high standard and quality of living; however, social integration remains a challenge in the city. Affordable housing is its most pressing issue.

The two approaches presented in this chapter have been developed in the context of junior research group Jufo Salus (2013). The group started a project titled 'Cities as healthy place to live, regardless of social inequality' in 2013. The central aim of the project is to address the goal of creating healthy cities by integrating the concepts of sustainability and health in the urban planning processes of German municipalities. In this context, our approaches focus on exploring and understanding health inequalities in the urban area, including both the salutogenic (green parks, forests) and pathogenic (air pollution, noise pollution) factors of the environment and their unequal distribution across various social vulnerability factors.

Two German cities, Dortmund and Munich, were chosen as case study cities in the Jufo-Salus research project. Both cities, as members of the WHO Healthy Cities Network, differ significantly in their spatial structure and socio-demographic factors. The city of Dortmund is located in the western part of Germany in the highly urbanised region of the Ruhr. It has a long history of coal mining, steel production and beer brewing, which produced a booming economy for the city from the early 19th to the mid-20th centuries. Its history of industrialization and subsequent deindustrialization, leading to long-lasting economic transformation, has had a great impact on the city: unemployment rates are high compared to German averages. Recent research points to the existence of large socio-economic disparities in the city, with significant differences in environmental burdens and benefits (Shrestha et al., 2016). In addition to strong social and ethnic segregation, Dortmund is characterised by a spatial structure that is typical for many other cities in the Ruhr, with better-off neighbourhoods in the south and disadvantaged neighbourhoods in the north (Flacke et al., 2016). There is also a significant difference in health outcomes in the northern and southern neighbourhoods. The average life expectancy in 2011 ranged from 66.3 years in the northern district of Nordstadt to 76.3 years in one of the southern districts of Dortmund.

The city of Munich, located in the south of Germany, is the capital and the most populated city in the German state of Bavaria. Munich is the third-largest city in Germany, with a population of around 1.5 million inhabitants. The city is a major centre for industrial production, services and education in Germany and enjoys a very high standard and quality of living. However, social integration remains a challenge in the city, and affordable housing is its most pressing issue (Cucca and Ranci, 2016). Studies have also shown that environmental inequalities exist with respect to neighbourhood, socio-economic position and negative environmental exposure in the city (Schüle et al., 2016; Schüle et al., 2017).

6.4 METHODOLOGY

BOX 6.2   Methods Applied in the Chapter

This chapter describes a conceptual framework for engaging stakeholders in the process of knowledge co-production and social learning that is mediated by an interactive GIS-based approach drawing upon two empirical approaches. The ISUSS approach combines interactive mapping and rich picture drawings. In the interactive mapping, indicator maps are prepared as online web services and accessed as dynamic maps. Rich pictures use large vertical pieces of paper to allow iconographic drawing. The Interactive-CuBA approach uses the CuBA method, which uses a grid-based spatial resolution, a relative procedure for normalization/standardization of indicators and integrates both environmental burdens/benefits with social vulnerability factors using a simple aggregation method. The CuBA method is then adapted to produce a flexible and auditable model. Both approaches use a MapTable as a shared workspace and are supported with a facilitator. The secondary data set, which was used to produce indicator maps, comprises environmental factors, social vulnerability factors and administrative boundaries supplied by municipalities. Primary data was collected for the analyses of knowledge co-production, and social learning processes happened during the workshops supported with these approaches. These data were obtained from workshop transcripts, observers' notes, screen capture of MapTable, voice recordings, debriefing notes, questionnaires and maps. In both approaches, practitioners and researchers were involved, but they could also be adapted to involve citizens, who bring their own tacit knowledge with them. The approaches can only be applied in face-to-face workshops with small numbers of participants around a single MapTable.

Figure 6.1 shows the general methodology that was used in the development of our interactive GIS-based approaches. We used a System Development Life Cycle (SDLC) (Zhang et al., 2004), which can be divided into four sequential steps: development of conceptual framework, operationalization of framework, implementation and evaluation (Carugati, 2008).

Accordingly, we first defined the conceptual framework for our two approaches of interactive GIS-based support systems. The conceptual frameworks depict the way interactive GIS-based support systems facilitate knowledge co-production, social learning and shared understanding in the context of environmental health issues.

Operationalization of the framework in both approaches involved selecting the platform and user interface, designing the tool/model specific to the context and defining the facilitation task. In both approaches we used a MapTable as an interactive medium for user engagement. A MapTable is a large-scale, horizontal interactive display that shows digital contents in terms of maps and allows users to interact with the content through touching and gestures. The current ISUSS approach uses Phoenix 1.1 software™ (Geodan), and the Interactive-CuBA approach uses CommunityViz

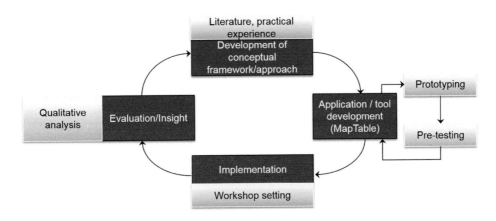

FIGURE 6.1   General methodology used in the development of interactive GIS-based approaches.

Scenario 360™ (CommunityViz Scenario 360). These approaches were first prototyped and pre-tested with researchers from the Jufo Salus research group and MSc students from TU Dortmund University before being used with stakeholders in Dortmund and Munich.

The approaches were implemented during face-to-face workshops organised in both cities. However, because of the preparation needed to set up the workshops and the required harmonization of these workshops with the others of the research group, we only conducted the ISUSS workshop in Dortmund on 26 October 2014. The two workshops for Interactive-CuBA were conducted in Dortmund on 29 October 2015 and in Munich on 7 November 2015. The selection of stakeholders for both was based on their professional roles, expertise and knowledge of environmental health issues in Dortmund/Munich. For the ISUSS workshop, a total of six individuals participated in the workshop – two municipal staff members (one each from the health and urban planning departments), three practitioners with backgrounds as NGOs/social entrepreneurs (one in a neighbourhood association, one in childcare and one from Local Agenda 21) and one researcher in urban planning. For the Interactive-CuBA workshops, a total of five individuals participated in Dortmund and seven individuals in Munich. Participants in the Dortmund workshop included one municipal staff member from the urban planning department, three practitioners associated with NGOs/social entrepreneurs (one each from Local Agenda 21, a tenant association and a youth organization) and one researcher in public health. The Munich workshop included three municipal staff members (one each from the social planning department, the environmental department and the health department), one practitioner associated with an environmental NGO, one representative from a health insurance company, one representative from a social housing company and one researcher in urban planning.

Evaluation of the workshops included the collection of data obtained by screen capture of the MapTable visualizations during the mapping sessions, voice recordings, observers' notes, a post-session questionnaire and maps produced by the participants on the MapTable. In the ISUSS workshop, a short debriefing of each participant and a rich picture drawing produced additional data. Voice recordings were later transcribed and translated into English from German with time coding added. Micro-analysis of the workshop transcripts was done in order to gain insights into the participants' interactions with visual artefacts such as models and interactive maps as well as among each other, facilitated by the use of technology.

6.5   INTERACTIVE SPATIAL UNDERSTANDING SUPPORT SYSTEM APPROACH (ISUSS)

The main goal of the ISUSS approach is to support knowledge co-production and integration by engaging stakeholders from a variety of relevant sectors in order to develop a shared understanding of a locally specific problem. Typically, this is done during an early phase of health-related planning processes.

6.5.1   CONCEPTUAL FRAMEWORK OF THE ISUSS APPROACH

Figure 6.2 depicts the conceptual framework of the ISUSS approach. The framework acts as a platform for enabling the integration of explicit knowledge (such as data, model, indicators) and stakeholders' tacit knowledge (experiences, expertise) during a collaborative workshop by combining two specific methods – interactive mapping and rich picture drawing.

Interactive mapping uses dynamic maps supported by interactive interfaces that facilitate the exploration of problems and solutions in geographic space and support the integration of spatial information during the reasoning, deliberation and communication activities of stakeholders. A rich picture is a graphical, 'cartoon-like' representation of a problematic situation. The rich picture is drawn using roughly sketched symbols, for example stick figures with short text, and provides a strong visual expression of features of interest. It also acts as a tool for encouraging conversation

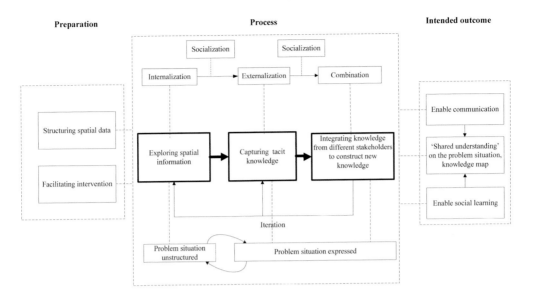

FIGURE 6.2 Conceptual framework of the ISUSS approach. (Shrestha et al., 2017).

and learning. A detailed description of these two methods and the framework is to be found in Shrestha et al. (2017).

The ISUSS framework consists of three phases: preparation phase, process phase and intended outcome phase. The preparation phase involves structuring spatial data and determining facilitating interventions. Structuring spatial data essentially means the collecting, processing, arranging and visualizing of spatial information as a meaningful indicator at an appropriate scale and resolution. To do so, stakeholders could be involved beforehand in the identification of relevant factors and indicators for the context. Depending on the availability of data, the available resolution and the requirements, these indicators may need to be further refined. The facilitation interventions involve determining the tasks for a facilitator. The facilitator moderates workshops by guiding the process, asking the right questions, making sure that participants remain engaged and actively participate and ensuring that no participants dominate the discussions. Moreover, the participatory activity that uses technology also requires tool-related facilitation (Pelzer et al., 2015). Therefore, the facilitator (referred to as the chauffeur) is entrusted with the task of helping the participants with the tool/model implemented in MapTable. The process phase consists of three iterative steps that the stakeholders need to carry out with the support of interactive maps and rich picture drawings. Using interactive maps, the ISUSS approach first asks stakeholders to explore and understand the spatial information in order to get insights into the context. Next, it guides stakeholders to articulate their views, experiences and concerns as a part of the problem definition step that is to be captured spatially on interactive maps. To get a complete picture of the problem, stakeholders collaborate to combine their knowledge and construct a holistic picture of the problematic situation by using the rich picture. In this way, the ISUSS approach intends to enable communication and social learning throughout the process and so develop a shared understanding of the problematic situation in its specific context.

6.5.2 THE ISUSS APPROACH FOR DORTMUND

The ISUSS approach for the Dortmund workshop was developed using data provided by the city of Dortmund. Therefore, the indicator maps listed in Table 6.1 were prepared based on issues discussed during earlier stakeholder workshops of Jufo Salus.

TABLE 6.1
List of Indicator Maps

Factors	Indicators	Description
	Green and water areas	According to RVR mapping
Environmental factors	Noise level from industries; noise level from rail transport: railway, tram, road traffic	24 h level in db, 2007
	Total Particulate Matter (PM_{10})	Annual average in g/m^3, 2000–2012, modelled in 2013
	Total Nitrogen Dioxide (NO_2)	
	Average Daily Traffic	Volume DTV in vehicles/24 h, 2007
Social factors	Social benefits: SGB II	Percentage of working age population (15–65 years) receiving SGB II at statistical sub-district, 2012
	Basic security in old age, disability, assistance for livelihood share: SGB XII	Percentage of basic security receiver per statistical district 2011
	Children up to 6 years	Percentage of children per statistical district, 2012
	Children and adolescents (>6–14 years)	
	People with migration background	Percentage of people with a migration background per statistical sub-district, 2012
Other factors	Administrative boundary	Dortmund city districts (12), statistical districts (62), statistical sub-districts (170)
	Land use plan	Existing/planned land use zoning, 2004
	Real land use: commercial and industrial use; residential and mixed use	According to RVR mapping
	Point locations: schools, kindergartens, playgrounds, hospitals, nursing homes	According to RVR mapping, 2003–2008
	Powerline distribution	in KV, 2004

The indicator maps were prepared as online web services with varying levels of detail when zoomed in – city district, district, sub-district and building block level – depending on the availability of data for each indicator. Phoenix software was used to access the indicator map and use it as an interactive map implemented in MapTable, thus enabling user engagement; see Figure 6.3b. To explore the interactive maps, stakeholders can use functions such as zoom in/out or pan for navigation, or use map structuring and overlaying to explore the maps in multiple combinations and view the area in 3D. To support capture of tacit knowledge spatially, functions such as annotating with text, sketching using coloured lines and polygon shapes or representing issues with iconography were provided on the interactive maps. Rich pictures were administered by allowing stakeholders to draw vertically with multi-coloured pens on a large piece of paper hung on the wall; see Figure 6.3a.

6.5.3 FINDINGS FROM THE ISUSS WORKSHOP IN DORTMUND

The use of interactive maps implemented in MapTable acted as a platform for supporting exploration of spatial information. Spatial visualization of indicators was reported to be particularly useful for seeing the connection between the places stakeholders know and the respective spatial information, as well as seeing the effect of combining various driving factors. User-friendly interfaces, together with the functionality to zoom in, pan and produce overlays of different layers, supported stakeholders while they were exploring various combinations of the environmental and social indicators. Additionally, structuring data to visualise indicators at a disaggregated level was noted to be important for understanding spatial inequalities by comparing specific areas to the rest of the city.

a)   b)

FIGURE 6.3   a) Stakeholders drawing a rich picture b) Stakeholders engaging with an interactive map implemented on the MapTable

Likewise, it also provided a good platform for 'triggering' discussions and 'capturing' stakeholders' tacit knowledge. Participants found that with the support of spatial visualization, and annotation and sketching functionalities, on interactive maps they were able to draw their 'ideas and scenarios quickly and discuss them further with others'. Stakeholders therefore added their knowledge at the neighbourhood scale, as shown in Figure 6.4a. The participants were further supported in elaborating their issues and concerns through the use of pictograms on the rich picture and by making links to other views that enabled them to construct an integrated picture of environmental health in the study area, as shown in Figure 6.4b.

Communication among the participants was observed throughout the entire workshop. Spatial visualization of interactive maps implemented in MapTable appeared to support communication of information and knowledge among the participants in an interactive and dynamic way. The MapTable appeared to be a useful medium for bringing people together. Being guided through structured facilitation, the participants were able to work actively and were encouraged to 'show what they meant'. In addition, combined use of interactive maps and the rich picture supported the articulation, sharing and integration of both local spatial knowledge and non-spatial, or yet to be spatialised, knowledge in the discussion. As such, social learning was clearly apparent at various moments during the workshop. For instance, by overlaying several environmental and social indicators (e.g. share of children under 6 years overlaid with noise levels from streets) using interactive maps implemented in MapTable, participants commented that the children's routes to schools are in some parts of the city greatly affected by noise and air pollution. Participants also sought to change their existing knowledge through reasoning and interaction. For instance, a strikingly high share of

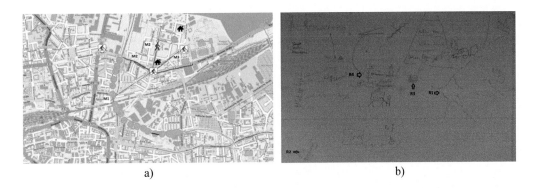

a)   b)

FIGURE 6.4   (a) Knowledge added using interactive maps, and (b) elaboration of issues on a rich picture.

recipients of unemployment aid (SGBII) along one street forced the participants to think in depth about the potentially underlying reasons; elsewhere, the participants realised that PM_{10} levels on a specific street were still problematic, but better than assumed. Likewise, development of a shared understanding was observed while the participants were engaged in constructing shared meanings and identifying shared interests. For instance, while considering accessibility to green areas participants discussed whether the focus should be on time and distance or on safety of the access route. Consensus was reached on assessing the quality of the access from a safety point of view. Similarly, improving accessibility to green areas was identified as important for promoting health in the case study area.

6.6 INTERACTIVE CUMULATIVE BASED ASSESSMENT APPROACH (INTERACTIVE-CuBA)

The main goal of the Interactive-CuBA approach is to facilitate stakeholder dialogue in order to collaboratively assess cumulative burdens and in doing so engender knowledge co-production and social learning of the stakeholders.

6.6.1 CONCEPTUAL FRAMEWORK OF THE INTERACTIVE-CuBA

Figure 6.5 depicts the conceptual framework that shows the way Interactive-CuBA supports knowledge co-production and social learning. The details of the framework have been described in Shrestha et al. (2018). The framework describes three attributes necessary in Interactive-CuBA: a flexible and auditable model; an interface-driven shared workspace for user engagement; and skilled facilitation. By providing two types of process support – communication support and information support – the approach contributes to the intended outcomes of knowledge co-production and social learning.

A flexible and auditable model allows the stakeholders to freely select indicators or change certain assumptions relevant for their CuBA. By explicitly giving users an opportunity to work with a flexible and auditable model, it may enable them to perform the calculations in a way that is transparent, relevant and easy to comprehend. Additionally, an interactive interface-driven shared workspace provides opportunities for interactive run times with the model, which is necessary to maintain user engagement. This implies providing a 'dialogue space' where stakeholders can interact with the model by providing their own input or changing assumptions in the model and receiving feedback in real time through a tangible user interface.

User engagement further requires skilled facilitation to maximize the goals of the participatory activity. In doing so, the approach aims at providing communication support by enabling active dialogue among the stakeholders, allowing them to question underlying assumptions made in the model and encouraging them to exchange their perspectives. Likewise, the approach aims at providing information support by allowing the stakeholders to explore the information dynamically and spatially using a shared spatial language. It also enables elicitation and integration of both tacit and explicit knowledge that yields an understanding greater than either could have produced separately.

6.6.2 INTERACTIVE-CuBA APPROACH FOR DORTMUND AND MUNICH

The interactive-CuBA approach adopts an indicator-based method of cumulative burden assessment that has been discussed in Shrestha et al. (2016). The approach uses a grid-based, small-scaled spatial resolution to represent indicators on both environmental and social vulnerability factors. A relative procedure for normalization and standardization of indicators using environmental standards and city-wide averages, respectively, is adopted in order to integrate both environmental burdens/ benefits into an index together with social vulnerability factors.

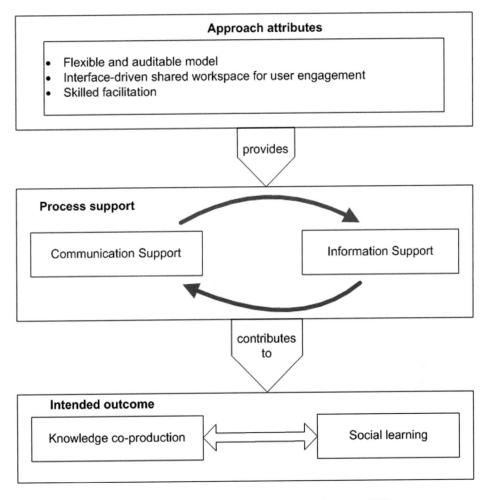

FIGURE 6.5 Conceptual framework of an Interactive-CuBA. (Shrestha et al., 2018).

Accordingly, the CuBA model developed for Dortmund and Munich used a 125 m × 125 m grid as small-scaled spatial resolution for both individual environmental indicators and the cumulative environmental index. The list of indicators used in the CuBA model for Dortmund and the CuBA model for Munich is presented in Table 6.2.

City-wide data on air quality was acquired in 125 m × 125 m resolution for Dortmund, whereas air quality data for Munich was available in the form of line data measured along the roads. Therefore, the indicator and index on air quality for Munich were prepared as line data. For both models, noise nuisance data were interpolated from point data to a 125 m × 125 m resolution. using the Inverse Distance Weighted function in GIS (Farcaş and Sivertun, 2010). Indicators on access to green areas (parks and forests) for both cities were prepared by calculating 'crow flight' distance for each 125 m grid cell to any of the nearest green areas. All the green and forest areas greater than 1 ha were taken into consideration. Environmental standards were used to normalise the environmental data and city-wide averages were used to standardise social vulnerability. Additive aggregation as described by Vlachokostas et al. (2014) was used to construct a cumulative environmental index, and a simple additive aggregation as provided by Shrestha et al. (2016) was used for the social vulnerability index.

The CuBA model was adapted using CommunityViz Scenario 360 to make it flexible and auditable and then was provided with interactive tools and dynamic visualization. In this form,

TABLE 6.2

List of Indicators Used in the CuBA Models for Dortmund and Munich

Dimension	Domain	Description of Indicators
Environmental burdens	Air quality	Annual average concentration of PM_{10} ($\mu g/m^3$)
		Annual average concentration of NO_2 ($\mu g/m^3$)
		Number of days PM_{10} exceeds the limit of 40 $\mu g/m^3$ (d/a)
	Noise nuisance	Noise levels from individual sources (industries, street and tram) measured in decibels (dbA)
Environmental benefits	Green spaces	Accessibility to green areas >1 ha in size within walking distance
		Accessibility to forest areas of >1 ha in size within walking distance
Social vulnerability	Sensitive population	Number of people aged between 6–11 (persons/625 m^2)
		Number of older adults aged 65 years and over (persons/625 m^2)
	Social and economic	Number of people with migration background (persons/625 m^2)
		Number of people receiving SGB II (persons/625 m^2)
		Number of people receiving SGB XII (persons/625 m^2)

stakeholders can overlay individual social vulnerability indictors on various environmental indica-tors and create either a separate index for environmental burdens and social vulnerability or an integrated index combining both indices. The interactive tools also allow stakeholders to select/deselect indicators to be combined into the index or change the threshold values of environmental standards in the case of environmental indicators. The effect of these changes with regard to distri-bution of cumulative burdens can then be viewed dynamically in real time and spatially as 2D maps. Likewise, to characterise and visualise the area being exposed to various levels of environmental burdens and social vulnerability, the tool uses relative scale visualization. Such visualization is derived by comparing the indicator value with a set of standards or city-wide averages and then rep-resented in a relative scale as shown in Figure 6.6a for the environmental indicator/index, Figure 6.6b for the social vulnerability indicator/index, and Figure 6.6c for the integrated environmental and social vulnerability index.

Interactive-CuBA uses a MapTable to provide an interactive interface-driven shared workspace, and the models for Dortmund and Munich were implemented in MapTable. Two facilitators – a process facilitator and a tool facilitator (also known as the chauffeur) – provided facilitation simi-lar to the ISUSS approach.

6.6.3 Findings from the Interactive-CuBA Workshops in Dortmund and Munich

In general, the Interactive-CuBA approach was able to provide communication support in assessing cumulative burdens in both workshops. Active dialogue was observed to be mediated through the use of the MapTable, which provided a shared workspace in which indicator maps were used as a common language. The use of the flexible and auditable model provided opportunities for partici-pants to interact with the model, for example by changing the threshold values of environmental indicators or by selecting the set indicators and viewing the results of the change instantaneously, which prompted the participants to question the underlying assumptions used in the model. In this way, the model became transparent as well as open to scrutiny. Likewise, the model implemented in MapTable acted as supporting material for the various participants to talk with each other and thereby exchange different perspectives on the same topic, as well as on information that was acknowledged to be important but not included in the model.

The Interactive-CuBA approach provided information support in assessing cumulative burdens in both workshops by enabling participants to explore the indicators dynamically, using interactive

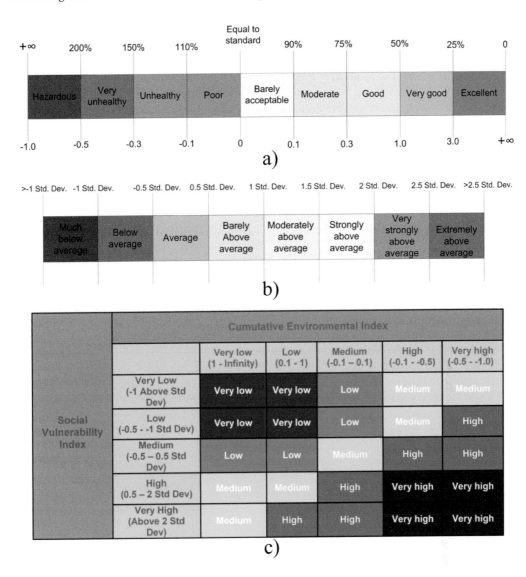

FIGURE 6.6 Relative scale representation of (a) the environmental indicator/index, (b) the social vulnerability indicator/index, and (c) the integrated environmental and social vulnerability index.

tools such as zooming in/out and selecting/deselecting. They were, therefore, able to explore the indicators at various scales – street-level to city-wide – and in various ways – individually, combining indicators to form an index, overlapping one indicator with another, changing the threshold. The use of small-scaled spatial resolution further supported deliberation about cumulative burdens at the local level. Similarly, by becoming engaged in exploring the model implemented in MapTable, participants' tacit knowledge seemed to be elicited, which they used to make others understand the information presented through the indicators, even highlighting relevant information not captured in the model.

Instances of co-production of knowledge and social learning were clearly observed during the workshops. For instance, the use of 70 dBA as the threshold for acceptable noise pollution was decided to be very high, and the use of a logarithmic scale was acknowledged to be necessary to remove the compensatory effects of various noise indicators. The use of 1 ha as the smallest size for parks was identified to filter out important parks in the city centre. Shortcomings and difficulties in combining environmental indicators into the index were also jointly discussed. Likewise, the importance of including the subjective perception of residents in relation to noise pollution and the quality of green areas to promote a healthy environment was also established. Several problems and opportunities to assess cumulative burdens for resource allocation were jointly identified. For instance, cumulative burdens are generally not confined within fixed administrative boundaries, and as a result several administrative units and boards are responsible, which ultimately makes it difficult to assess and plan interventions. On the other hand, the participants perceived the opportunities for different departments to work together using the model implemented in MapTable, as these departments would be able to integrate data from other relevant departments (e.g. health, environment, social, education) and together generate meaningful information.

6.7 SUPPORTING KNOWLEDGE CO-PRODUCTION AND SOCIAL LEARNING BY MEANS OF INTERACTIVE GIS-BASED APPROACHES

Figure 6.7 provides the conceptual framework of an interactive GIS-based approach for supporting social learning and knowledge co-production through participatory activity. The framework: (A) comprises attributes to be included in the design of an interactive GIS-based approach; (B) sheds light on the process components and process attributes of the general framework of social learning that in principal need to be addressed through the approach; and (C) elaborates on those process attributes that can be specifically strengthened through the approach.

The first part of the framework (A) describes the attributes that are necessary for such an interactive GIS-based approach. These approach attributes include flexibility and dynamics in tools/models; interface-driven shared workspace for user engagement; and skilled facilitation. These three elements are considered essential for the development of any interactive GIS-based approach. Depending on the context and the task to be supported, the specifications of the tools/models, the

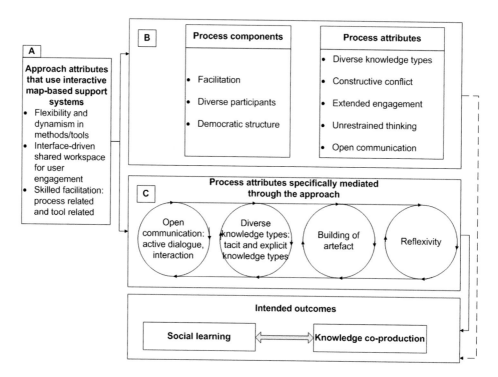

FIGURE 6.7 Conceptual framework of an interactive GIS-based approaches to support social learning and knowledge co-production.

type of the hardware and software to augment the tools/models with flexibility and dynamics, the type of shared workspace and details of the facilitation tasks all need to be decided. In the ISUSS approach, the use of the interactive maps of varying levels of detail, combined with rich picture drawings, provided the flexibility and dynamism. In the case of Interactive-CuBA, the opportunities to explore the indicators/index individually, or in combination, and to select the indicator and subsequently construct an index preferred by the participants ensured flexibility and dynamism in the process. In both approaches, the MapTable acted as the shared workspace, and skilled facilitation was provided by appointing a facilitator to stimulate interaction and commitment among stakeholders and a chauffeur to provide guidance on the use of the tools.

The second part of the framework (B) shows the process components and process attributes synthesised from studies on social learning. These components and attributes are considered prerequisites for successful participatory activities that are aimed at engendering knowledge co-production and social learning, including the activity supported by means of an interactive GIS-based approach. Having a diverse group of participants, for instance, ensures access to a greater breadth of worldviews, mental models and experiences during the process. Similarly, ensuring democratic structures provides opportunities for participants to contribute to the process. Approaches that incorporate these process components are considered to generate process attributes, which are diverse knowledge types, constructive conflicts, extended engagement, unrestrained thinking and open communication. Open communication enables participants to engage with various knowledge types and differing opinions. Similarly, ensuring a mixture of diverse knowledge types and encouraging unrestrained thinking about those types can lead to learning across traditional disciplinary or professional boundaries. Yet, conflicts may arise about contested values and worldviews. If managed properly, constructive conflicts can foster new understanding. Finally, extended engagement beyond a single participatory activity is needed to foster social learning. Some of these process components and process attributes could be clearly identified in our interactive GIS-based approaches; see Shrestha (2018) for more details.

The third part of the framework (C) shows the attributes that could particularly be strengthened by the interactive GIS-based approach. These attributes include open communication and inclusion of diverse knowledge types taken from social learning studies. Next to these, two new attributes – building of artefact and reflexivity – were identified in our study. An interactive approach could support open communication among participants by providing a shared workspace with a flexible and dynamic method/tool, such as interactive maps or a CuBA model implemented in MapTable. Working together in a small group around the table, and viewing the same information, could provide a 'dialogue space' for people to actively engage and talk with each other; this was observed in both the ISUSS and Interactive-CuBA approaches. Inclusion of diverse knowledge types, in the form of both tacit and explicit knowledge, is another attribute that could be well supported with an interactive GIS-based approach. Explicit knowledge could be included in the form of interactive maps, as in the ISUSS approach, or as indicators maps or environmental standards, as in the CuBA model. Tacit knowledge, on the other hand, needs to be elicited and articulated. This could be encouraged by allowing stakeholders to interact with a flexible and dynamic method/tool. During this activity, participants tend to relate their knowledge specifically to geographic reference points. Moreover, by allowing stakeholders to mark and annotate areas while adding their knowledge or drawing a rich picture as in the ISUSS approach, support could be provided further in capturing tacit knowledge. Furthermore, participants could be enabled to build visual artefacts in the participatory activity. Visual artefacts are human-created visual representations of objects or entities that take form through constant negotiation of meaning among participants during participatory work (Singh, 2011). When the participants are enabled to build a visual artefact, for example drawing a rich picture as in the ISUSS approach or constructing an index as in the Interactive-CuBA approach, this could provide them with an opportunity to get hands-on experience (i.e. learning-by-doing). In doing so, building the artefact could also help by provoking participants to reflect on the topics. For instance, when participants noticed a strikingly high proportion of unemployment aid recipients along a certain street during the ISUSS workshop, they were drawn to discuss the potential causes of the situation. Focusing on concrete experiences and locally relevant information is also considered to be a basis for reflecting, thinking and acting (Wibeck, 2014). As such, an interactive GIS-based approach has the potential to draw participants into a deeper exploration of the issues, their own knowledge and their experience by providing the means for them to visualise information at various scales and levels of detail. In general, these attributes are implanted dynamically rather than linearly, thereby fostering social learning and knowledge co-production.

6.8 CONCLUSIONS AND OUTLOOK

In the current practice of understanding environmental health issues, GIS-based tools and analytical methods are primarily used to analyse the socio-spatial distribution of inequalities of environmental burdens and benefits, thereby providing information for health-related planning practices. Despite claim from several researchers for transdisciplinary research practice by involving both researcher and practitioners on an equal footing as co-creators of knowledge in a mutual learning process, the practice thereof is often limited. This can often be attributed to simplistically equating knowledge with information that can be transferred through a one-way process and considering that learning can happen when people are just brought together. Scholars have argued, however, that for social learning to occur, nurturing of learning opportunities is required. People learn and knowledge is co-produced when their tacit knowledge and perspectives are confronted and sufficiently challenged with explicit knowledge and the perspectives of others. In this regard, the development of approaches as presented in this chapter reflects the possibilities for supporting collaboration among and between researchers and practitioners that is geared towards involving stakeholders in the process of knowledge co-production and social learning.

Although the ISUSS and the Interactive-CuBA approaches have been implemented with practitioners and researchers from Dortmund and Munich, experiences with their application in real-world planning contexts are still inconclusive, and the results are not yet exhaustive. Therefore, further research is called for. In doing so, the conceptual GIS-based framework we have described in this chapter could be adopted as the basis for engaging multiple stakeholders in knowledge co-production and social learning.

REFERENCES

Abernethy, P. (2016). Bridging conceptual "silos": Bringing together health promotion and sustainability governance for practitioners at the landscape scale. *Local Environment, 21*(4), 451–475. doi:10.1080/13549 839.2014.968841

Arciniegas, G., Janssen, R., & Rietveld, P. (2013). Effectiveness of collaborative map-based decision support tools: Results of an experiment. *Environmental Modelling & Software, 39*(Suppl C), 159–175. doi:10.1016/j.envsoft.2012.02.021

Armitage, D., Marschke, M., & Plummer, R. (2008). Adaptive co-management and the paradox of learning. *Global Environmental Change, 18*(1), 86–98. doi:10.1016/j.gloenvcha.2007.07.002

Carugati, A. (2008). Information system development activities and inquiring systems: An integrating framework. *European Journal of Information Systems, 17*(2), 143–155. doi:10.1057/ejis.2008.23.

Chakraborty, J., Maantay, J. A., & Brender, J. D. (2011). Disproportionate proximity to environmental health hazards: Methods, models, and measurement. *American Journal of Public Health, 101*(S1), S27–S36. doi:10.2105/AJPH.2010.300109

CommunityViz Scenario 360. Available online. Retrieved from http://communityviz.city-explained.com/communityviz/scenario360.html

Cucca, R., & Ranci, C. (2016). *Unequal Cities: The Challenge of Post-Industrial Transition in Times of Austerity.* New York: Taylor & Francis.

Dana, G. V., & Nelson, K. C. (2012). Social learning through environmental risk analysis of biodiversity and GM maize in South Africa. *Environmental Policy and Governance, 22*(4), 238–252. doi:10.1002/eet.1587

Diez Roux, A. V., & Mair, C. (2010). Neighborhoods and health. *Annals of the New York Academy of Sciences*, *1186*(1), 125–145.

Farcaş, F., & Sivertun, Å. (2010). Road traffic noise: GIS tools for noise mapping and a case study for Skåne region. *The International Archives of the Photogrammetry, Remote Sensing and Spatial Information Sciences*, *34*.

Fazey, I., Evely, A. C., Reed, M. S., Stringer, L. C., Kruijsen, J., White, P. C., … Phillipson, J. (2013). Knowledge exchange: A review and research agenda for environmental management. *Environmental Conservation*, *40*(01), 19–36. doi:10.1017/S037689291200029X

Flacke, J., Schüle, S. A., Köckler, H., & Bolte, G. (2016). Mapping environmental inequalities relevant for health for informing urban planning interventions – a case study in the city of Dortmund, Germany. *International Journal of Environmental Research and Public Health*, *13*(7), 711. doi:10.3390/ijerph13070711

Geertman, S., & Stillwell, J. (2003). *Planning support systems in practice* (1st ed.). Heidelberg: Springer.

Geertman, S., & Stillwell, J. (2009). *Planning Support Systems Best Practice and New Methods* (Vol. 95). Dordrecht: Springer Science & Business Media.

Geodan. Phoenix: Touch table application for spatial cooperation. Retrieved from https://www.geodan.com/solutions/phoenix/

Goodspeed, R. (2013). *Planning Support Systems for Spatial Planning Through Social Learning*. (PhD Thesis), Massachusetts Institute of Technology. Retrieved from http://hdl.handle.net/1721.1/81739

Head, B. W., & Xiang, W.-N. (2016). Why is an APT approach to wicked problems important? *Landscape and Urban Planning*, *154*, 4–7. doi:10.1016/j.landurbplan.2016.03.018

Huang, G., & London, J. K. (2012). Cumulative environmental vulnerability and environmental justice in California's San Joaquin Valley. *International Journal of Environmental Research and Public Health*, *9*(5), 1593. doi:10.3390/ijerph9051593

Huang, G., & London, J. K. (2016). Mapping in and out of "messes": An adaptive, participatory, and trans-disciplinary approach to assessing cumulative environmental justice impacts. *Landscape and Urban Planning*, *154*, 57–67. doi:10.1016/j.landurbplan.2016.02.014

Innes, J. E., & Booher, D. E. (2016). Collaborative rationality as a strategy for working with wicked problems. *Landscape and Urban Planning*, *154*, 8–10. doi:10.1016/j.landurbplan.2016.03.016

Jerrett, M., Burnett, R. T., Kanaroglou, P., Eyles, J., Finkelstein, N., Giovis, C., & Brook, J. R. (2001). A GIS–environmental justice analysis of particulate air pollution in Hamilton, Canada. *Environment and Planning A*, *33*(6), 955–973. doi:doi.org/10.1068/a33137

Johnson, K. A., Dana, G., Jordan, N. R., Draeger, K. J., Kapuscinski, A., Schmitt Olabisi, L. K., & Reich, P. B. (2012). Using Participatory Scenarios to Stimulate Social Learning for Collaborative Sustainable Development. doi:10.5751/ES-04780-170209

Jufo Salus. (2013). Junior research group: The city as healthy living environment independent of social inequalities. Retrieved from http://www.jufo-salus.de/cms/en/Welcome/index.html

Kaiser, D. B., Weith, T., & Gaasch, N. (2017). Co-production of knowledge: A conceptual approach for integrative knowledge management in planning. *Transactions of the Association of European Schools of Planning*, *1*(1), 18–32.

Kreuter, M. W., De Rosa, C., Howze, E. H., & Baldwin, G. T. (2004). Understanding wicked problems: A key to advancing environmental health promotion. *Health Education & Behavior*, *31*(4), 441–454. doi:10.1177/1090198104265597

Kühling, W. (2012). Mehrfachbelastungen durch verschiedenartige Umwelteinwirkungen. *Umweltgerechtigkeit Chancengleichheit bei Umwelt und Gesundheit: Konzepte, Datenlage und Handlungsperspektiven*; Hans Huber: Bern, Switzerland, 135–150.

Lakes, T., Brückner, M., & Krämer, A. (2014). Development of an environmental justice index to determine socio-economic disparities of noise pollution and green space in residential areas in Berlin. *Journal of Environmental Planning and Management*, *57*(4), 538–556. doi:10.1080/09640568.2012.755461

Marmot, M., Friel, S., Bell, R., Houweling, T. A., & Taylor, S. (2008). Closing the gap in a generation: Health equity through action on the social determinants of health. *The Lancet*, *372*(9650), 1661–1669. doi:10.1016/S0140-6736(08)61690-6

Morello-Frosch, R., Zuk, M., Jerrett, M., Shamasunder, B., & Kyle, A. D. (2011). Understanding the cumulative impacts of inequalities in environmental health: Implications for policy. *Health Affairs*, *30*(5), 879–887.

Pearce, J. R., Richardson, E. A., Mitchell, R. J., & Shortt, N. K. (2010). Environmental justice and health: The implications of the socio-spatial distribution of multiple environmental deprivation for health inequalities in the United Kingdom. *Transactions of the Institute of British Geographers*, *35*(4), 522–539.

Pelzer, P., & Geertman, S. (2014). Planning support systems and interdisciplinary learning. *Planning Theory & Practice, 15*(4), 527–542. doi:10.1080/14649357.2014.963653

Pelzer, P., Geertman, S., van der Heijden, R., & Rouwette, E. (2014). The added value of planning support systems: A practitioner's perspective. *Computers, Environment and Urban Systems, 48*, 16–27. doi:10.1016/j.compenvurbsys.2014.05.002

Pelzer, P., Goodspeed, R., & te Brömmelstroet, M. (2015). Facilitating PSS workshops: A conceptual framework and findings from interviews with facilitators. In J. Ferreira, R. Goodspeed, & J. Stillwell (Eds.), *Planning Support Systems and Smart Cities* (pp. 355–369). Cham: Springer.

Pretorius, C. (2017). Exploring procedural decision support systems for wicked problem resolution. *South African Computer Journal, 29*(1), 191–219. doi:10.18489/sacj.v29i1.448

Rittel, H. W., & Webber, M. M. (1973). Dilemmas in a general theory of planning. *Policy sciences, 4*(2), 155–169. doi:10.1007/BF01405730

Sadd, J. L., Pastor, M., Morello-Frosch, R., Scoggins, J., & Jesdale, B. (2011). Playing it safe: Assessing cumulative impact and social vulnerability through an environmental justice screening method in the South Coast Air Basin, California. *International Journal of Environmental Research and Public Health, 8*(5), 1441.

Schüle, S. A., Fromme, H., & Bolte, G. (2016). Built and socioeconomic neighbourhood environments and overweight in preschool aged children. A multilevel study to disentangle individual and contextual relationships. *Environmental Research, 150*, 328–336. doi:10.1016/j.envres.2016.06.024

Schüle, S. A., Gabriel, K. M., & Bolte, G. (2017). Relationship between neighbourhood socioeconomic position and neighbourhood public green space availability: An environmental inequality analysis in a large German city applying generalized linear models. *International Journal of Hygiene and Environmental Health*. doi:10.1016/j.ijheh.2017.02.006

Schulz, A. J., & Northridge, M. E. (2004). Social determinants of health: Implications for environmental health promotion. *Health Education & Behavior, 31*(4), 455–471. doi:10.1177/1090198104265598

Schulz, A. J., Kannan, S., Dvonch, J. T., Israel, B. A., Allen III, A., James, S. A., … Lepkowski, J. (2005). Social and physical environments and disparities in risk for cardiovascular disease: The healthy environments partnership conceptual model. *Environmental Health Perspectives, 113*(12), 1817.

Schusler, T. M., Decker, D. J., & Pfeffer, M. J. (2003). Social learning for collaborative natural resource management. *Society & Natural Resources, 16*(4), 309–326. doi:10.1080/08941920390178874

Seto, E. Y. W., Holt, A., Rivard, T., & Bhatia, R. (2007). Spatial distribution of traffic induced noise exposures in a US city: An analytic tool for assessing the health impacts of urban planning decisions. *International Journal of Health Geographics, 6*(1), 24. doi:10.1186/1476–072X-6-24

Sexton, K. (2012). Cumulative risk assessment: An overview of methodological approaches for evaluating combined health effects from exposure to multiple environmental stressors. *International Journal of Environmental Research and Public Health, 9*(2), 370.

Shrestha, R. (2018). *Interactive Map-Based Support Systems: Supporting Social Learning and Knowledge Co-production On Environmental Health Issues*. (PhD Thesis), University of Twente, Forthcoming.

Shrestha, R., Flacke, J., Martinez, J., & Van Maarseveen, M. (2016). Environmental health related socio-spatial inequalities: Identifying 'hotspots' of environmental burdens and social vulnerability. *International Journal of Environmental Research and Public Health, 13*(7), 691. doi:10.3390/ijerph13070691

Shrestha, R., Köckler, H., Flacke, J., Martinez, J., & Van Maarseveen, M. (2017). Interactive knowledge co-production and integration for healthy urban development. *Sustainability, 9*(11), 1945. doi:10.3390/su9111945

Shrestha, R., Flacke, J., Martinez, J., & van Maarseveen, M. (2018). Interactive cumulative burden assessment: Engaging stakeholders in an adaptive, participatory and transdisciplinary approach. *International Journal of Environmental Research and Public Health, 15*(2), 260. doi:10.3390/ijerph15020260

Siebenhüner, B. (2005). The role of social learning on the road to sustainability. In U. Pestschow, J. Rosenau, & E. v. W. Ulrich (Eds.), *Governance and Sustainability: New Challenges for States, Companies and Civil Society* (1st ed., Vol. 1, pp. 85–99). London: Routledge.

Singh, A. (2011). Visual artefacts as boundary objects in participatory research paradigm. *Journal of Visual Art Practice, 10*(1), 35–50. doi:10.1386/jvap.10.1.35_1

Te Brömmelstroet, M. (2017). PSS are more user-friendly, but are they also increasingly useful? *Transportation Research Part A: Policy and Practice, 104*(Suppl C), 96–107. doi:10.1016/j.tra.2016.08.009

Tippett, J., Searle, B., Pahl-Wostl, C., & Rees, Y. (2005). Social learning in public participation in river basin management – early findings from HarmoniCOP European case studies. *Environmental Science & Policy, 8*(3), 287–299. doi:10.1016/j.envsci.2005.03.003

Vlachokostas, C., Banias, G., Athanasiadis, A., Achillas, C., Akylas, V., & Moussiopoulos, N. (2014). Cense: A tool to assess combined exposure to environmental health stressors in urban areas. *Environment International, 63*, 1–10. doi:10.1016/j.envint.2013.10.014

Vonk, G., Geertman, S., & Schot, P. (2005). Bottlenecks blocking widespread usage of planning support systems. *Environment and Planning A, 37*(5), 909–924. doi:10.1068/a3712

Wibeck, V. (2014). Enhancing learning, communication and public engagement about climate change – some lessons from recent literature. *Environmental Education Research, 20*(3), 387–411. doi:10.1080/13504 622.2013.812720

Xiang, W.-N. (2013). Working with wicked problems in socio-ecological systems: Awareness, acceptance, and adaptation. *Landscape and Urban Planning* (110), 1–4. doi:10.1016/j.landurbplan.2012.11.006

Zhang, P., Carey, J., Te'eni, D., & Tremaine, M. (2004). *Integrating human–computer interaction development into the systems development life cycle: A methodology.* Paper presented at the Proceedings of the Americas Conference on Information Systems, New York.

Zhang, X., Lu, H., & Holt, J. B. (2011). Modeling spatial accessibility to parks: A national study. *International Journal of Health Geographics, 10*(1), 31. doi:10.1186/1476-072X-10-31

7 Role of Public Spaces in Promoting Social Interaction in Divided Cities

The Case of Nicosia, Cyprus

Marija Kukoleca, Ana Mafalda Madureira, and Javier Martinez

CONTENTS

7.1 Introduction 103
7.2 Public Spaces as Places of Social Interaction in
    Divided Cities 104
    7.2.1 Divided Cities 104
    7.2.2 Public Spaces 105
    7.2.3 Social Interaction 105
7.3 Methodology 106
    7.3.1 Case Study 106
    7.3.2 Data and Methods 110
7.4 Results 112
7.5 Discussion and Conclusion 114
References 119

7.1 INTRODUCTION

Nicosia is the last divided capital city in the world, being both the capital city of the Republic of Cyprus and the capital city of the Turkish Republic of Northern Cyprus. A UN-administered buffer zone still divides the island of Cyprus and passes through the centre of Nicosia. This buffer zone, also referred to as the 'dead zone', can only be crossed through checkpoints. Currently there are three crossing points in Nicosia, around which several public spaces exist, ranging from commercial streets to parks and markets. The aim of this chapter is to analyse opportunities for social interaction in a divided city by focusing on the potential of public spaces in Nicosia to enable this social interaction among the city's divided communities.

Divided cities originate as a result of dividing a nation in two (Kliot and Mansfeld, 1999). Examples are the formerly divided cities of Berlin and Belfast and Nicosia (Kliot and Mansfeld, 1999). The spatial division of a wall can create a distinct urban environment on either side of it. This division of space has a physical function that is significant as it affects on administrative and political structures, thereby creating an imbalanced distribution of resources and opportunities (Abdelmonem and McWhinney, 2015). It also becomes a negative feature that deters mobility, interaction and social cohesion. The built-up fabric of the city becomes an object of remembrance that is paradoxical and contested, with different meanings and connotations (Bevan, 2007).

However, in divided cities where the citizens are allowed to move across borders, public spaces provide shared spaces where people can interact (Pullan et al., 2012b). For residents of both sides, these public spaces can become a common space for shared functions and social activities, and important catalysts for change that allows people to experience life on the other side and bond with their neighbours. This interaction discourages stigmatization of 'the other' through shared experiences that promote mutual trust and respect. Hence, public spaces become places of exchange with a significant social role as complex systems of open socio-spatial engagement (Marcus and Francis, 1998). Even streets can act as shared spaces and binding factors within divided cities (UN Habitat, 2013). Any public space, if accessible to both conflicted communities, could contribute to renewing past relationships and memories of what was once a unified area.

Researchers have long been interested in divided cities and on the effect this has had on public spaces within those cities, particularly focusing on political or historical aspects of the division (Till et al., 2013; Öngül, 2012). Others stress the importance of public spaces as binding mechanisms in divided cities, analysing the physical aspects and the ways in which people use the space (Nagle, 2009). In locations where a physical barrier that divides the city sends a clear message of exclusion, public spaces contest this notion by presenting a shared space where everyone is welcome. Gaffikin et al. (2010) analyse the concept of public spaces in divided cities from the perspective of urban design and the role those spaces play in that context, concluding that public spaces provide an opportunity for social contact, which can lead to social interaction. Other researchers have observed how people interact with the public space, as well as how they interact within one another (Abu-Ghazzeh, 1999; Talen, 1999).

We developed an index that measures the potential of the public spaces in Nicosia to promote social interaction. Experts were then interviewed about the index outcomes. These experts were people involved in bi-lateral projects at both the municipality and the NGO level. The index illustrates those dimensions of public spaces that have the greatest potential to enable social interaction. The interviews with experts were used to validate the index and to contextualise its results.

Following this introduction, in this chapter we discuss what divided cities are and how public spaces can enable social interaction within them. We then introduce the case study of Nicosia, briefly describe the public spaces that we analysed, explain how the data were collected and analysed and explain how an index of social interaction was built. The results focus on public spaces that scored highest and lowest in the index, with expert interviews providing a context for the results. The discussion and conclusions are presented in the final section.

7.2 PUBLIC SPACES AS PLACES OF SOCIAL INTERACTION IN DIVIDED CITIES

7.2.1 Divided Cities

A divided city has been defined as a city comprising two or more distinct entities that have to be spatially separated, mutually exclusive and relatively homogeneous enclaves (Nagle, 2009). Van Kempen (2007) and Marcuse (1993) consider highly fragmented cities to be divided cities, whereas O'leary (2007) refers to this phenomenon as a political partition, where an entity becomes divided by a barricade in the form of a wall, fence or another type of physical obstacle.

The physical borders dividing cities have a practical and a symbolic function. They represent 'infrastructures of conflict' (Till et al., 2013), such as walls, barricades and buffer zones, which are not only a physical but also a social and symbolic divide (Pullan et al., 2012a). The symbolic divide may be the deepest. It causes stigmatization, feelings of insecurity and disassociation from the population on the other side (Pullan et al., 2012a). Symbols of divisions can also become 'infrastructures of peace' (Till et al., 2013), as is, for example, the 'Home for Cooperation', a revitalised building in the buffer zone of Cyprus that is located on neutral ground and administered by both Greek and Turkish Cypriots. Such symbols send a message of unity and cooperation, turning an artefact of conflict into a promise of peace (Till et al., 2013).

Even in cases where members of the divided communities are free to move across the divide, they often opt not to do so, either due to personal beliefs or out of principle (Pullan et al., 2012b).

In Nicosia, some of the Greek Cypriot population consider that by crossing the divide and showing their identification at the border they acknowledge the legitimacy of the Turkish Republic of Northern Cyprus (Webster and Timothy, 2006). However, opening the border has helped the economic development of both communities in Nicosia and revived public spaces in the area (Gaffikin et al., 2010).

7.2.2 Public Spaces

Any space accessible to the general public and part of the built environment could be considered a public space: squares, streets, gardens, cafés and markets, among others, qualify as public spaces. Dymnicka (2010) argues that a public space is formed by the interactions of its users, while at the same time shaping their interactions. These are social spaces of public life, whereas public life is an interaction between life itself and a public space (Gehl and Svarre, 2013). Nevertheless, a public space is defined by a set of rules and restrictions that applies to it and determines who can use it and how.

Public spaces have been studied based on their functions at the city level, their use and the roles of different stakeholders in shaping them (Madanipour, 2010). In a physical sense, public spaces are urban generators that stimulate communication channels with morphological, environmental and aesthetic values (Lynch, 1960; Marcinczak and Sagan, 2011; Woolley, 2003). Culturally, they are common points of convergence that host numerous traditional functions that allow for symbolic value embedded in identity and sense of place. The cultural and political meanings of these spaces are vital in day-to-day life, where meanings emerge through social interactions (Low, 2000). Politically, they are spaces for demonstrations, an arena for negotiations of conflict and for political action (Van Deusen, 2002; Low, 2000; Mitchell, 1995). They can host clashes among opposing groups, but social tensions can change over time and the resulting effect can be witnessed in shared spaces.

The functions that public spaces allow diverge extensively, and so do their users. Activities in public spaces are determined by the socio-economic attributes of their users (Aziz et al., 2012; Aratani, 2010), as well as influenced by the political ideology of their users. Differences in personal beliefs can change the way people behave in these spaces. Public spaces are experimental environments that groups use to legitimise themselves through decisions about where to stay, gather and socialise (McCann, 1999). They gain genuine significance as locations in which groups can react to the condition of co-existence and overcome boundaries of division in a quest to build a consensus of shared living (Abdelmonem and McWhinney, 2015).

Amin (2002) argues that most public spaces are 'places of transit', where meaningful interactions among strangers are unlikely. There is, however, potential for chance encounters among strangers, and 'in such serendipities rests the opportunity for exchange and learning that can help break barriers' (Gaffikin et al., 2010, p. 498). In divided cities, the motivation for using public spaces may arise from daily tasks such as going to shops and parks and doing other activities, rather than a desire to integrate with neighbouring communities (Pullan et al., 2012b). Thus, the roles, function and even definition of public spaces can have varied connotations. For the purposes of this chapter, we consider a public space as an urban space, including but not limited to streets, squares, cafés and parks that are accessible to everyone pending a set of rules applied to all users, visible from street level and determined by and determining the interactions of its users.

7.2.3 Social Interaction

Social interaction is the contact that takes place between individuals, groups and environments (Talen, 1999). Individuals and groups feel the need for social interaction and find opportunities for it within public spaces (Drucker and Gumpert, 1998; Marcus and Francis, 1998). Social interaction can happen anywhere, between any two individuals or groups. Researchers often emphasise the importance of social interaction in public spaces on a local level, such as a neighbourhood or a residential complex. Talen (1999) found that public spaces that are attractive to visitors promote a sense of community. Hickman (2013) analysed the social interaction promoted by spaces such as cafés, parks and shops and found that much of neighbourhood interaction takes place in these spaces. One of their most important characteristics is their functional role, for example for acquiring goods, for entertainment or for recreation.

Researchers also underline the importance of 'place of contact' (Abu-Ghazzeh, 1999; Farida, 2013; Talen, 1999), i.e. locations where strangers come into contact with each other in a relatively impartial and casual manner. These places of contact have special significance in divided cities. Regardless of the level of social interaction such places produce, their absence can have great consequences in divided cities and on the future of social life in them (Pullan et al., 2012b).

There are different levels of social interaction, starting with very superficial contact that includes, among other things, observing other individuals and greeting them, to a high level of interaction, which can be found in neighbourhoods where individuals form communities and share emotional investments in the same things (Talen, 1999). In situations where communities are physically separated by barriers, any level of social interaction becomes relevant and contributes to the social life of a divided city.

7.3 METHODOLOGY

7.3.1 CASE STUDY

BOX 7.1   Case Study Area

Nicosia is the last divided capital city in the world, being both the capital of the Republic of Cyprus, where people identify as Greek Cypriot, and the Turkish Republic of Northern Cyprus, where the population identifies itself as Turkish Cypriot. The island of Cyprus has been the home of Greeks and Turks since the 16th century. Conflicts between the two nationalities began in the 20th century, and by 1958 its largest city, Nicosia, was divided by a wire border. As the conflict escalated, the Greek and the Turkish populations immigrated to the south and the north of the island, respectively. The military invasion by the Republic of Turkey was followed by an intervention by the United Nations, which set up a border, or buffer zone, in the city of Nicosia and the rest of the country in 1974 (Öngül, 2012). This division ended the conflict and the two parts of Nicosia continued to develop separately.

The UN-administered buffer zone still divides Cyprus. This strip of land, only 3.5 m wide at some points, and up to 5 km wide at others, passes through the centre of Nicosia (Grichting, 2014). The buffer zone is also referred to as the 'dead zone' since it was evacuated by the residents in the 1970s (Kliot and Mansfield, 1997). It divides Nicosia into a northern, Turkish Cypriot part, and a southern, Greek Cypriot part. The buffer zone can only be crossed through checkpoints. Currently, there are in total only seven checkpoints in Cyprus, three of which are located in Nicosia. The public spaces that were analysed in this research are located around two of these crossings: Ledra Palace and Ledra Street.

Our research focuses on the city of Nicosia (Cyprus), which is divided by an UN-administered zone, also known as the 'buffer zone' or the 'green line'. The buffer zone can only be crossed through checkpoints, of which there are currently three in the city. The *Ayios Dhometios* (Metehan) crossing is located in the western suburb of Nicosia and is mostly used by cars. The *Ledra Palace* crossing is located just outside the Venetian walls that circle the old town, which is at the same time the centre of the city. The *Ledra Street* crossing is located in the very centre of the old town. Ledra Palace and Ledra Street are intended for pedestrians. Our analysis focused on public spaces in the vicinity of these two crossings.

The public spaces selected for our study had to meet the following criteria: they would need to be freely accessible to everyone; visitors would have to occupy these spaces for an amount of time longer than that just needed for passing through; the space would need to be used daily, not just during special events; and the public space would need to be used by the members of both majority communities in Nicosia. Based on field observations and expert interviews, we selected seven public spaces for the study, three on the Greek Cypriot side of Nicosia (Ledra Street, Faneromeni Square and the Municipal Gardens), three on the Turkish Cypriot side (Lokmaci Street, Buyuk Han market and Bandabulya market) and one located entirely in the UN buffer zone (Markou Drakou Street) (Figure 7.1).

Ledra Street is a pedestrian area, except in the early morning hours, when delivery trucks have access to the many shops, cafés and restaurants there. A department of the University of Cyprus is also located there, which brings many students and young people to the area. Ledra Street is very popular with both Greek and Turkish communities. According to the experts interviewed, Ledra Street and the surrounding areas underwent a revival after the Ledra crossing opened in 2008 (Figure 7.2a).

Ledra Street became more alive, right now, after the borders were opened. It's more commercialized (than the northern side), […] When you walk from the north of Ledra Street to the south, you would

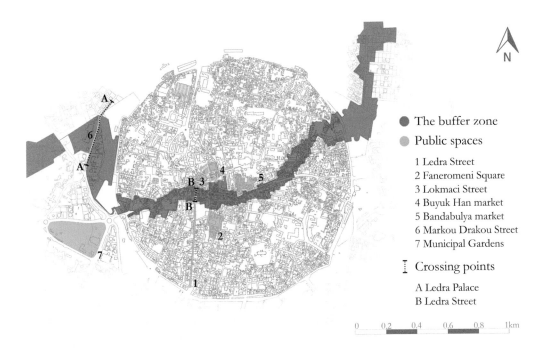

FIGURE 7.1 Location of crossings and public spaces analysed. (Base map source: University of Cyprus, 2016.)

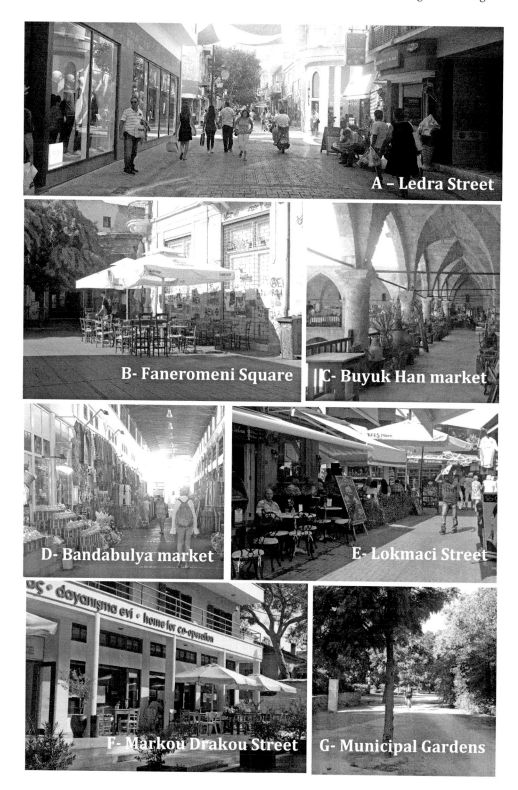

FIGURE 7.2 (a) Ledra Street, Greek side, 2016; (b) Faneromeni Square, Greek side, 2016; (c) Buyuk Han market, Turkish side, 2016; (d) Bandabulya market, Turkish side, 2016; (e) Lokmaci Street, Turkish side, 2016; (f) Markou Drakou Street, buffer zone, 2016; (g) Municipal Gardens, Greek side, 2016.

see that it's gradually getting more commercialized and more internationalized. So, in the north you would see the local shops and in the south you would see more of the global brands. But there are many communities, not just Greek Cypriots and Turkish Cypriots – but many tourists walking and visiting the place.

(Interviewee 7)

Faneromeni Square hosts one of the oldest schools on Cyprus, and a church, a mosque and a museum are in its vicinity. The area also includes a square and a network of alleys with cafés and shops and is very popular with both communities (Figure 7.2b).

Faneromeni is not being used because it's a square, but because it's part of a café. If cafés bring their chairs outside, it is still public in that sense. But if we didn't have the cafés there, we wouldn't have many people sitting there.

(Interviewee 1)

Faneromeni is a lovely place to sit and have coffee and chat, and I'm sure Turkish Cypriots come over and Greek Cypriots come over. And it's a much more pleasant area than this area, for example (Markou Drakou Street).

(Interviewee 10)

Historically, *Lokmaci Street* was the street of craftsmen and tradesmen, and this tradition lives on in the types of shops found there nowadays. There are also several local cafés. The area gained significance after the Ledra Street crossing was opened (Figure 7.2e).

This area emerged after the borders were opened. It was deserted before the checkpoints opened; now it has have come alive. It has attracted new and interesting uses, like cafés, cafés which breed respect for diversity. Places which are visited by all sorts of people. The mixing of different people creates a culture of diversity. If you look at pictures of the area of Lokmaci before they opened, it was deserted. And if you compare it with the picture of how it is now, it has come alive.

(Interviewee 11)

The *Buyuk Han market* originated in Ottoman times, when it was used as an inn. Nowadays it is used as a market and a public space. The open space on the ground floor has cafés, restaurants and several shops, locally owned, that sell hand-made products and typical Cypriot ornaments. On the first floor, shops coexist with artist's workshops and studios. Several festivals take place in Buyuk Han, as well as food markets and live music sessions (Figure 7.2c). Every Saturday, there is a regular meeting for Greek and Turkish Cypriots at one of the cafés, and everyone is welcome to join in.

Buyuk Han is a prime location, where Greek and Turkish Cypriots meet. For example, I go there every Saturday, and I meet my friends there.

(Interviewee 11)

The *Bandabulya market* is a municipal market dating back to 1939 and is located in the old town. Vendors, mostly Turkish Cypriots, sell fresh fruit and vegetables, meat, ornaments and souvenirs. There are art studios, cafés and restaurants. The market attracts tourists and the Greek and Turkish Cypriot population. The majority of people who sit in the cafés and spend time here, however, are members of the local, older population (Figure 7.2d).

Bandabulya is the old market. It's actually recently been renovated. And here you would see the Greek Cypriots also, but I don't know if you would see them spending a lot of time. They would go there for shopping, I would say. Again, it's local people from the north selling. They also have somethings unique to Cyprus, like ornaments and souvenirs, etc.

(Interviewee 6)

Markou Drakou Street, located entirely in the buffer zone, is neutral ground for Greek and Turkish Cypriots, with many meetings and political discussions taking place there. International and non-governmental organizations are situated there, mainly in the 'Home for Cooperation'. This building is an educational centre, in which events, dialogues, workshops and Greek and Turkish language courses are held. In the building's café, many entertainment events are organised, such as bi-communal music nights, dance lessons and festivals (Figure 7.2f). The experts interviewed identified Markou Drakou Street as a place frequented by Greek and Turkish Cypriots, in most part thanks to the 'Home for Cooperation' and the events that take place there. Other experts observed that visitors to Markou Drakou Street form a select group who participate in the events held there and engage in social interaction. Several private businesses are also located there, as are educational centres and a restaurant, where many diplomatic events take place:

This space here (Markou Drakou Street), OK, it's controlled by the United Nations, but it's usually very quiet unless somebody arranges an event or something. You don't see people hanging out here, there's nothing to do here, […] The 'Home for Cooperation' is very active, […]. But you always see the same people who're going to come and interact here. It's always the same people.

(Interviewee 10)

The *Municipal Gardens* are a large green area, one of the few in Nicosia. Besides seating benches, a children's playground and one café, there are no other activities here (Figure 7.2g). According to some experts, these gardens are used mainly by minority communities during weekends. Experts from the Turkish Republic of Northern Cyprus, who are not very familiar with this area, suggested that it is used by cyclists from the Turkish side, while others claim that the park is frequented during events.

The Municipal Gardens are frequented by Sri Lankans and Filipinos. The Asian minorities use it more than Greek and Turkish Cypriots.

(Interviewee 13)

Some of my friends are passing [through] and riding bikes in this park. If it is this park. But I am not sure about it. But I know some, as I said, Turkish Cypriots will pass from the checkpoint and go to the park on the Greek side, and they're riding bikes and spending time there.

(Interviewee 4)

7.3.2 DATA AND METHODS

BOX 7.2   Methods Applied in the Chapter

Data were collected from expert interviews and fieldwork observations. The interviews were analysed with Atlas.ti™ software. The map used for quantitative analysis was in *.dwg* format, and the analysis was done with AutoCAD™ Map 3D software. The map includes all the public spaces, the location of the border and details of the network of roads and streets. The road network was used to calculate the distance from the public space to the nearest border crossing point. Network analysis was performed in AutoCAD Map 3D.

The indicators for the index of social interaction were selected after review of the relevant literature and further aligned with expert interviews and fieldwork observations. Each of the indicators was calculated in its own unit and then standardised to a value ranging from 0 to 1. The same formula was used for standardizing most indicators, considering that they all present benefits:

$$\frac{\text{The value of 'X' − Minimum value}}{\text{Maximum value − Minimum value}} = \text{Standardized value of 'X'.}$$

Where 'X' stands for the value of an indicator for a public space, while the minimum and maximum values represent the lowest and highest values of the public spaces for that indicator.

Data for the index of social interaction was collected through expert interviews and fieldwork observations. The semi-structured interviews were set up to obtain information about the social interaction between the two major communities in Nicosia and the public spaces in the city, the way those spaces were used, what affects the social interaction in those public spaces and how the public spaces influence social interaction. A total of 13 interviews were conducted from 16 September to 18 October 2016 with Turkish and Greek Cypriot experts and international experts involved in bi-communal* work who have first-hand knowledge of the public spaces and bi-communal dynamics of the city. The interviews were analysed with the Atlas.ti software.

Quantitative analysis started by georeferencing the map of Nicosia provided by the University of Nicosia. The map was in .*dwg* format and the analysis was done with AutoCAD Map 3D software. The geographic coordinate system that was selected was the WGS 84 in the Universal Transverse Mercator zone 36N projection.

The map did not cover the entire city of Nicosia, but all of this study's public spaces were included, as were the location of the border and details of the network of roads and streets. The road network was used to calculate the distance from the public space to the nearest border crossing point. Analysis was performed in AutoCAD Map 3D. For each of the public spaces, possible routes from the nearest crossing point to the closest point of the public space were also calculated. The distances were measured taking into consideration all possible combinations with every street (for vehicles or pedestrians) located outside of the UN buffer zone, i.e. accessible to civilians. The closest distance was chosen as the best result.

The index of social interaction was designed as a tool for quantifying the potential ability of public spaces to host and facilitate social interaction. The indicators were selected through a review of the relevant literature and, based on the expert interviews and fieldwork observations, further aligned to our research aims. Each of the indicators was calculated in its own unit and then standardised to a value from 0 to 1. The same formula was used for standardizing most indicators, considering that they all present benefits:

$$\frac{\text{The value of 'X' − Minimum value}}{\text{Maximum value − Minimum value}} = \text{Standardized value of 'X'.}$$

---

* Bi-communal refers to 'involving or including two distinct communities of people: of relating to or being a society composed of two distinct or separate communities, often with conflicting interests' (Merriam Webster, 2017).

Where 'X' stands for the value of an indicator for a public space, while the minimum and maximum values represent the lowest and highest values of the public spaces for the indicator in question.

There are two exceptions to this. (i) The indicator 'Presence of different age groups in the public space' was given a value 1 if members of all age groups were present during the counting sessions and a value 0 if any age group was not represented. (ii) In the category 'Number of events taking place in the public space', the data was divided into three categories. The first category (< 2 events/month) was given the value 0, the second category (3–8 events/month) was given a value of 0.5 and the last category, (> 8 events/month) was given a value of 1.

The 'cafés and restaurants' sub-indicator was given a sub-weighting of 0.2, while the other sub-indicators were given a sub-weighting of 0.1, adding up to a total of 1; this was done because this activity was mentioned during the expert interviews as more likely to promote social interaction.

To determine the weights for each indicator, the indicators prioritised were those indicated by experts to have the biggest influence on social interaction.

Table 7.1 presents the overview of the indicators used, how the data was collected, their rationale and their weight.

7.4   RESULTS

The main feature of the index of social interaction is a score for each of the public spaces, representing its potential ability to promote and facilitate social interaction between the Greek Cypriot and the Turkish Cypriot communities. The results for each public space, as well as its position relative to the border and distance from the nearest crossing point, are shown in Figure 7.3. The public spaces are presented on the map as shaded circles, with darker shading representing higher index scores and the lighter shading lower scores.

Figures 7.4 through 7.6 show the results of the standardised values, from 0 to 1, for each of the indicators, grouped in dimensions of public spaces that influence their ability to promote social interaction: physical, social and activities/events. Figure 7.4 illustrates the physical and social dimensions of occupant density. The public space with the highest standardised value of the surface-area indicator has the lowest occupant density, both for walking and static visitors.

Figure 7.5 shows the extent to which the public spaces are visited by members of both communities and members of all age groups.

Figure 7.6 presents activities/events in the public spaces: the number of different types of activities, the number of each of type of activity and the number of events.

Closer examination of the public spaces with the highest and lowest index scores – Ledra Street and the Municipal Gardens, respectively – provides a better understanding of the relationship between the index and the situation in the public spaces. The results for Ledra Street are shown in Figure 7.7. The score is dominated by the fact that the area has the highest number of shops (both local and branded), cafés and restaurants among all the public spaces analysed. Also, many people occupy this public space, standing or sitting, considering its size; see Figure 7.2a. This can be related to the high concentration of cafés and restaurants, most of which have a sitting area outside; this therefore makes the outside area of the café a part of the public space in a functional sense, even though we do not consider it to be a public space in full, because one needs to pay for/consume something in the café or restaurant in order to sit on its sidewalk. The results of the index of social interaction for the Municipal Gardens are shown in Figure 7.8. In most of the indicators, these public gardens score 0. However, this does not mean that the occupant density of visitors that are walking or spending time in the Municipal Gardens is 0 people/m^2: it simply means that the value of these indicators was the lowest among all the public spaces before they were standardised. Some of the experts mentioned events that are organised in the Municipal Gardens, but due to the low frequency of these events, this indicator also had a low score.

TABLE 7.1

Indicators of the Index of Social Interaction and Their Rationale

Indicator	Data Collection and Measurement	Rationale	Weight of Indicator
Surface area (m²)	Extracted from the map of Nicosia. The surface area of any structure that was not a public space was deducted from the total surface area.	The surface area of a public space determines its capacity and the number of different activities that can be located there. More activities make the public space potentially more attractive for visitors and generate more opportunities for social interaction (Van Deusen, 2002).	0.1
Accessibility[a]	The road network was used to calculate the distance of each public space to the nearest border crossing point. The analysis was performed in AutoCAD Map 3D. Distances were measured considering all possible combinations with every street. The shortest distance was chosen as the best result.	A public space must be accessible to the population it serves (Whyte, 1988). Public spaces that are easily reachable and at a convenient location have more visitors (Shaftoe, 2008).	–
No. of different types of activities taking place	Guided observation: activities were counted and divided into groups: shopping; cafés and restaurants; education; art and culture; religion; private businesses; beauty services; non-governmental organizations; and international organizations.	The variety of activities attracts more visitors to an area. Activities such as cafés and restaurants, shopping, entertainment, culture, etc., can induce communities to cross divides and participate in social interaction (Pullan et al., 2012b).	0.1
No. of each type of activity taking place[b]	Guided observation.	A higher number of activities (shops, etc.) of the same type offers visitors a wider choice, increasing the chances of getting visitors interested in spending time in the public space.	0.1
No. of events taking place	Information about events in the public spaces came from websites of different organizations, expert interviews and social media. Grouped as 2 or less per month; 3–8 per month; more than 8 per month.	Events such as festivals or concerts in public spaces have the ability to build solidarity among participants (Shaftoe, 2008). Events also attract more visitors and make a public space more interesting (Holland et al., 2007).	0.15
Occupant density, walking	Guided observation: counting the number of people in a public space and then dividing by the surface area of the public space. The unit of this indicator is people/m². Public spaces were divided into sections of approximately the same size via imaginary lines, called 'gates', which were observed in five-minute counting sessions. Only those people who crossed the imaginary line were counted. The counting sessions were repeated at each gate at different times of the day (Grajewski and Vaughan, 2001). Counting was done at three different times of the day, morning (from 8 am to 10 am), lunch time (from 12 noon to 2 pm) and afternoon (from 4 pm to 6 pm).	The more visitors that circulate through a public space, the higher the opportunity for social interaction. The occupant density can also indicate the popularity of the public space.	0.15

(Continued)

TABLE 7.1 (CONTINUED)

Indicators of the Index of Social Interaction and Their Rationale

Indicator	Data Collection and Measurement	Rationale	Weight of Indicator
Occupant density, static	Guided observation: Counted in a similar way as 'Occupant density, walking', i.e. the number of people who were standing or sitting in the public space inside each section. The people who were sitting in cafés and restaurants were only taken into account if they were sitting outdoors, located in the public space. Average number of visitors divided by surface area, providing an average occupant density for static users of the public space.	The more visitors that spend time in the public space, the higher the opportunity for social interaction. The occupant density can also indicate the popularity of the public space.	0.15
Presence of different age groups	Guided observation: the age-group classes were: 0–14; 15–24; 25–44; 45–64; 65 and older (Provisional guidelines on standard international age classifications, 1982).	Accommodating the needs of all age groups increases the sense of community (UN Habitat, 2013) and social interaction. Different age groups visit the public spaces for different reasons. Presence of all age groups in a public space indicates that it meets diverse needs and that it is safe and inviting.	0.05
Presence of members of both communities	Determined during expert interviews: interviewed experts were asked to identify whether members of both communities visited the public space or not. The testimonies of experts were counted as 1 if the answer was positive and then added up, giving each of the public spaces a maximum possible score of 13 (= total number of interviews).	In divided cities, residents have the option of whether or not to cross the dividing barrier (Pullan et al., 2012b). If both communities are present in public spaces on either side of the divide, the chances of interaction between them increase.	0.2

[a] The indicator of accessibility was not relevant in the case of Nicosia, because all the public spaces identified were equally accessible and close to one another. Thus, it would not have a significant effect on the outcome of the index.

[b] During the expert interviews, it was determined that locally owned shops and cafés have a greater impact on social interaction. Thus, locally owned activities were given a higher weighting.

7.5   DISCUSSION AND CONCLUSION

This chapter analyses social interaction in the public spaces of a divided city, Nicosia. An index of social interaction was designed and used to quantify the potential for public spaces to promote social interaction, and to offer an approach for discussing the social interaction that occurs there. Different public spaces have different characteristics and qualities that attract visitors, thus making comparison difficult. The results of this index were matched against the opinions of experts to gain a better understanding of the characteristics of the public spaces analysed.

Qualities that influence the potential for a public space to encourage social interaction can be incorporated in three dimensions: physical, social and activities in the public space. It is the representation of these dimensions and their combinations that determines the potential of public spaces to promote social interaction. Physical size was considered as beneficial due to the increased surface area available for hosting various activities, such as shops, cafés, events and festivals – hence

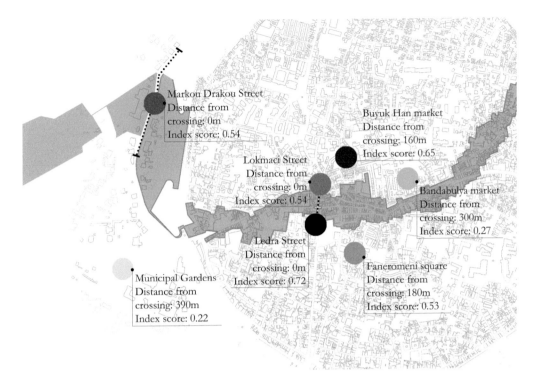

FIGURE 7.3 Map of selected public spaces (shaded circles) and their score on the social interaction index; the darker the shading, the higher the index score. (Base map source: University of Cyprus, 2016.)

more people for interaction. Nonetheless, the public space with the largest surface area was the one with the lowest score on the index of social interaction. This finding concurs with the study of Abu-Ghazzeh (1999), who found that, in the context of residential areas, smaller open spaces felt more inviting and intimate for residents. Residents also preferred smaller public spaces because these were more recognizable: residents knew who was using the place. Although Abu-Ghazzeh's study applies to public spaces in residential areas, it lends some insight as to why, in the context of

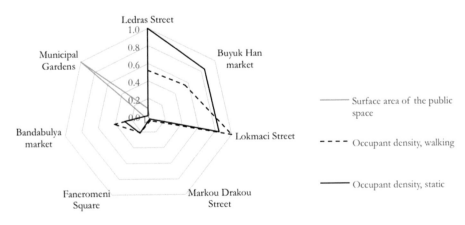

FIGURE 7.4 Standardised values of the indicators for the physical size and social dimensions of occupant density of public spaces.

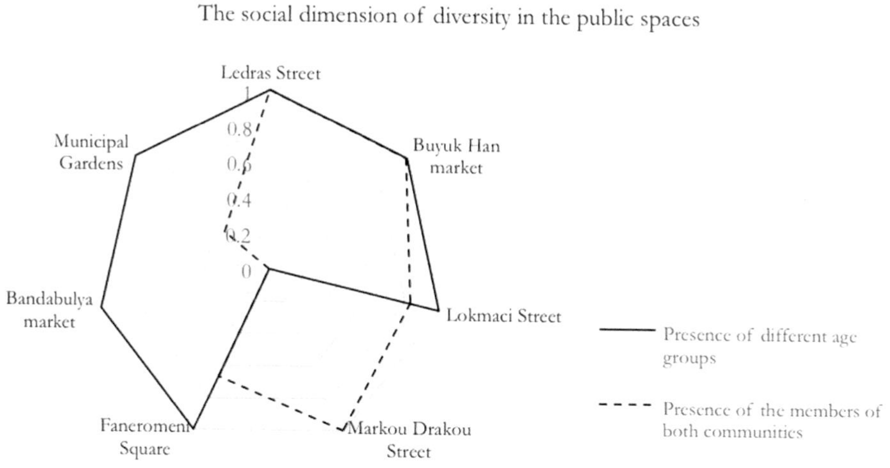

The social dimension of diversity in the public spaces

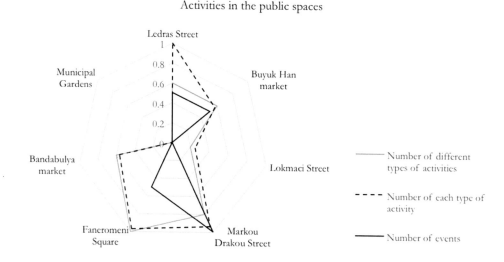

FIGURE 7.5   Standardised values of the indicators for the social dimension of diversity in public spaces.

our study, it is the smaller public spaces that have a bigger role in promoting social interaction and instigating stronger connections between their users.

There are other physical characteristics of public spaces that might also play a role in how they perform with regards to their potential to attract people to use them: namely their shape, exposure to weather, presence of urban furniture and proximity to major access roads or green infrastructure (Lynch, 1960; Gehl and Svarre, 2013; Marcus and Francis, 1998). We did not take these aspects into account in the development of our index. This is a limitation that is likely to be addressed in future research, judging by the wide range of work in urban design linked with the ability of places to promote walkability. Ewing and Handy (2009), for example, focused on measuring the subjective qualities of the urban street environment.

The social dimension of public spaces included the indicators of occupant density (walking and static) and diversity (of age groups and communities). The public spaces with the highest score are, apart from Markou Drakou Street, those with the highest occupant density and presence of different age groups.

Activities in the public spaces

FIGURE 7.6   Standardised values of the indicators for the activities/events dimension of public spaces.

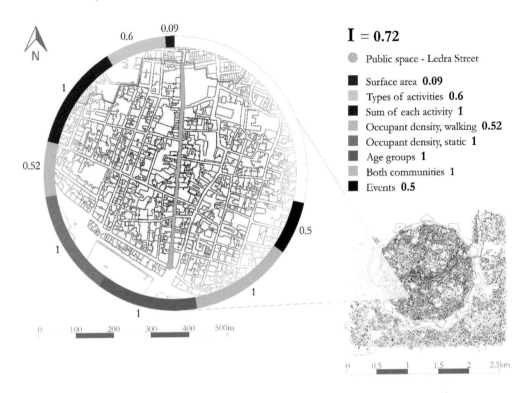

FIGURE 7.7   Index of social interaction for Ledra Street. (Base map source: University of Cyprus, 2016.)

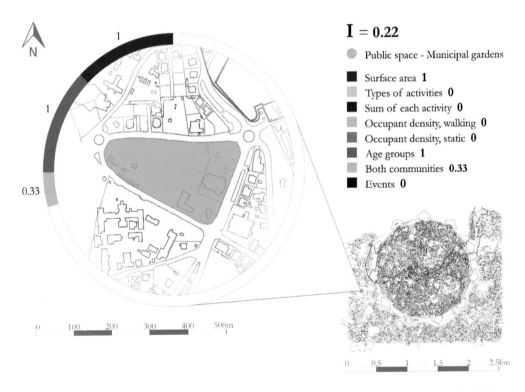

FIGURE 7.8   Index of social interaction for the Municipal Gardens. (Base map source: University of Cyprus, 2016.)

The dimension of activities included the number of different types of activities, their individual number and the number of events in the public space, all important aspects in Nicosia's relationship with public spaces according to the expert interviews. Markou Drakou Street has the single highest score for the indicator of events in public spaces and significantly high scores for the number of different types of activities and numbers of each type of activity.

Social interaction in Nicosia is place-dependent. Most experts agree that the form of interaction between the Greek and Turkish Cypriots varies depending on the place where it is happening. It is, however, always present to a certain degree, especially in cafés and in areas with high concentrations of both communities.

> Recently there is a couple of new bars that opened in the north, and Greek Cypriot youth, […] they are going there frequently. Not necessarily because they have friends or whatever, but they feel so comfortable to cross and go and sit there and have a drink […] you would say that older people and maybe middle-aged people, they were so much more comfortable interacting because they knew people, and they knew why they were doing it. Whereas now, through kind of very casual occurrences, young people start to do the same.
>
> **(Interviewee 2)**

According to expert testimonies, the role of public spaces as places of social interaction is significant because these are the rare spaces where the two communities come into contact and have the opportunity to interact. All public spaces we analysed were publicly owned areas and properties with free access, located within and around the walled city. Until the opening of the Ledra Street crossing, that area was heavily militarised. Even though the area had been used, it became significantly busier and livelier after the checkpoint was opened in 2008; of the public spaces we analysed, Ledra Street was the one with the highest potential for promoting social interaction. Some of the public spaces in this area underwent a revival after 2008.

For the newer generations of Cypriots, frequenting and sharing these spaces with the community from the other side of the dividing buffer zone is still a relatively new phenomenon. People are going back to using, or learning to use, the area of the walled city and the public spaces within for social and civic life, as opposed to letting their militarised, border-zone character prevail. 'Infrastructures for peace', such as the 'Home for Cooperation', are being built around and replacing 'infrastructures of conflict' (Till et al., 2013), i.e. militarised zones. One of the main streets and public spaces in Nicosia in the pre-conflict era was the Ermou Street. Now, the UN buffer zone passes through most of Ermou Street, restricting access to it. The citizens of Nicosia nevertheless still remember the spaces of the city for cooperation and sharing, and they commemorate it through a festival called 'Ermou 1900', organised by the Centre of Visual Arts and Research. The festival is an example of events organised in public spaces that evoke memories of conflict and division, but also of reunion and peace.

> Ermou 1900, every Saturday before Christmas, Greek and Turkish Cypriot craftsmen, peddlers, come here in Ermou Street, and we dress up people as they were in the 1900s, and they sell their products, as they used to in the 1900s at Ermou Street, which was the most commercial street in the city, it was the hub, the heart of the city.
>
> **(Interviewee 12)**

Ermou Street is an example of a public space that vanished with the division of the city, although others emerged from it. Markou Drakou Street would not be a public space if it were not for the 'Home for Cooperation'. This street was also identified by the experts as one of the public spaces with the highest levels of interaction. This suggests that the city is acquiring new 'places of contact' (Abu-Ghazzeh, 1999; Farida, 2013; Talen, 1999) between the two communities. The challenge now is how to turn these casual interactions into occasions that overcome the city's partition.

I think that public spaces help people to come together and be active, instead of being individuals. They become part of a team. It's a place to connect. [...] I think it's a great way to have people in an open space and react with each other [...] This is what I think people use them for, having fun. That is the main use of public spaces.

(Interviewee 9)

Expert testimony on the use of public spaces and their role in Nicosia acknowledge that these are not used to their full potential, perhaps because of people's perceptions of the role of the streets and where people traditionally spend their free time, where they socialise with neighbours and friends.

The Cypriot culture has never favoured active public spaces. Maybe because Cypriots were more home-based. I think the role of public spaces is not very prominent in the Cypriot culture. They're not particularly visited or populated. It's not a major element in the city. Of course it's there, and people put out chairs and they sit there and they have coffee. But people go to the restaurants, they go to cafés. If you go to any part of Nicosia, you find people in cafés, rather than in a square.

(Interviewee 11)

There is a tendency for inhabitants to use public spaces where semi-public activities develop outside, in the street or on a square. This is the case for cafés and restaurants, via their sidewalk areas, but also for shops or events that 'spill over' into the public streets. The public spaces that had a larger number of such activities had a higher potential for promoting social interaction, albeit not completely for free. Focusing on public spaces mainly located around the two pedestrian crossings in the city, we also observed that an area marked by infrastructures of conflict (Till et al., 2013), such as the UN- administered area and the dividing wall, is slowly evolving into an area where the two communities come together to interact, sometimes more and sometimes less intensively (Talen, 1999), offering a hopeful image of new places for exchange and interaction between the two sides of the city.

REFERENCES

Abdelmonem, M. G., & McWhinney, R. (2015). In search of common grounds: Stitching the divided landscape of urban parks in Belfast. *Cities, 44*, 40–49. doi.org/10.1016/j.cities.2014.12.006

Abu-Ghazzeh, T. M. (1999). Housing layout, social interaction, and the place of contact in Abu-Nuseir, Jordan. *Journal of Environmental Psychology, 19*(1), 41–73. doi.org/10.1006/jevp.1998.0106

Amin, A. (2002). Ethnicity and the multicultural city: Living with diversity. *Environment and Planning A, 34*(6), 959–980. doi.org/10.1068/a3537

Aratani, Y. (2010). Public housing revisited: Racial differences, housing assistance, and socioeconomic attainment among low-income families. *Social Science Research, 39*(6), 1108–1125. doi.org/10.1016/j.ssresearch.2010.04.007

Aziz, A. A., Ahmad, A. S., & Nordin, T. E. (2012). Flats outdoor space as a vital social place. *Asian Journal of Environment-Behaviour Studies, 3*(7), 13–24. doi.org/10.21834/aje-bs.v2i5.221

Bevan, R. (2007). *The destruction of memory: Architecture at war.* London, UK: Reaktion Books.

Bicommunal. (2017). Retrieved February 12, 2017, from https://www.merriam-webster.com/dictionary/bi communal

Bollens, S. A. (2006). Urban planning and peace building. *Progress in Planning, 66*, 67–139.

Drucker, S., & Gumpert, G. (1998). Public spaces and the right of association. *Free Speech Yearbook, 36*(1), 25–38. doi.org/10.1080/08997225.1998.10556223

Dymnicka, M. (2010). The end of place as we know it? Attempts at conceptualization. *Human Geographies – Journal of Studies and Research in Human Geography, 4*(1), 53–65.

Ewing, R., & Handy, S. (2009). Measuring the unmeasurable: Urban design qualities related to walkability. *Journal of Urban Design, 14*(1), 65–84. doi.org/10.1080/13574800802451155

Farida, N. (2013). Effects of outdoor shared spaces on social interaction in a housing estate in Algeria. *Frontiers of Architectural Research, 2*(4), 457–467. doi.org/10.1016/j.foar.2013.09.002

Gaffikin, F., Mceldowney, M., & Sterrett, K. (2010). Creating shared public space in the contested city: The role of urban design. *Journal of Urban Design, 15*(4), 493–513. doi.org/10.1080/13574809.2010.502338

Gehl, J., & Svarre, B. (2013). Public space, public life: An interaction. In J. Gehl & B. Svarre (Eds.), *How to study public life* (pp. 1–8). London, UK: Island Press.

Grajewski, T., & Vaughan, L. (2001). Space syntax observation manual. In *'Space syntax' observations procedures manual* (pp. 1–18).

Grichting, A. (2014). Cyprus: Greening in the Dead Zone. In K. G. Tidball & M. E. Krasny (Eds.), *Greening in the Red Zone: Disaster, Resilience and Community Greening* (pp. 429–443). Dordrecht, the Netherlands: Springer Netherlands. http://doi.org/10.1007/978-90-481-9947-1_33

Hickman, P. (2013). 'Third places' and social interaction in deprived neighbourhoods in Great Britain. *Journal of Housing and the Built Environment*, 28(2), 221–236. doi.org/10.1007/s10901-012-9306-5

Holland, C., Clark, A., Katz, J., & Peace, S. (2007). *Social interactions in urban public places*. Bristol, UK: Policy Press.

Kliot, N., & Mansfeld, Y. (1999). Case studies of conflict and territorial organization in divided cities. *Progress in Planning*, 52(3), 167–225.

Low, S. M. (2000). *On the plaza: The politics of public space and culture*. Austin: University of Texas Press.

Lynch, K. (1960). *The image of the city* (1st ed.). Massachusetts: Joint Center for Urban Studies.

Madanipour, A. (2010). *Whose public space?: International case studies in urban design and development*. London, UK: Taylor & Francis.

Marcinczak, S., & Sagan, I. (2011). The socio-spatial restructuring of Lodz, Poland. *Urban Studies*, 48(9), 1789–1809. doi.org/10.1177/0042098010379276

Marcus, C. C., & Francis, C. (1998). *People places: Design guidelines for urban open space* (2nd ed.). New York: John Wiley & Sons, Inc.

Marcuse, P. (1993). What's so new about divided cities? *International Journal of Urban and Regional Research*, 17(3). doi:10.1111/j.1468-2427.1993.tb00226.x

McCann, E. J. (1999). Race, protest, and public space: Contextualizing Lefebvre in the U.S. city. *Antipode*, 31(2), 163–184. doi:10.1111/1467-8330.00098

Mitchell, D. (1995). The end of public space? People's park, definitions of the public, and democracy. *Annals of the Association of American Geographers*, 85(1), 108–133.

Nagle, J. (2009). Sites of social centrality and segregation: Lefebvre in Belfast, a 'divided city'. *Antipode*, 4(2), 326–347. doi:10.1111/j.1467-8330.2009.00675.x

O'Leary, B. (2007). Analysing partition: Definition, classification and explanation. *Political Geography*, 26, 886–908. doi.org/10.1016/j.polgeo.2007.09.005

Öngül, Z. (2012). Analysing the City Identity of Nicosia from a Historical Perspective: External Effects, Solutions Proposed. *Procedia - Social and Behavioral Sciences*, 35, 284–292. http://doi.org/10.1016/j.sbspro.2012.02.090

United Nations. (1982). Provisional guidelines on standard international age classifications. New York.

Pullan, W., Anderson, J., Dumper, M., & O'Dowd, L. (2012a). Rethinking conflict infrastructure: How the built environment sustains divisions in contested cities. *Conflict in Cities*, 21(2). Retrieved from: https://www.urbanconflicts.arct.cam.ac.uk/downloads/briefing-paper-2

Pullan, W., Anderson, J., Dumper, M., & O'Dowd, L. (2012b). Sharing space in divided cities: Why everyday activities and mixing in urban spaces matter. *Conflict in Cities*, 21(4). Retrieved from: http://www.conflictincities.org/PDFs/Briefing%20Papers/Briefing%20Paper%204.pdf

Shaftoe, H. (2008). *Convivial Urban Spaces, Creating Effective Public Places*. Sterling, VA: Earthscan.

Talen, E. (1999). Sense of community and neighbourhood form: An assessment of the social doctrine of new urbanism. *Urban Studies*, 36(8), 1361–1379. doi.org/10.1080/0042098993033

Till, K. E., Sundberg, J., Pullan, W., Psaltis, C., Makriyianni, C., Zincir Celal, R., ... Dowler, L. (2013). Interventions in the political geographies of walls. *Political Geography*, 33, 52–62.

UN-Habitat. (2013). *Streets as public spaces and drivers of urban prosperity*. Retrieved from: http://doi.org/978-92-1-132590-4

Van Deusen, R. (2002). Public space design as class warfare: Urban design, the 'right to the city' and the production of Clinton Square, Syracuse, NY. *GeoJournal*, 58(2), 149–158. doi.org/10.1023/B:GEJO.0000010834.17907.5e

Van Kempen, R. (2007). Divided cities in the 21st century: Challenging the importance of globalisation. *Journal of Housing and the Built Environment*, (22), 13–31. doi.org/10.1007/s10901-006-9064-3

Webster, C., & Timothy, D. J. (2006). Travelling to the 'other side': The occupied zone and Greek Cypriot views of crossing the green line. *Tourism Geographies*, 8(2), 162–181. doi.org/10.1080/14616680600585513

Whyte, W. H. (1988). *City: Rediscovering the center*. Philadelphia: University of Pennsylvania Press.

Woolley, H. (2003). *Urban open spaces*. London: Spon Press.

8 Spatial Variability of Urban Quality of Life in Kirkos Sub-City (Addis Ababa)

Elsa Sereke Tesfazghi, Javier Martinez, and Jeroen Verplanke

CONTENTS

8.1 Introduction 121
8.2 Domains of Quality of Life and Their Attributes 122
8.3 Methodology 124
    8.3.1 Study Area 124
    8.3.2 Data Set and Sampling Procedure 125
    8.3.3 Data Analysis 127
        8.3.3.1 Relation Between Quality of Life and Satisfaction with Domains of Life 127
        8.3.3.2 Variability of Levels of Satisfaction Across Domains of Life 127
        8.3.3.3 Factor Analysis of Quality of Life 128
8.4 Results and Discussion 128
    8.4.1 Descriptive Statistics on Quality of Life and Satisfaction in Domains of Life 128
    8.4.2 Variability in Levels of Satisfaction across Life Domains 130
    8.4.3 The Relation between Quality of Life and Satisfaction in Domains of Life 132
    8.4.4 Factor Analysis of Quality of Life 134
    8.4.5 Context and Policy Implications 137
8.5 Conclusions 137
References 138

8.1 INTRODUCTION

Quality of life (QoL) is a broad term that encompasses notions of a good life, a valued life, a satisfying life and a happy life (McCrea et al., 2006). For our study, we adopted a definition from Foo (2000) that defines urban QoL as the overall satisfaction of an individual with life. There is a consensus among researchers, policy-makers and planners on the need to study QoL in urban areas. Such studies are important, for instance, when policy-makers are deciding on the most effective means of improving standards of living. Outcomes of QoL studies may help city planners understand and prioritise the problems a community faces. QoL research in urban areas has, therefore, received increasing attention (Bonaiuto et al., 2003; Brereton et al., 2008; Lee, 2008; Møller, 2007; Moro et al., 2008).

There are several domains of life that may determine overall QoL. The types of domains used in other, previous studies vary since they depend on the objective of the particular study and the socio-economic setting of the study area. Housing, family income and social connectedness are among the domains often taken into account in QoL studies. The causality direction between overall QoL and specific domains of life is currently under debate. Most studies assume that satisfaction with domains of life causes an overall satisfaction with QoL (bottom-up model) (Lee, 2008), while others attribute domain satisfaction (Schyns, 2001) to a top-down effect of life satisfaction. This controversy has arisen because satisfaction with domains of life can be treated as causes for satisfaction with QoL as well as a consequence of QoL satisfaction. A detailed explanation of the applications of these two approaches can be found in Headey, Veenhoven and Wearing (1991) and Saris (2001). Rojas (2006) argues that it is important to study the nature of the relation between life satisfaction and satisfaction in domains of life, favouring a bottom-up argumentation in which not all domains are equally important. Our study follows a bottom-up model and, without claiming direct causality, identifies the specific domains that have a greater impact on the QoL.

Most QoL studies have been carried out in cities in the Global North, which indicates that the theories and empirical studies of QoL mostly have their origins in Western society. As a result, the main factors that affect the QoL of individuals in the cities of many countries in the Global South have not been clearly identified. Thus, the applicability of the methodologies that originate from the Global North have not convincingly demonstrated what the main factors are that affect the QoL of individuals in cities of other regions. The main purpose of our study is to demonstrate the applicability of the methodology of QoL assessment in a non-Western, non-affluent urban setting and, in particular, to investigate the relative importance of various domains of urban life that contribute to the QoL in the Kirkos sub-city of Addis Ababa (Ethiopia) and their variability at local scale. With this empirical information and some new insights, policy-makers may be able to formulate and implement QoL-enhancing policies in the sub-cities of Addis Ababa. This study can improve understanding of urban QoL and its domains, as well as understanding of the methods for measuring it in cities of similar geographic and socio-economic settings.

This chapter is structured as follows. After this introduction (Section 8.1), Section 8.2 discusses relevant literature on QoL domains and attributes, as well as providing our justification for the domains of life identified in this study. Section 8.3 considers methodological aspects, in particular the methods of data collection and analysis. The results of the study, such as descriptive statistics on domain satisfaction, spatial variability of satisfaction levels for different domains and a multiple regression analysis of the relation between QoL and domains, are presented in Section 8.4. Context and policy implications are also presented in Section 8.4. Finally, conclusions are presented in Section 8.5.

8.2   DOMAINS OF QUALITY OF LIFE AND THEIR ATTRIBUTES

The literature on the domains of life states that life can be approached as an aggregate construct of many specific domains, and life satisfaction (QoL) can be understood as the result of satisfaction in the domains of life (Rojas, 2008). On the other hand, satisfaction in the domains of life (e.g. housing) can be understood as the result of satisfaction in attributes of each respective domain (e.g. housing conditions, number of rooms) (Grzeskowiak et al., 2003; Lee, 2008).

The domains of life that are relevant for studying QoL are often determined by research objectives. There is no consensus on which domains should be included when studying QoL. However, Rojas (2008) argues that the number of domains of life must be manageable. Lee (2008) used five domains of life (i.e., civic services, neighbourhood satisfaction, community status, neighbourhood environmental assessment, and local attachment) when studying QoL in the city of Taipei (Taiwan) and discovered that community status is the domain that best predicts QoL. Foo (2000) compared two studies that used 14 and 18 domains of life for the city–state of Singapore and discovered that in both cases the most important domains were health and family life.

In our study, the domains of life and their specific attributes that may provide relevant information for urban planning were identified based on relevant literature. Personal experience with the study area and local knowledge were also instrumental in this regard. We chose eight domains of life: housing, built-environment, neighbourhood safety, neighbourhood sanitation, quality of public services, access to public services, social connectedness and family income. Table 8.1 shows the domains of life identified in this study and references to relevant literature.

Housing is a basic need and, as such, can affect the QoL in urban areas (Zebardast, 2009). In the absence of appropriate housing, people cannot meet their basic needs. Factors that affect

TABLE 8.1
Domains of Life Chosen to Measure QoL

Domains of Life	Attributes	Studies
Housing	Level of home ownership Crowding in a dwelling Housing condition Number of rooms Housing utilities affordability	Foo (2000), Turksever and Atalik (2001), Ibrahim and Chung (2003), McCrea et al. (2006), Santos and Martins (2007), Li and Weng (2007), Das (2008), Zebardast (2009)
Built-environment	Living place attractiveness Noise pollution Garbage collection Suitability for raising children Neighbourhood congestion	Turksever and Atalik (2001), Li and Weng (2007), Das (2008)
Neighbourhood safety	Crime rate Road safety Police protection in the neighbourhood Nearness to police stations	Foo (2000), Turksever and Atalik (2001), Ibrahim and Chung (2003), Santos and Martins (2007), Das (2008)
Neighbourhood sanitation	Neighbourhood sanitation	Richards et al. (2007), Das (2008)
Quality of public services	Quality of primary schools Quality of health facilities Quality of recreational facilities Garbage collection	Foo (2000), Ibrahim and Chung (2003), Rojas (2008), Das (2008)
Access to public services	Access to primary schools Access to secondary schools Access to health facilities Accessibility to public transport Access to main shopping area Access to sporting and recreational facilities	McCrea et al. (2006), Brereton et al. (2008)
Social connectedness	Social connectedness	Foo (2000), Turksever and Atalik (2001), Ibrahim and Chung (2003), Bonaiuto et al. (2003), Richards et al. (2007), Lee (2008)
Family income	Family income	Foo (2000), Brereton et al. (2008), Rojas (2008)

appropriate housing (attributes of the housing domain) are housing condition, overcrowding and home ownership or tenure.

The built-environment of urban areas can affect the degree to which people appreciate where they live, which in turn can affect their QoL. For instance, the more attractive, less congested and more suitable the built-environment is for raising children, the more satisfied will be individuals who live there.

Neighbourhood safety greatly affects how individuals view their living environment and, thus, their QoL. Unsafe environments can be described in terms of the level of crime and road traffic conditions, as well as perceived levels of safety. McCrea et al. (2006), for example, indicated that perception of crime rates is one of the most important predictors of neighbourhood satisfaction in Brisbane (Queensland, Australia).

Neighbourhood sanitation is often a major concern in cities of developing countries. The status of neighbourhood sanitation can be evaluated based on how the disposal of solid and liquid wastes is managed, as well as the adequacy and quality of water supply.

The quality of public services, such as education, healthcare and recreational facilities, can also affect QoL. It is assumed that the better the quality and coverage of these services are, the higher the QoL will be. Access to these services may also determine the QoL of individuals. It is also assumed that the nearer public services are, the better the QoL. In Genesee County (Michigan, USA), Grzeskowiak et al. (2003) found that the greater the satisfaction with individual community services, the greater the overall satisfaction with life.

Social connectedness in our study is defined as the degree of social interaction in the neighbour-hood. Social connectedness can indicate the strength of interaction in a community. The strength of interaction within a community increases the sense of belonging for residents in the community and, thus, can affect QoL. Grzeskowiak et al. (2003) mentioned that the greater the resident's social ties in the community, the greater the resident's overall satisfaction with the community. In Ethiopia, there are traditional ways by which people enhance interaction within their community, including *Idir* (a burial association for mutual support with deaths and funerals), *Iquib* (a rotating savings and credit group) and *Mahiber* (a religious association through which people gather to celebrate rituals). When there are social or financial problems, individuals in Ethiopia depend on their neighbours, friends and relatives through traditional mechanisms such as *Idir, Mahiber* and *Iquib* (Kebede and Butterfield, 2009).

Family income is the last domain of life considered in this study. It is assumed that the higher family income is, the better QoL will be. A strong relation between income and QoL is commonly observed, and this becomes more pronounced in economically weak societies (Rojas, 2004). This relation implies that increase in family income can make a greater difference in improving QoL in poor areas than in high-income neighbourhoods.

8.3 METHODOLOGY

8.3.1 STUDY AREA

Kirkos sub-city is one of the ten sub-cities (districts) of Addis Ababa, the capital city of Ethiopia. As Figure 8.1a shows, Kirkos sub-city is located in the centre of Addis Ababa; it is divided into 11 Kebeles (Figure 8.1b), the smallest administrative unit used in Ethiopia. As shown in Table 8.2, Kirkos covers an area of 1,472 hectares and is one of the four most densely populated sub-cities of Addis Ababa.

National sport and cultural facilities such as the Addis Ababa stadium and Meskel square are located in Kirkos. It is also host to the international offices of the Organization for African Union (OAU) and the United Nations Economic Commission for Africa (ECA).

BOX 8.1 Case Study Area

Addis Ababa is the largest city and commercial hub of Ethiopia and covers an area of approximately 524 km². The city is divided into ten administrative districts or 'sub-cities', which are further subdivided into Kebeles. Since 2005, the city has undergone physical transformation and investments in subsidised housing ('condominiums') for low- and middle-income residents have been made. Nevertheless, the housing offered is not affordable for 20% of the city's residents, whose income is below the poverty line (UN-HABITAT, 2017).

The study area, Kirkos, is located in the centre of Addis Ababa. It covers an area of 1,472 ha and

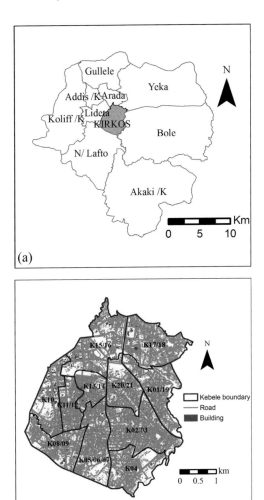

FIGURE 8.1 Study area (a) Addis Ababa sub-city boundaries; (b) map of buildings and roads in Kirkos.

a population of 220,991 inhabitants. Kirkos is one of the four most densely populated sub-cities of Addis Ababa. National sport and cultural facilities such as the Addis Ababa stadium and Meskel square are located in Kirkos. It is also host to the international offices of the Organization for African Union (OAU) and the United Nations Economic Commission for Africa (ECA). The area was selected for this case study because of its diversity in conditions affecting quality of life and for the high concentration of urban renewal initiatives (UN-HABITAT, 2017) being undertaken there.

8.3.2 DATA SET AND SAMPLING PROCEDURE

The data used in the study were collected through a household survey conducted in the 11 Kebeles of Kirkos in 2008. Some 607 households were selected for the survey using a stratified and systematic sampling approach.

TABLE 8.2

Characteristics of the Ten Sub-Cities of Addis Ababa

Name of Sub-City	Population (2007)	Sub-City Area (ha)	No. of Kebeles	Population Density (persons/ha)
Arada	212,129	954	10	222
Addis Ketema	254,972	742	9	344
Lideta	201,613	1,100	9	183
Kirkos	220,991	1,472	11	150
Bole	308,714	11,970	11	26
Nifas Silk Lafto	315,134	5,851	10	54
Yeka	345,807	8,190	11	42
Akaki Kaliti	183,288	12,470	8	15
Kolfe Keranio	428,219	6,325	10	68
Gullele	267,381	3,334	10	80
Totals, Addis Ababa	2,738,248	52,408	99	52.25

Sources: Addis Ababa City administration, 2008; Central Statistical Agency of Ethiopia, 2009 (Retrieved January 15 2009, from http://www.csa.gov.et).

For the stratified sampling, the 11 Kebeles of Kirkos were subdivided into smaller homogenous strata that shared similar characteristics, such as residential structure, social connectedness (determined by a transect walk and informal communication with residents) and access to public services. Knowledge of the study area, supported by QuickBird™ (2002) satellite imagery and Google Earth™, was helpful in identifying these homogenous areas. Classifying Kebeles into strata helped ensure spatial diversity in the selection of households. Within each selected household, an adult of at least 18 years of age was interviewed. In some cases, the household selected did not have any member available to respond to the questionnaire. In such instances, a nearby household was surveyed.

Respondents were asked to indicate their level of QoL satisfaction for each domain and attribute. The QoL score was measured as a response to the question *How do you feel about your life as a whole?* The domain satisfaction score was measured as a response to the question *What is your level of satisfaction for the following domains of life?* The level of satisfaction score for each specific attribute was measured as a resident's response to the question *What is your level of satisfaction for the following attributes?*

Quality of life domains and attributes are often measured using a Likert scale even though there is no commonly defined range. For instance, Foo (2000) applied a 5-point Likert scale ranging from a 'very dissatisfied' to a 'very satisfied', while Santos and Martins (2007) used a 6-point Likert scale, and Brereton et al. (2008) used a 7-point Likert scale. The 6-point Likert scale used in our study allowed respondents to choose the following levels of satisfaction: (1) completely dissatisfied; (2) very dissatisfied; (3) dissatisfied; (4) satisfied; (5) very satisfied; (6) completely satisfied. Considering the context of the study, we assumed there would be no neutral responses. The two end categories – completely dissatisfied and completely satisfied – were included to represent extreme levels of satisfaction or dissatisfaction. This classification is typical for bipolar scales that have been used in similar studies (e.g. Cummins, 1995; Oktay and Rustemli, 2010; Santos and Martins, 2007).

8.3.3   DATA ANALYSIS

8.3.3.1   Relation Between Quality of Life and Satisfaction with Domains of Life

The conceptual model for the Detroit Area Study (DAS), as adapted from Lee (2008), was used in this study (Figure 8.2). The DAS model is based on the relation between the satisfaction of individuals with various domains of life and their QoL (Campbell et al., 1976; Marans, 2003) and as such can provide policy-makers and planners with empirical findings upon which to base their decision-making.

Ferrer-i-Carbonell and Frijters (2004) stated that assuming either cardinality or ordinality of answers to the general question of satisfaction is relatively unimportant in its effect on the results: i.e. they show that there is no substantial difference when satisfaction is treated either as a cardinal or an ordinal variable. Quality of life variables are commonly considered cardinal, and statistical methods, such as descriptive statistics and multiple regression analysis, are commonly applied for such variables (Pacione, 2003; Richards et al., 2007; Turksever and Atalik, 2001). Norman (2010) also justified the use of various parametric methods, for example regression and correlation, with Likert scale data without concern for 'getting the wrong answer' (Norman, 2010, p. 625). With this in mind, in this research we applied descriptive statistics to measure overall QoL and individual domain satisfactions at the Kirkos level, while multiple regression analysis was used to identify the relative importance of satisfaction for various domains on QoL in Kirkos.

A correlation matrix was computed for the raw data as a preliminary assessment of the relation between domain satisfactions. Multiple regression assumptions (e.g. multicollinearity, normal distribution non-zero variance and independent errors) were checked. Regression analysis was carried out between domain satisfaction and QoL scores and between domain satisfaction scores and their respective attributes. The regression coefficients indicated the relative contribution of each domain and the respective effect of attributes on QoL.

8.3.3.2   Variability of Levels of Satisfaction Across Domains of Life

A coefficient of variation (CV) was applied to study the variability of satisfaction in domains of life at the Kebele level. The CV is computed as the standard deviation of satisfaction with domains of life scores divided by the mean of the scores, and as such it shows an absolute variability. A Geographic

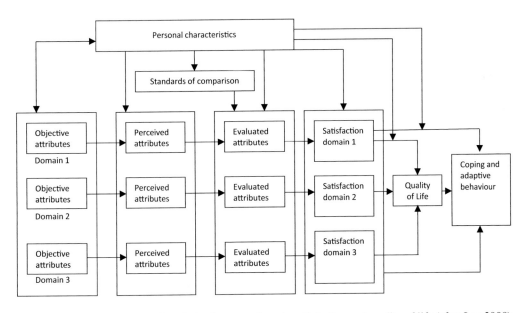

FIGURE 8.2   Model showing the relation between domain satisfaction and quality of life (after Lee 2008).

Information System (GIS) was applied to visualise the variability and identify the spatial grouping of high and low variability by mapping the calculated CV for each Kebele.

8.3.3.3 Factor Analysis of Quality of Life

Prior to conducting the household survey, the literature was consulted to identify perceived attributes of each domain of life. Factor analysis helps identify domains of life by grouping QoL attributes into factors and helps to identify whether the classification of the attributes to respective domains is correct or if some modification is required for future work. It also helps to assess whether there is interaction across domains.

Before applying factor analysis, the suitability of the data was checked according to Kaiser–Meyer–Olkin (KMO) and Bartlett's tests (Li and Weng, 2007). To determine the number of factors extracted, both scree plot and eigenvalue criteria were applied.

In our study, orthogonal varimax rotation was applied to ensure that the attributes were maximally correlated with only one factor and to facilitate the interpretation of the factors. To ensure that the factor scores were uncorrelated, the Anderson–Rubine method was applied to identify the factor-score coefficients. The internal scale reliability of the attributes was checked using Cronbach's alpha (Field, 2013).

BOX 8.2   Methods Applied in the Chapter

Simple GIS methods were used in this study to visualise and communicate statistical results on variations in quality of life. Kebeles, which are the smallest administrative unit in Ethiopia, are also used as spatial elements for analysis of Kirkos sub-city. Descriptive statistics, factor analysis, multiple regression and GIS operations were applied to analyse the QoL and visualise the results.

Geospatial data such as building footprints, land-use maps, road maps, administrative boundaries, distribution of public services (i.e. health and education) and home ownership attribute data were compiled from the municipal records of Addis Ababa. Primary data were collected from a survey of the perception of residents on various attributes of QoL.

To date, GIS in quality of life studies has not been well explored, although spatial factors are expected to affect QoL and the mapping and communication of QoL results are meaningful to policy-makers. The analysis illustrated in this chapter shows how even simple GIS operations can produce input data for QoL studies, as well as facilitate communication to policy-makers by providing a spatial distribution and representation of QoL conditions. These methods can be applied in any study that requires the mapping of indicators that are disaggregated at a scale that is meaningful for planners and policy-makers.

8.4 RESULTS AND DISCUSSION

8.4.1 Descriptive Statistics on Quality of Life and Satisfaction in Domains of Life

Respondents were asked about their QoL in Kirkos sub-city. As Table 8.3 shows, the majority of respondents (62%) expressed some level of dissatisfaction, while only 4% expressed complete satisfaction. The mean score (3.06) indicates that, on average, the respondents in the sub-city were dissatisfied with their QoL.

The score for domain satisfaction was measured as the percentage of respondents to the question *What is your level of satisfaction with the following domains of life?* The percentage of respondents for each level of satisfaction for each domain and the mean and standard deviation of that domain in Kirkos are shown in Table 8.4. Of the eight measured domains, more than half of the respondents

TABLE 8.3

Descriptive Statistics of Levels of Quality of Life in Kirkos

Level of Quality of Life	Proportion (%)	Cumulative Proportion (%)
Completely dissatisfied	15	15
Very dissatisfied	16	31
Dissatisfied	31	62
Satisfied	26	88
Very satisfied	8	96
Completely satisfied	4	100
Mean (Likert)	3.06	
Standard deviation	1.3	

felt dissatisfied or worse in five of them: housing, built-environment, neighbourhood sanitation, quality of public service and family income. A minority of respondents felt dissatisfied or worse with regard to neighbourhood safety, access to public services and social connectedness.

As shown in Table 8.4, the mean score for satisfaction in each domain varied. On average, the least favourably evaluated domain was family income, followed by neighbourhood sanitation. The most favourably evaluated domain in terms of the mean score was social connectedness, followed by neighbourhood safety. This result may be due to the presence of local institutions such as *Idir*, which promotes social interaction among neighbours in Ethiopia (see Ellis and Woldehanna, 2005).

Other QoL studies show variations in domain satisfaction that are, to a degree, similar to our study in relation to levels of satisfaction for social connectivity and safety. Richards et al. (2007) reported that there is a higher level of satisfaction with social connectivity in the informal settlements of the cities of Buffalo and Durban in South Africa. Ibrahim and Chung (2003) reported that respondents were most satisfied with public safety and least satisfied with the environment in the industrial areas of Singapore. Das (2008) reported that respondents in Guwahati City (India) were most satisfied with housing conditions and least satisfied with traffic conditions. These differences can be attributed to differences in context, measurement and the socio-economic settings of the regions in which these studies were conducted.

TABLE 8.4

Statistics for Satisfaction in Domains of Life in Kirkos Sub-city

Level of Satisfaction	Domain of Life (%)							
	HH	BE	SF	SN	PS	AC	SC	IN
Completely dissatisfied	18.0	5.5	3.8	18.1	4.3	2.3	2.5	23.4
Very dissatisfied	12.0	20.1	7.2	23.6	16.5	15.5	3.8	18.8
Dissatisfied	23.6	26.7	14.3	32.0	31.5	24.4	11.9	29.2
Satisfied	25.2	28.7	37.7	17.1	33.4	40.4	41.4	17.6
Very satisfied	10.7	10.5	20.1	4.9	9.6	12.0	22.7	7.2
Completely satisfied	10.5	8.7	16.8	4.3	4.8	5.4	17.8	3.8
Mean (Likert)	3.30	3.45	4.10	2.80	3.40	3.60	4.31	2.78
Standard deviation	1.54	1.31	1.28	1.30	1.14	1.12	1.16	1.37

HH = housing; BE = built-environment; SF = neighbourhood safety; SN = neighbourhood sanitation; PS = quality of public service; AC = access to public service; SC = social connectedness; IN = family income.

8.4.2 Variability in Levels of Satisfaction across Life Domains

Variation of satisfaction with QoL domains within a specific administrative unit is relevant for the implementation of geographically targeted policies, such as area-based policies. However, in the implementation of such policies, aggregated values must be interpreted carefully. The spatial variability of QoL within each Kebele is large (Tesfazghi et al., 2010).

Social connectedness is the domain with the highest level of satisfaction in the study area; the mean score was 4.31(Table 8.4). Spatial variation in satisfaction with social connectedness does, however, occur within each Kebele of the study area, although the level of variation is not the same across Kebeles. Figure 8.3 shows that the satisfaction of residents with degree of social connectedness is highly variable in Kebeles 20/21 and Kebeles 05/06/07, as the high CV indicates. The variability of satisfaction of residents for this domain is lowest in Kebeles 08/09 and 11/12.

Furthermore, family income presents the highest level of dissatisfaction, and its variation is larger than that of social connectedness. Figure 8.4 shows a high level of variation in satisfaction with family income in Kebeles 08/09, located in the southwestern part of the sub-city. Neighbouring Kebeles (10 and 11/12) present less variation (are more homogenous) in family income, suggesting that income differences may vary significantly over small distances.

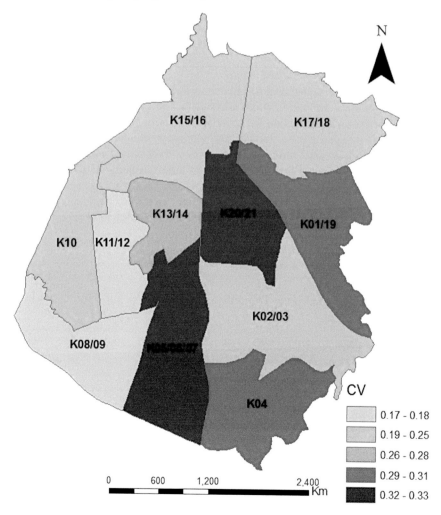

FIGURE 8.3 Variability in social connectedness expressed as a coefficient of variation (CV). Note that high values indicate high variability.

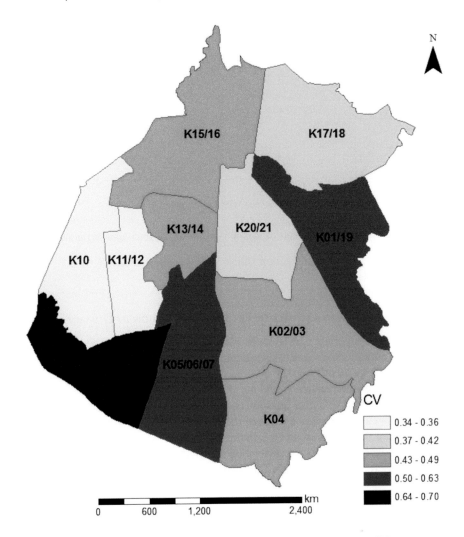

FIGURE 8.4 Variability in family income expressed as a coefficient of variation (CV).

The relationship between the mean and CV of satisfaction with family income is shown in Figure 8.5. The mean score and CV of this domain correlate negatively. This shows that the higher the satisfaction with family income in a particular area, the less variable the family income is. Those Kebeles with high levels of satisfaction with family income are relatively homogeneous in terms of their residents' family income. In contrast, Kebeles with residents having low levels of satisfaction with this domain are relatively heterogeneous in terms of family income.

The exponential relationship between the mean satisfaction score for family income and the CV value is defined as:

$$CV = 130.04e^{-0.365M}$$

where:

CV = the coefficient of variation

M = the mean score of satisfaction with family income

e = the base of the natural logarithm, which is equivalent to 2.718. The equation fitted with $R^2 = 0.54$.

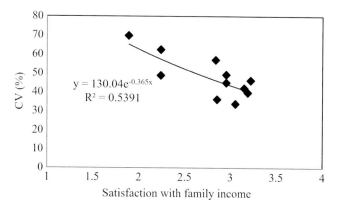

FIGURE 8.5 Relation between coefficients of variation and mean satisfaction with family income.

The plot of mean and CV of satisfaction shows strong scatter (Figure 8.6). As such, there is a weak relation between the two.

8.4.3 THE RELATION BETWEEN QUALITY OF LIFE AND SATISFACTION IN DOMAINS OF LIFE

Equally important to policy-makers is to know what types of domains of life highly influence life satisfaction. One of our objectives was to investigate the relation between QoL and satisfaction with various domains of life by identifying the dominant domains that affect QoL in the sub-city. From this point of view, a relevant question would be: *Which domains of life have the highest impact on the QoL in Kirkos?* The answer to this question could be used to prioritise intervention strategies in terms of domain-related sectors. Knowledge of the relative impact of domains will also contribute to the design of future studies on urban QoL in the region.

For preliminary analysis, Pearson's correlation coefficient was calculated to examine the relation between the different domains of life. Table 8.5 shows the correlation matrix, with statistically significant correlations between the domains of life surveyed in our study, with the level of significance set at 0.05. Multiple regression analysis (quality of life as a dependent variable) was applied to identify the dominant domains of life (Table 8.6). In the analysis, no intercept was applied because the respondents expressed their QoL after considering their satisfaction with each of the specific domains of life. As such, the QoL was assumed to have a score of 1 when all domain scores had a value of 1, which means complete dissatisfaction in all domains results in complete dissatisfaction for an individual's QoL. The relation between QoL and satisfaction in the domains of life in the sub-city takes the following form:

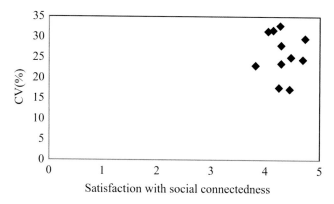

FIGURE 8.6 Relation between coefficients of variation and mean satisfaction for social connectedness.

TABLE 8.5

Correlation Matrix between Satisfaction in Surveyed Domains of Life

	HH	BE	SF	SN	PS	AC	SC	IN
HH	1							
BE	0.375[a]	1						
SF	0.220[a]	0.526[a]	1					
SN	0.217[a]	0.344[a]	0.259[a]	1				
PS	0.167[a]	0.262[a]	0.341[a]	0.162[a]	1			
AC	0.204[a]	0.228[a]	0.252[a]	0.052	0.459[a]	1		
SC	0.080[b]	0.208[a]	0.246[a]	0.103[b]	0.192[a]	0.198[a]	1	
IN	0.494[a]	0.284[a]	0.233[a]	0.277[a]	0.177[a]	0.175[a]	−0.057[a]	1

HH = housing; BE = built-environment; SF = neighbourhood safety; SN = neighbourhood sanitation; PS = quality of public service; AC = access to public service; SC = social connectedness; IN = family income.

[a] Correlation is significant at the 0.01 level (2-tailed).

[b] Correlation is significant at the 0.05 level (2-tailed).

TABLE 8.6

Multiple Regression Analysis for Quality of Life and Satisfaction in Surveyed Domains of Life

Domains of Quality of Life	B	Std. Error	t	Sig.
Level of satisfaction with housing	0.12	0.03	4.27	0.000
Level of satisfaction with built-environment	0.01	0.04	.29	0.775
Level of satisfaction with neighbourhood safety	0.09	0.04	2.58	0.010
Level of satisfaction with neighbourhood sanitation	0.09	0.03	2.96	0.003
Level of satisfaction with quality of public services	0.10	0.04	2.68	0.008
Level of satisfaction with access to public services	0.03	0.04	.89	0.374
Level of satisfaction with social connectedness	0.06	0.03	2.16	0.031
Level of satisfaction with family income	0.46	0.03	14.23	0.000

$R^2 = 0.93$.

$$QoL = 0.12HH + 0.01BE + 0.09SF + 0.09SN + 0.10PS + 0.03AC + 0.06SC + 0.46IN \qquad (8.1)$$

where:

HH = housing
BE = built-environment
SF = neighbourhood safety
SN = neighbourhood sanitation
PS = quality of public service
AC = access to public service
SC = social connectedness
IN = family income

The total variance of the QoL explained by the model in Equation 1 is 93% (significant at 0.05). The coefficients in the model indicate the relative impact of each domain on the QoL. All domains are positively related to QoL, which is to be expected since the higher the satisfaction levels for these domains are, the better an individual's QoL is expected to be. However, the contribution of each domain to QoL is not equal.

In Kirkos, satisfaction with family income (0.46) had the strongest impact on the QoL of the respondents. To check whether this finding was influenced by a large number of respondents with low incomes, an analysis was performed that excluded respondents earning low incomes (i.e. < 500 Birr/month). The results showed that family income has the strongest impact on the QoL for all income groups, which implies that, compared to each of the domains of life, family income is the highest contributor to the level of QoL in the sub-city. The domain that had the second strongest impact on QoL was satisfaction with housing (0.12). Satisfaction with neighbourhood safety, quality of public services, neighbourhood sanitation and social connectedness all had roughly the same impact on QoL in the sub-city, while satisfaction with the built-environment (0.01) had the least impact all. These findings imply that any effort to improve the QoL in the area should give priority to the domains that have the highest impact: family income and housing.

Lee (2008) reported community status as having the strongest causal impact in determining the QoL for Taipei (Taiwan). However, that study did not include family income as a domain of life, so a direct comparison cannot be made with the findings of our study.

Figure 8.7 shows the relation between domain satisfaction and its respective perceived attributes, which were also measured on a 6-point Likert scale. Home ownership had a dominant impact on housing satisfaction, followed by the number of rooms in a house. Housing condition and crowding in a dwelling had the least impact on housing satisfaction. Noise pollution was found to be the most important predictor of built-environment satisfaction. Crime rates in the neighbourhood were likewise found to be a better predictor of neighbourhood safety than the other attributes of safety. The quality of primary schools had a strong impact on satisfaction with the quality of public services. Health service accessibility was also shown to have a strong impact on satisfaction.

8.4.4 FACTOR ANALYSIS OF QUALITY OF LIFE

Factor analysis was applied using 27 perceived QoL attributes that were obtained from the household survey (Table 8.1). The KMO value for this study was 0.67, and Bartlett's test had a significant level of about 0.0, suggesting that the data were suitable for factor analysis.

One of the challenges in using factor analysis is how to interpret the loadings in each factor and attach physical meaning to the factors. Comrey and Lee (1992), cited in Li and Weng (2007), recommended the following: a loading ≥ 0.71 indicates the relation between the factors and variables is excellent; 0.63 indicates a very good relation; 0.55 indicates a good relation; 0.45 indicates a fair relation; and 0.32 indicates a poor relation.

The results of the factor analysis are shown in Table 8.7. Ten factors were extracted by the eigenvalue criterion. However, for ease of interpretation, and considering the total number of attributes, seven factors were finally extracted using a scree plot. These seven factors explain 57% of the total variance:

Factor F1 shows the highest loadings on the attributes related to the built-environment.
Factor F2 shows the highest loadings on the attributes related to neighbourhood safety.
Factor F3 shows the highest loadings on the attributes of housing – physical aspects.
Factor F4 shows the highest loadings on the attributes of access to public services.
Factor F5 shows the highest loadings on the attributes of housing – economic aspects.
Factor F6 shows the highest loadings on the attributes of public service quality.
Factor F7 shows the highest loadings on the attributes related to sport and recreational facilities.

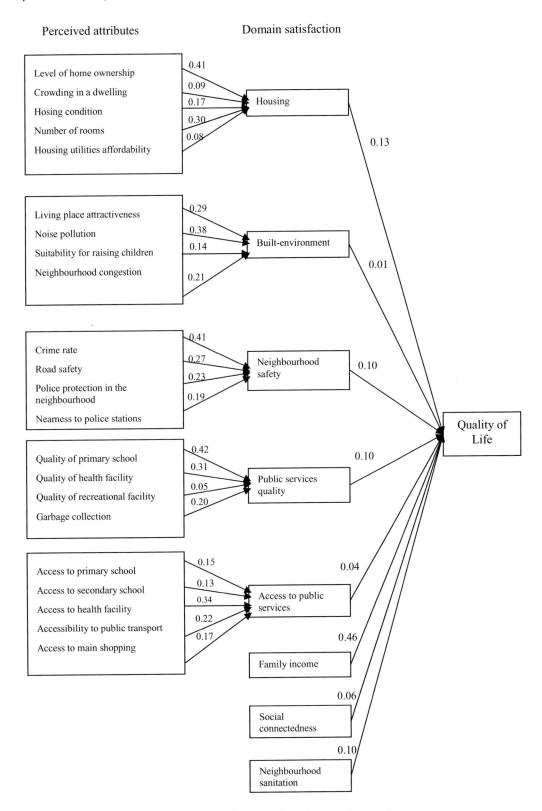

FIGURE 8.7 Relation between quality of life and its domains and attributes in Kirkos.

TABLE 8.7

Factor Loading Matrix for Quality of Life Attributes

Perceived Attributes	Factors						
	F1	F2	F3	F4	F5	F6	F7
1. Neighbourhood congestion	0.678						
2. Living place attractiveness	0.668						
3. Noise pollution levels	0.661						
4. Garbage collection	0.559						
5. Suitable for raising children	0.516						
6. Police protection in the neighbourhood		0.724					
7. Crime rate		0.630					
8. Road safety		0.556					
9. Nearness to police stations		0.386					
10. Number of rooms			0.767				
11. Housing condition			0.736				
12. Crowding in a dwelling			0.719				
13. Access to primary schools				0.818			
14. Access to secondary schools				0.796			
15. Access to health facilities				0.563			
16. Access to main shopping area				0.407			
17. Accessibility of public transport					0.626		
18. Housing utilities affordability					0.544		
19. Level of home ownership					0.523		
20. Family income					0.410		
21. Housing affordability					0.341		
22. Quality of health facilities						0.779	
23. Quality of primary schools						0.765	
24. Quality of secondary schools						0.482	
25. Road maintenance						0.317	
26. Access to sport and recreational facilities							0.750
27. Quality of sport and recreational facilities							0.674
Eigenvalue	4.08	3.15	2.05	1.84	1.51	1.38	1.34
Proportion of variance explained (%)	9.45	9.17	9.06	8.51	7.45	6.92	6.33
Proportion of total variance explained (%)	57						

Extraction method: principal component analysis.

Rotation method: varimax with Kaiser normalization.

The internal scale reliability of the attributes was checked using Cronbach's alpha. The degree of reliability for QoL attributes was 0.74, which implies that, on the whole, the 27 attributes measured the same construct (QoL).

The factor analysis identified five of the eight surveyed domains (i.e. built-environment, neighbourhood safety, housing, access to public services and quality of public service), which suggests that the domain of housing can be divided into physical and economic aspects. We note that there are no attributes related to neighbourhood sanitation and social connectedness; as such, the factor analysis did not identify these domains, while, furthermore, it incorporated family income into the economic aspect of housing. However, factor analysis identified a new domain: that of sport and recreation. Therefore, depending on the context, future studies in this study area should consider this domain.

8.4.5   CONTEXT AND POLICY IMPLICATIONS

In previous sections, we have shown that the residents of Kirkos expressed dissatisfaction with most domains of life and overall QoL. There are some ongoing projects in the sub-city that are aimed at improving residents' quality of life. These projects are either being implemented by the Addis Ababa city administration or the Kirkos sub-city or Kebele administrations. For instance, to improve housing conditions in Kirkos, the Addis Ababa city administration has undertaken several upgrading and renewal projects (e.g. construction of condominiums) with priority for low-income households. These apartments are highly subsidised and the city is unable to satisfy the high demand for housing by low-income households, which account for over 80% of the city's population (Addis Ababa City Government, 2011). Considering the low level of satisfaction with housing in Kirkos sub-city indicated by our study, the city administration should ensure the continuation of the existing activities and look for housing strategies for low-income households.

Kirkos is one of the sub-cities in Addis Ababa that has a poor built-environment. The poor quality of roads, shortage of pedestrian walkways and poor construction of housing make the sub-city undesirable to live in. The upgrading of roads, construction of condominiums, protection and rehabilitation of green areas and construction of playgrounds are part of the city's current master plan. However, our study shows that the built-environment has the least effect on quality of life (0.01). Other domains should therefore receive more attention if the quality of life is to be improved.

Out of the solid waste generated daily in Addis Ababa, less than 68% is collected (Addis Ababa City Government, 2011). The remainder is simply dumped in drainage channels, rivers and open areas and even on the streets. Kirkos is one of the sub-cities where sanitation problems associated with poor waste collection and disposal practices are widespread. Nevertheless, Kebele administrations are organizing unemployed individuals to collect dry waste from households. The involvement of individuals and the private sector in waste collection could be an option for improving neighbourhood sanitation and, consequently, satisfaction with the QoL in the sub-city.

To improve the quality of public services, renewal and upgrading activities are being undertaken by the sub-city administration. For instance, to improve the quality of government and public schools, the sub-city is upgrading existing schools by constructing additional classrooms and providing better equipment.

Providing access to savings and credit facilities is one of the activities carried out by Kebele administrations to improve the income of residents. However, access to sufficient credit for the poor seems to be limited. As expected, there is a low level of satisfaction with income in Kirkos, which highlights the importance of creating and strengthening income generation activities, such as micro- and small enterprises.

As shown in this study, the respondents of the sub-city expressed satisfaction with neighbourhood safety. The presence of strong social connectedness may be making respondents feel safe in their neighbourhood. Also, Kebele administrations appoint individuals to safeguard the neighbourhood. In Ethiopia, there are several types of local institutions (e.g. *Idir*, *Iquib* and *Mahiber*) that promote social connectedness, and the sub-city should support and cooperate with such local organizations.

The analysis of the domains and QoL in Kirkos sub-city undertaken in this study can be used by policy-makers in Addis Ababa to prioritise their development activities. The outcome of our study clearly shows that when it comes to enhancing the residents' level of satisfaction with their QoL, family income and the housing situation should receive priority over other domains.

8.5   CONCLUSIONS

The main purpose of our study was to identify the relationship between QoL and satisfaction with domains of life and to visualise its variability. We found that, on average, individuals in Kirkos are dissatisfied with their QoL. The source of this dissatisfaction was assessed by studying certain domains of life (i.e. housing and family income) and their relation to the QoL.

Factor analysis reduced 27 attributes of QoL to seven factors, of which six were related to the QoL domains chosen in our study. The newly identified domain (sport and recreation) should be considered in future studies as a new domain. Although sport and recreation are often considered a domain in studies that are carried out in Western countries (e.g. Bonaiuto et al., 2003; Oktay and Rustemli, 2010), it has been mostly ignored in studies conducted in developing countries (e.g. Møller and Saris, 2001). Though most studies consider housing as a single domain, the results of our factor analysis indicate that it can be split into two domains: physical aspects and economic aspects.

The QoL in the sub-city was generated from different domains than those commonly employed in Western contexts. In Kirkos sub-city, the respondents expressed satisfaction with only three domains of life: neighbourhood safety, access to public service facilities and social connectedness. Our result has some similarities with Richards et al. (2007) and Ibrahim and Chung (2003), who reported satisfaction with social connectivity and safety for informal settlements in South Africa and for an industrial area in Singapore. Contrary to our results, Das (2008) reported that respondents in Guwahati City (India) were most satisfied with housing conditions. The differences in these results show the importance of the context, measurement and socio-economic setting of the region in which such studies are carried out.

The degree to which the domains affect the QoL in Kirkos was identified through multiple regression analysis. The domain of family income had the strongest impact on QoL, while that of the built-environment had the weakest impact. This result is similar to the findings of Saris (2001) for a low-income area in Russia. However, this result contrasts with most studies that are performed in Western countries, where income is shown to have less impact on the QoL because most people in Western countries have sufficient income to cover their basic needs (Headey et al., 1991).

The results of our study show, therefore, that the relation between the QoL and its domains is different in the Global South than in Western countries. The findings of this study can serve as a platform to understand and further analyse the QoL in this region, as well as other sub-cities of Addis Ababa. The findings of the study also show the importance of mapping the spatial variation of QoL satisfaction since within some domains of life (i.e. income) neighbouring areas show contrasting levels of heterogeneity.

REFERENCES

Addis Ababa City Government. (2011). *Revision of Addis Ababa Master Plan*. Addis Ababa: Office for the Revision of Addis Ababa Master Plan. Retrieved from http://www.telecom.net.et/~aamp/Housing.htm (last access: February 20, 2011).

Bonaiuto, M., Fornara, F., & Bonnes, M. (2003). Indexes of perceived residential environment quality and neighbourhood attachment in urban environments: A confirmation study on the city of Rome. *Landscape and Urban Planning, 65*(1–2), 41–52.

Brereton, F., Clinch, J. P., & Ferreira, S. (2008). Happiness, geography and the environment. *Ecological Economics, 65*(2), 386–396.

Campbell, A., Converse, P., & Rodgers, W. (1976). *The quality of American Life: Perception, evaluation and satisfactions*. New York: Russell Sage Foundation.

Comrey, A. L., & Lee, H. B. (1992). *A first course in factor analysis* (2nd ed.). Hillsdale, N.J.: L. Erlbaum Associates.

Cummins, R. A. (1995). On the trail of the gold standard for subjective well-being. *Social Indicators Research, 35*(2), 179–200. doi:10.1007/bf01079026

Das, D. (2008). Urban quality of life: A case study of Guwahati. *Social Indicators Research, 88*, 297–310.

Ellis, F., & Woldehanna, T. (2005). *Ethiopia participatory poverty assessment 2004–05*. Addis Ababa: Ministry of Finance and Economic Development.

Ferrer-i-Carbonell, A., & Frijters, P. (2004). How important is methodology for the estimates of the determinants of happiness? *Economic Journal, 114*(497), 641–659.

Field, A. P. (2013). *Discovering statistics using IBM SPSS statistics: and sex and drugs and rock 'n' roll* (4th ed.). London: SAGE.

Foo, T. S. (2000). Subjective assessment of urban quality of life in Singapore (1997–1998). *Habitat International*, *24*(1), 31–49.

Grzeskowiak, S., Sirgy, M. J., & Widgery, R. (2003). Residents' satisfaction with community services: Predictors and outcomes. *Journal of Regional Analysis and Policy*, *33*(2), 1–36.

Headey, B., Veenhoven, R., & Wearing, A. (1991). Top-down versus bottom-up theories of subjective well-being. *Social Indicators Research*, *24*(1), 81–100. doi:10.1007/bf00292652

Ibrahim, M., & Chung, W. (2003). Quality of life of residents living near industrial estates in Singapore. *Social Indicators Research*, *61*, 203–225.

Kebede, W., & Butterfield, A. K. (2009). Social networks among poor women in Ethiopia. *International Social Work*, *52*(3), 357–373. doi:10.1177/0020872808102069

Lee, Y. J. (2008). Subjective quality of life measurement in Taipei. *Building and Environment*, *43*(7), 1205–1215. doi:10.1016/j.buildenv.2006.11.023

Li, G., & Weng, Q. (2007). Measuring the quality of life in city of Indianapolis by integration of remote sensing and census data. *International Journal of Remote Sensing*, *28*(2), 249–267.

Marans, R. W. (2003). Understanding environmental quality through quality of life studies: The 2001 DAS and its use of subjective and objective indicators. *Landscape and Urban Planning*, *65*(1–2), 75–85. doi:10.1016/S0169-2046(02)00239-6

McCrea, R., Shyy, T.-K., & Stimson, R. (2006). What is the strength of the link between objective and subjective indicators of urban quality of life? *Applied Research in Quality of Life*, *1*(1), 79–96. doi:10.1007/s11482-006-9002-2

Møller, V. (2007). Quality of life in South Africa: The first ten years of democracy. *Social Indicators Research*, *81*(2), 181–201.

Møller, V., & Saris, W. E. (2001). The relationship between subjective well-being and domain satisfactions in South Africa. *Social Indicators Research*, *55*(1), 97–114. doi:10.1023/a:1010851412273

Moro, M., Brereton, F., Ferreira, S., & Clinch, J. P. (2008). Ranking quality of life using subjective well-being data. *Ecological Economics*, *65*(3), 448–460.

Norman, G. (2010). Likert scales, levels of measurement and the 'laws' of statistics. *Advances in Health Sciences Education*, *15*(5), 625–632. doi:10.1007/s10459-010-9222-y

Oktay, D., & Rustemli, A. (2010). Measuring the quality of urban life and neighbourhood satisfaction: Findings from Gazimagusa (Famagusta) Area Study. *International Journal of Social Sciences and Humanity Studies (IJ-SSHS)*, *2*(2), 27–37.

Pacione, M. (2003). Urban environmental quality and human wellbeing – a social geographical perspective. *Landscape and Urban Planning*, *65*(1–2), 19–30.

Richards, R., O'leary, B., & Mutsonziwa, K. (2007). Measuring quality of life in informal settlements in South Africa. *Social Indicator Research*, *81*, 375–388.

Rojas, M. (2004). *The complexity of well-being: A life satisfaction conception and a domains-of-life approach*. Paper presented at the International Workshop on Researching Well-being in Developing Countries, Delmenhorst near Bremen, Germany July 2–4.

Rojas, M. (2006). Life satisfaction and satisfaction in domains of life: Is it a simple relationship? *Journal of Happiness Studies*, *7*(4), 467–497. doi:10.1007/s10902-006-9009-2

Rojas, M. (2008). Experienced poverty and income poverty in Mexico: A subjective well-being approach. *World Development*, *36*(6), 1078–1093.

Santos, L. D., & Martins, I. (2007). Monitoring urban quality of life: The Porto experience. *Social Indicators Research*, *80*(2), 411–425. doi:10.1007/s11205-006-0002-2

Saris, W. E. (2001). What influences subjective well-being in Russia? *Journal of Happiness Studies*, *2*(2), 137–146. doi:10.1023/a:1011556429502

Schyns, P. (2001). Income and satisfaction in Russia. *Journal of Happiness Studies*, *2*(2), 173–204. doi:10.1023/a:1011564631319

Tesfazghi, E., Martinez, J., & Verplanke, J. (2010). Variability of quality of life at small scales: Addis Ababa, Kirkos Sub-City. *Social Indicators Research*, *98*(1), 73–88. doi:10.1007/s11205-009-9518-6

Turksever, A. N., & Atalik, G. (2001). Possibilities and limitations for the measurement of the quality of life in urban areas. *Social Indicators Research*, *53*, 163–187.

UN-HABITAT. (2017). *The State of Addis Ababa 2017: The Addis Ababa we want*. Nairobi: UN-Habitat.

Zebardast, E. (2009). The housing domain of quality of life and life satisfaction in the spontaneous settlements on the Tehran metropolitan fringe. *Social Indicators Research*, *90*, 307–324.

CONTENTS

9.1 Theoretical Framework 142
    9.1.1 Introduction 142
    9.1.2 City Signature and Identity 142
        9.1.2.1 Actors 142
        9.1.2.2 Foundation 143
        9.1.2.3 Driving Forces 143
        9.1.2.4 Linkages between Actors, Foundation and Driving Forces 144
    9.1.3 Transport Planning, Urban Planning and Urban Design 144
    9.1.4 Gender-Related Issues in Urban Transport and Planning for Urban Safety 144
    9.1.5 Filling the Gaps 145
    9.1.6 Approaches for Linking Urban Design and Transport Planning 145
        9.1.6.1 Space Syntax 145
        9.1.6.2 Individual Movements 146
9.2 Study Area and Data Collection 147
    9.2.1 Study Area 147
    9.2.2 Fieldwork and Data Compilation 149
9.3 Methodology 149
    9.3.1 Urban Morphology Indicators 149
    9.3.2 Attractiveness Index and SMCE 152
9.4 Results and Discussion 153
    9.4.1 Survey Results and Analysis 153
        9.4.1.1 Demographics 153
        9.4.1.2 Travel Distance and Purpose 154
        9.4.1.3 Time Spent per Type of Activity 154
    9.4.2 Sampling Daily Routines and Tracing of Activities 154
    9.4.3 Indicator Results 155
        9.4.3.1 Activity Locations and Accessibility Effect on Travel Patterns 155
        9.4.3.2 Socio-Cultural Variances and Urban Structure in Relation to Activity
                Patterns and Travel Behaviour 157
        9.4.3.3 How Can Urban Form Be Defined and Quantified? 157
        9.4.3.4 How Do Women's Activities and Cultural Aspects Affect Urban
                Structure in the City? 157
        9.4.3.5 How Does the Street Network Shape Movements in Cities? 157
        9.4.3.6 What Barriers Do Women Face on Their Daily Trips? 159
9.5 Discussion 159
9.6 Conclusion 159
References 160

9.1 THEORETICAL FRAMEWORK

9.1.1 INTRODUCTION

Where gaps exist between urban design practices and urban transport planning, problems are created for travellers, especially for vulnerable groups of travellers such as women. These gaps evolve particularly when in urban design the focus is only on the physical structures involved, thereby ignoring social needs and those of individuals in design and planning. As such, mobility behaviours and preferences are also ignored. To tackle this issue, we developed a theoretical framework that addresses the city's physical and social elements with respect to the travel behaviour of women. Aided by participatory GIS, the framework we describe in this chapter combines and links various theories from urban design, urban morphology, transport planning and the behavioural sciences (focusing on gender-sensitive perceptions). These various dimensions are applied in a case study of several neighbourhoods in Istanbul, Turkey.

9.1.2 CITY SIGNATURE AND IDENTITY

A city is a complex entity (Batty, 2007) that is characterised by (1) its residents' social diversities and (2) its physical structure (i.e. built environment), which is enriched dynamically by (3) its citizens' activities, thus creating a unique signature and identity for every city. To integrate all these dimensions, it is essential to find the *linkages* between them instead of focusing on one dimension. The theoretical structure of our framework focuses on three core elements that define urban complexity, which we expand upon in the remainder of this subsection:

1. Actors (human beings and their social dimension);
2. Foundation (urban morphology and geometry);
3. Driving forces (human activities and mobility), as the link between actors and foundation.

9.1.2.1 Actors

All social groups live within the same urban space of the city. Each individual in this unit is active and has different needs (e.g. playgrounds for children, ramps for disabled people, street lighting for women, especially during the night) (Beall, 1996). Together, women, men, the elderly and children all shape a city based on their experiences, and all participate in different ways in forming the image of the city. Changes to a city are not simply the result of shaping its physical structure; rather, they arise out of the role humankind lives out in the city (Vance, 1977).

In the matter of urban morphology and movement, women and men have different needs, demands and perceptions (UN-Habitat, 2008), and therefore gender plays an important role in urban environments (Tacoli and Satterthwaite, 2013). In traditional societies (e.g. Turkey and other countries in the Middle East), women are often more connected with the family and indoor activities, while men are more connected with outdoor activities and jobs (Küçük, 2013). Both women and men are concerned about safety. Women, however, have a different perception than men and often feel less safe (Crabtree and Nsubuga, 2012). For example, women are more concerned about their personal security, while men are more concerned about road safety (FHWA, 1996). Key issues for understanding the complexity of urban security are individual local needs and gender-sensitive perceptions, protection from sexual violence, and avoidance of public spaces tinged with fear 'through the unmasking of diverse masculinities and femininities' (Feuerschutz, 2012).

A city's transport system is based in some way on its structure, while its society is founded on gender-related variations in terms of responsibilities, activities and behaviour (UN-Habitat, 2008). Thus, travel patterns and safe mobility are among the most important gender-related issues in urban design (Wachs, 1996).

9.1.2.2   Foundation

Cities are the outcome of an interaction between movement technology (mobility and accessibility), communication and land values (Frey, 1999). The physical outcome of this interaction is called urban morphology, which considers the advanced description and classification of urban form (Pacione, 2005).

Conzen (1960) divided the urban landscape into three main elements: (1) the town plan or city layout; (2) building form; and (3) land use. This classification was based on the degree of changes in these elements, i.e. land use is mainly subject to change, and buildings are physically long-lasting, while their use might change over time. City layouts are the most permanent elements in the urban geometry. Similarly, Lynch (1960) addressed the spatial interaction between humans and the urban environment by comparing the city structure as seen by its citizen (i.e. a mental map) against an actual map. Lynch listed five major elements of (physical) *urban structure* that were identified by inhabitants: (1) landmarks, (2) districts, (3) paths, (4) nodes and (5) edges. They are descriptive in terms of *urban spatial arrangements*. In other words, urban morphology is the physical structure of the city.

Traditionally, designers and planners tried to improve the physical form of urban edges, ignoring the physical and social impact of standardised street patterns on the urban environment (Gökariksel and Mitchell, 2005). Street patterns, however, have a large effect on location suitability. Similarly, Southworth and Ben-Joseph (2003) linked the *spatial configuration* of a city with suitability for particular types of land use. For example, they found that a grid-structure configuration in a neighbourhood makes it suitable for high density building and commercial use; most likely, that neighbourhood will be a city centre. This structure makes an area visually highly permeable and accessible for pedestrians. A neighbourhood with a curvilinear structure, according to Southworth and Ben-Joseph (2003), is highly suitable for single-row housing and other residential use. Following the classification of Conzen (1960) and Lynch (1960), our study focused on *street and building layouts* as the main elements of urban morphology.

9.1.2.3   Driving Forces

People's need for mobility is one of the strongest forces that shapes a city's physical structure, with people's activities the real drivers behind that force, making mobility the pulse of the city (Johnson, 1967; Latham et al., 2009). People move to meet their social and economic needs (Algers et al., 2005; Meyer and Miller, 2001; Vance, 1977). This movement is called a 'journey', and the connections between different destinations – themselves called 'functional cluster of activities' – are called 'linkages' (Latham et al., 2009; Vance, 1977). Transport has a direct effect on households and on an individual's daily activity schedules, as well as on the type and quality of activities (Fox, 1995).

People's movement, or mobility, is an essential issue for both men and women. However, women are often restricted in their mobility by issues of access, opportunity, family commitment and customs (Pacione, 2005). Women's trip patterns and frequency are, as a result, commonly more complex than those of men (Hasson and Polevoy, 2012). Women often have more responsibilities than men in terms of household duties, family commitments and work and hence tend to combine different activities in one trip (work, shopping, taking children to/from school, etc.) (Mauch and Taylor, 1997).

The image of a neighbourhood has an effect on determining an individual's mobility and activity patterns. For example, fear of crime in a neighbourhood may encourage an inhabitant to withdraw physically and socially from the community and its activities, thus producing a decline in their mobility and participation in society (Wesly, 1986). Fear of being attacked or harassed, in particular, results in women losing their freedom to be or mingle in public places (Phadke, 2012). This especially applies to leisure-related trips that take place in evening hours.

To conclude, women's perception of travel is related to trip motivation, access, opportunity, family obligations, customs and personal security in public areas. This varies from one area/country to another, in many cases based on customs and women's roles in the society.

9.1.2.4   Linkages between Actors, Foundation and Driving Forces

Transport planning typically builds on the three elements of actors, foundation and driving forces. Given the symbiotic relation between people (the actors), transport (as a driving force) and urban morphology (as a foundation) described above, the ultimate goal is to integrate the objective (land use, building, street layout, etc.) and subjective (individual's previous experiences, daily behaviour, lifestyle, etc.) aspects – in other words, by considering 'the city on the ground and the city in the mind' (Pacione, 2005 p.541).

9.1.3   Transport Planning, Urban Planning and Urban Design

Urban planners focusing on urban change often consider cultural, social and economic factors, while human interaction with the surrounding environment is forgotten. Tradition and social customs shape the (power) relations between men and women. Societies treat women and men differently and expect different contributions and forms and levels of participation from them. In transport planning practice, women tend to be treated similarly to men – sometimes women's position/roles are even completely ignored – even though women and men have different transport needs (Beall, 1996).

Nenci and Troffa (2006) differentiated the way women choose their routes from men. In their study, women tended to choose longer routes, motivated by aesthetics, in particular the pleasantness and panoramic features of the selected route, while men chose shorter routes, motivated by the time to destination or route complexity. Thus, urban designers and architects are left with a complex task to link and fulfil three main challenges: provide infrastructure for the organization of human life, give shape to a city's structure and ensure the functioning of a diverse collection of activities, while considering the different needs of women, children and men.

9.1.4   Gender-Related Issues in Urban Transport and Planning for Urban Safety

Although women play an important role in urban development and often actively participate in developing their neighbourhood, in the past not enough attention has been given to their specific needs (Rakodi, 1991). For example, in urban design and architecture human perception is often forgotten, especially differences in gender-related perception. van Nes and Nguyen (2009) observed that gender differences in perceptions of urban public spaces are stronger at night when shops are closed. In many Middle Eastern societies, women cannot go out alone: they use the streets as a means of passage and go out into public spaces only in male company (Sadiqi and Ennaji, 2013).

In social sciences, women's and men's safety is related to the space in which they carry out their activities. In general, a woman is more vulnerable than a man, which makes her feel more exposed to risk in public spaces, e.g. when using public transport services (Kunieda and Gauthier, 2007). Aspects of urban safety include road safety but also perceptions of personal security. Road safety is related to public transport services, road speed, chances of deaths and road accidents, and chances of injuries from cars for pedestrians and cyclists. Men pay more attention to these aspects of safety (FHWA, 1996). For a woman, however, perceptions of personal security receive more attention, as personal security is related to physical strength and an environment that protects her against verbal and physical harassment (Kunieda and Gauthier, 2007).

In Istanbul, women's perception of security differs from place to place. Secor (2002) and Gökariksel and Mitchell (2005) addressed perception of women's personal security in public spaces

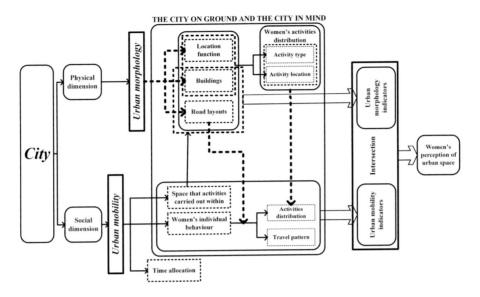

FIGURE 9.1 Conceptual framework.

in Istanbul. They found that women, whether they wear a veil or not, often face obstacles in their daily mobility, depending on varying community perceptions and degrees of secularism.

9.1.5 FILLING THE GAPS

In order to fill the gap between the disciplines of transport and urban planning/design, we present in this chapter a conceptual framework (Figure 9.1) that links these disciplines while, at the same time, focusing on the perceptions and needs of one social group, that of women. The framework combines the plans of individuals (knowledge and information about route, time and locations) extracted from questionnaires, mental maps and reference maps that have been input into a GIS. Urban morphology is then linked with urban mobility using spatial multi-criteria evaluation (SMCE) and GIS techniques to generate a spatial distribution of trip patterns that provide the basis for our analysis.

To address the identified gap, several approaches have been used, combining urban planning/design techniques such as space syntax and mental (or sketch) mapping (Kim and Penn, 2004; Long et al., 2007) with transport planning techniques such as time-use surveys and street interviews (Bhat and Koppelman, 2003). This provides the means for linking human activities and travel patterns through activity mapping, travel behaviour analysis and urban morphology analysis (Balmer et al., 2009; Nagel and Flötteröd, 2009).

9.1.6 APPROACHES FOR LINKING URBAN DESIGN AND TRANSPORT PLANNING

A link between urban design and transport planning can be established using techniques that relate human visual interaction with the environment, which links morphology and mobility.

9.1.6.1 Space Syntax

Urban design studies address the interaction of humans with their surrounding environment through vision, where a memory is formed about a place, the time and experiences that took place. Addressing the issue of spatial interaction between human beings and their urban environment, Lynch (1960) compared a city's structure as seen by its citizens (through mental maps) and a site map of the city. By exploring the city, Lynch identified five major elements of urban structure that can be used for describing spatial arrangements (see Subsection 9.1.2).

By addressing the question 'how does human living and culture connect through space?', Hillier (1999) introduced the 'space syntax approach' to measure the relations between spatial patterns in urban space. Based on network topology, his approach focuses on measuring the movements of pedestrians and vehicles and the influence of urban patterns on them. Graph theory provides a means for investigating the abstract topological relationships between network elements. In space syntax, the output of the street graph is not planar, thus it cannot be properly called a graph map. Instead, it has been called an axial map and its lines axial lines (Batty, 2004). The establishment of the space syntax concept is supported by two theoretical concepts called 'natural movements' and 'movement economy'.

The natural movement concept proposes a homogenous distribution of urban form in order to attract and generate movements within an area. This gives spatial arrangements of networks the opportunity to explain movements within that space, based on considering movements as linkages between buildings in the system. Likewise, the movement economy concept proposes – using the gravity concept – that there is a 'central driving force' behind the spatial growth of a city. To activate all those concepts, space syntax uses three indexes: connectivity, global integration and local integration. *Connectivity* measures direct point–line or line–line contact, depending on the direct links between elements (Batty, 2004). The more line-to-line connections that exist, the more attractive the area is. The *global integration* index is the sum of the minimum number of dominant lines that must be used in whole, or in part, to go from one line to another (Hillier, 1999). With limited logic on how the city is structured and how it functions, this index still gives good insights into the macro-scale functioning of a city. The *local integration* index is the sum of the minimum number of intervening lines that must be used in whole, or in part, to go from one line to another. Local integration has the spatial capability of explaining movement patterns for both pedestrians and vehicles (UN-Habitat, 2008). The local integration index describes local indicative movement patterns, connecting specific areas with other parts of the city. It presents a more detailed picture of movement structures, showing the real busy routes within the city.

By adopting Hillier's approach, the relation between a city's structural elements (e.g. road networks and building layouts) and activity patterns (an individual's movement) is defined. The concept has been developed using human vision, by constructing axial lines within the street layout and between building edges, with a maximum length of 1500 m (the limit of human vision) (Bentley et al., 1985; de Smith et al., 2009; Moughtin, 2003).

9.1.6.2 Individual Movements

Hillier assumed that human movements in a city are based on visual steps. Women's perceptions of a city are connected to aesthetic and safety considerations (Kunieda and Gauthier, 2007; Nenci and Troffa, 2006). For example, women change their travel behaviour (routes and times of travel) if they feel unsafe (Kunieda and Gauthier, 2007). In order to link this concept to urban mobility, in addition to explaining women's activities and movement patterns, participatory mapping techniques and GIS-based spatial analysis are required (UN-Habitat, 2008).

Women are socially more active than men; they tend to walk more and use public transport more, especially if they are from a single-car family. And married women with children generally makes more complex trips (i.e. combining activities) than single women or men. Thus, research into human behaviour is important (Bentley et al., 1985; Hillier, 1999; Kim and Penn, 2004; Lynch, 1960) for being able to track behaviour, for which qualitative measurements are required. These measurements can be based on a number of indicators, such as visual permeability, variety of use, connectivity and accessibility. Bentley et al. (1985), for example, link the number of choices that a place offers to its permeability, relating the visual permeability of a place to its location inside the surrounding network of routes. Variety of use is the variety of choices that a place offers in terms of access or use (Bentley et al., 1985). Such places are highly attractive and permeable. As a place becomes highly attractive, the concept of natural movement applies to that place. And as the attraction increases, the place becomes a magnet and attracts more and more movements (Hillier and Iida, 2005). Accessibility, on the other hand, is a function that is connected to a place's structure patterns, connectivity and legibility (Mahdzar and Safari, 2014).

9.2 STUDY AREA AND DATA COLLECTION

9.2.1 STUDY AREA

The city centre of Istanbul (Turkey) was chosen as the study area (Figure 9.2b). The selection of the specifically surveyed areas within Istanbul was guided by advice from local experts from the Istanbul Metropolitan Planning and Urban Design Centre (IMP). We selected areas from two districts: Beyoğlu and Fatih (Figure 9.2d). Within Beyoğlu, the neighbourhood of *Cihangir* (Figure 9.2f) was chosen, and within Fatih, the neighbourhood of *Koca Mustafa paşa* (Figure 9.2e). The criteria for selecting the areas were presence of mixed uses (residential, commercial, educational, etc.), availability of public services (e.g. transport, education and religious services), transport mode availability and accessibility of the area.

The transport system in Istanbul faces a range of problems typical of large metropolitan cities, among them road congestion and insufficient capacity of road and public transport systems (trams, light railways and metros), which cannot cope with the demand for public transport (Hennig, 2011). The current transport policy in Istanbul is, however, more oriented towards sustainability, giving priority to improving public transport and pedestrian movement while reducing transport costs, pollution and congestion, as well as improving vehicle infrastructure and sea transport. An understanding of female perceptions of urban safety is important for making these services more sustainable and affordable.

BOX 9.1   Case Study Area

Istanbul is the largest city in Turkey, the largest city in Europe and the seventh largest city in the world. The city has around 14.7 million inhabitants within a total area of 5,343 km². Istanbul is located in Marmara region, where the Bosporus, connecting the Sea of Marmara and the Black Sea, divides the city into two parts (a European and an Asian part). The city is divided into 39 districts (Figure 9.2). The case study area is located in the city centre of Istanbul. The areas selected for our study were within the districts of Beyoğlu and Fatih. Within Beyoğlu, the neighbourhood *Cihangir* was selected, and within Fatih the neighbourhood of *Koca Mustafapaşa*. The survey included 97 trips and activities carried out by women. The destinations of these trips were located within the 462 neighbourhoods of Istanbul (Figure 9.2). Our specific focus was on women's experiences with and perceptions of urban design and mobility by tracing and mapping their mobility and activity patterns. The criteria for selecting the survey areas for our study were presence of mixed uses (residential, commercial, educational, etc.), availability of services (e.g. transport, education and religious services), transport mode availability, and accessibility of the area.

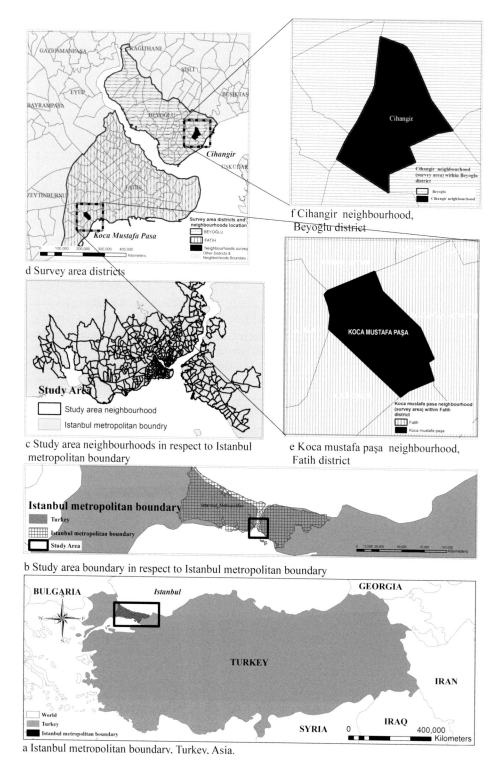

FIGURE 9.2   Istanbul study area with the selected neighbourhoods: (a) Istanbul metropolitan boundary; (b) study area boundary; (c) study area neighbourhoods; (d) survey area district; (e) Koca Mustafapaşa neighbourhood; (f) *Cihangir* neighbourhood.

TABLE 9.1
Research Database

Data Type	Location	Description	Source and Date	Data Condition
Spatial data	Fatih & Beyoğlu district	Time-use survey	Street survey, Istanbul citizens 2010	Paper questionnaire
	Fatih & Beyoğlu district	Interview locations	GPS 2010	GIS shape (vector)
	Istanbul metropolitan boundary	Activity locations	Sketch maps, Istanbul citizens 2010	Paper maps
	Istanbul metropolitan boundary	Bus, metro station	IMP, Istanbul 2006, 2010	GIS shape (vector)
	Istanbul metropolitan boundary	Building use	IMP, Istanbul 2010	
	Istanbul metropolitan boundary	Neighbourhood	IMP, Istanbul 2007	
	Istanbul metropolitan boundary	District	IMP, Istanbul 2007	
	Istanbul metropolitan boundary	Istanbul metropolitan boundary	IMP, Istanbul 2006	
	Istanbul metropolitan boundary	Road network & hierarchy	IMP, Istanbul 2006	

9.2.2 FIELDWORK AND DATA COMPILATION

The general focus of the survey was women's trips (from a multi-modal perspective) and activities, which were located within the 462 neighbourhoods of Istanbul (Figure 9.2). The survey's specific focus was on women's experiences and perceptions towards urban design and mobility by tracing and mapping their mobility and activity patterns. For the sampling, use was made of a combination of quota sampling, flow populations and the snowball method. A total of 97 women were surveyed. By means of a questionnaire and the use of geo-referenced sketch maps, women's interaction and experiences, areas of interests and the obstacles they faced during their daily trips were analysed.

The questionnaire was also designed to collect information on women's socio-economic situation, daily activities and movement patterns. Included in this was activity tracing, e.g. activity duration, start and end of the activity, daily schedule, preferred time to carry out daily activities, mode of transport used, daily routine and the combination of the activities.

In total, 97 mental maps were sketched during the interviews and a GIS database was provided by IMP; see Table 9.1. The questionnaire also allowed mapping of mobility and tracing the daily activity patterns of the interviewed women. Activity mapping was conducted by means of mental/sketch maps, for which the women were asked to point out the location of activities and routes selected. The activities categories we used were home, work, preferred shops, schools and visiting relatives but also included locations of interest and areas that were avoided. The route of each individual trip was accordingly sketched on a base map of Istanbul's streets (see, for example, Figure 9.3).

9.3 METHODOLOGY

9.3.1 URBAN MORPHOLOGY INDICATORS

The methodology builds on the theoretical framework (Figure 9.1), which establishes links between urban morphology and urban mobility. The three major outputs of the research are: (1) individual's

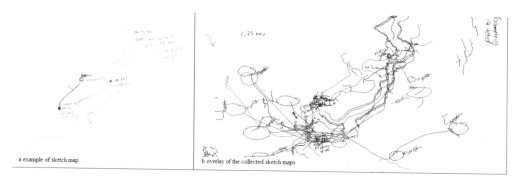

a example of sketch map | b overlay of the collected sketch maps

FIGURE 9.3   (a) Example of a sketch maps drawn by women interviewed; (b) the overlay of the sketch maps for the samples.

(women) daily trips through the city's physical structure; (2) an analysis of travel patterns of women in the city; and (3) women's activities in the city, which are combined with the city's morphological structure through GIS-based spatial analysis.

As described in Subsection 9.1.6 under *Individual Movements*, a number of indicators (visual permeability, variety of use, connectivity and accessibility) were established for measuring the quality aspects of women's travel behaviour. To observe the effect of urban morphology on urban mobility (Figure 9.1), it was essential to be able to provide a link between urban morphology and the elements of urban mobility that focus on human activities. Consequently, five urban morphology and mobility-related indicators were chosen (Table 9.2). The indicators for *urban morphology* fall into two main categories. The first set of indicators is based on the built environment, i.e. physical dimensions, while the second set is based on the function/use of the built environment (see Figure 9.4).

Physical dimensions are indicators that can be defined, described according to physical shape (i.e. linear, curved or irregular) and measured geometrically (i.e. width, length, area and volume). The integration of both physical configuration and geometrical measurements creates a system with predefined characteristics (i.e. grid, grid and square, radial, web, irregular or curvilinear; see *Foundation*, Subsection 9.1.2) and/or attributes (highly connected, globally/locally and highest/lowest level of integration in the case of a network), while functional indicators (i.e. commercial, mixed use, residential, etc.) are activity allocators and space attractions; see Figure 9.5b.

In the setting of physical and functional indicators (location-use measurement), an attractiveness index was developed that depends on the type of building use in a neighbourhood. Mixed use of buildings (i.e. residential, shops, commercial, social and institutional) attracted more people (see Figure 9.5b), while single use (i.e. industrial, commercial) decreased neighbourhood attractiveness. We assumed, therefore, that mixed use and high levels of permeability and accessibility would increase the number of women visiting an area.

Humans are the connection between urban mobility and urban morphology. Women and men differ in their requirements when pursuing activities, which are spatially distributed within an urban space. This can be referred to as the travel generator, as it motivates travel. Similarly, the connection between these activities is called a trip; trips are spatially distributed within the urban space.

There are two main categories of urban mobility indicators. One is based on activities carried out, while the other is related to daily travel distance and the spatial distribution of daily trips (see Figure 9.5a). Activity-based indicators deal with the spatial location of activities (carried out by the women within the city), showing the particular activity's spatial distribution and type.

In a final step, an index is created that combines all individual urban morphology and mobility indicators. The calculation and combination of all urban morphology indicators was done in ILWIS™, using its Spatial Multi Criteria Estimation (SMCE) tool (see Figures 9.8a, b, c, d), while the link between urban morphology and urban mobility was done in ArcGIS (attractiveness index; see Figure 9.8).

TABLE 9.2
Indicator Tree for Urban Morphology and Urban Mobility

Objective and Measurement	Figure	Measured Indicators	Sub-Categories	Indicators
The effect of infrastructure (in terms of urban street morphology) on urban mobility and activity patterns.	9.5c, 9.5d	Percentage of the built-up to the street area per neighbourhood	Physical	Urban morphology Figure 9.4
To model and map urban geometry of Istanbul, focusing on expressing urban street morphology.	9.7	Structural configuration (pattern & layout)/ neighbourhood		
Natural movement and movement economy.	9.5e	Network topology Connectivity		
To evaluate the influence of urban structure on movements within the built-up environment.	9.5f	Local integration		
Neighbourhood attractiveness in terms of location use. To identify dominant location use/neighbourhood.	9.5b	Location use	Functional	
Measure the effect of urban morphology's individual elements and urban morphology's cumulative index on the spatial distribution of activities. To measure the effect of urban morphology on the spatial distribution of activities and their types. To measure the effect of urban morphology's individual elements and the cumulative index effect on the spatial distribution of the activity types.	9.5a	Spatial activity distribution Activities spatial distribution (%) Activities type distribution (%)	Activities	Urban mobility Figure 9.4
Effects of attractiveness on trip destination. To measure the number of women's trips per urban morphology's attractive category.	9.5a	Daily travel activities of trip's spatial distribution (%)	Travel	

BOX 9.2   Methods Applied in the Chapter

This chapter presents a conceptual framework that combines surveying methods and techniques integrated into GIS, spatial analysis, spatial multi-criteria evaluation (SMCE) and statistical analysis to trace women's movements within the city's physical structure and the effect of these movements on their perception of fear. The movements of each woman interviewed were traced in a participatory GIS (PGIS, questionnaire mental map). The spatial data obtained were converted to shape files in ArcGIS™ by using Img2CAD™ 7.0 (for digitizing and geo-referencing). The resultant spatial data were used to carry out spatial analysis according to concepts of space syntax by using Axwomen™ 4.0 (GIS extension) (Jiang, 2015) to obtain connectivity and local and global integration indexes (see Figures 9.5e, f). The space syntax indexes showed the highest connected road in the road network at macro and micro levels. The demographic data collected were analysed using descriptive statistics. The secondary data supplied by **Istanbul** Metropolitan Planning and Urban Design Centre (IMP) were used to prepare indicators of neighbourhood physical structure (e.g. percentage of built-up street area of a neighbourhood, dominant activity types, structural configuration indicators), which were analysed by ILWIS SMCE. The final result relates the location of 'areas of fear, (as indicated by women) and the results of space syntax (connectivity) and SMCE maps (attractiveness index). This methodology was used to link urban morphology, urban mobility and women's perceptions.

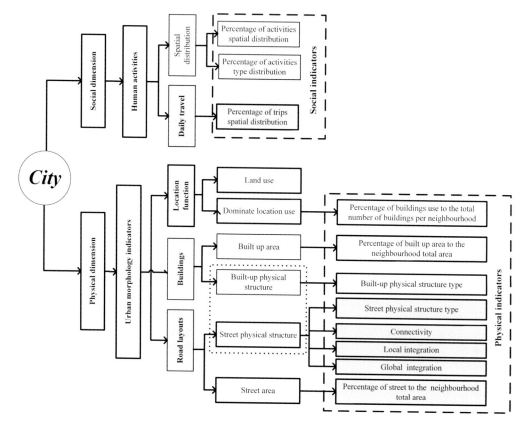

FIGURE 9.4   Design of urban morphology and mobility indicators.

9.3.2   ATTRACTIVENESS INDEX AND SMCE

ILWIS SMCE was used to combine the morphology indicators into an index of spatial attractiveness: the SMCE tool permits the creation of a multi-criteria tree for combining the indicators into a final (single) index. In this process, equal weights are used for the four main groups of morphological indicators:

1. Percentage of built-up/street area per neighbourhood (see Figures 9.5c, d);
2. Structural configuration (pattern) per neighbourhood (see Figure 9.7);
3. Network topology (connectivity and local and global integration; see Figures 9.5e, f);
4. Location use (building use; see Figure 9.5b).

The resulting index measures a neighbourhood's attractiveness in terms of location, variety of use, visual permeability, walkability, accessibility and connectivity.

The gap we have identified between the disciplines of urban planning/design and transport planning exists because it is difficult to link urban mobility to urban planning/design perspectives. By creating this link through an attractiveness index, our research connects women's subjective perception and behaviour (i.e. trip motivation and access choice) to the physical features of the trip (i.e. physical configuration of the road network and connectivity)

through the selection of the location of the activities, which differs for each individual.

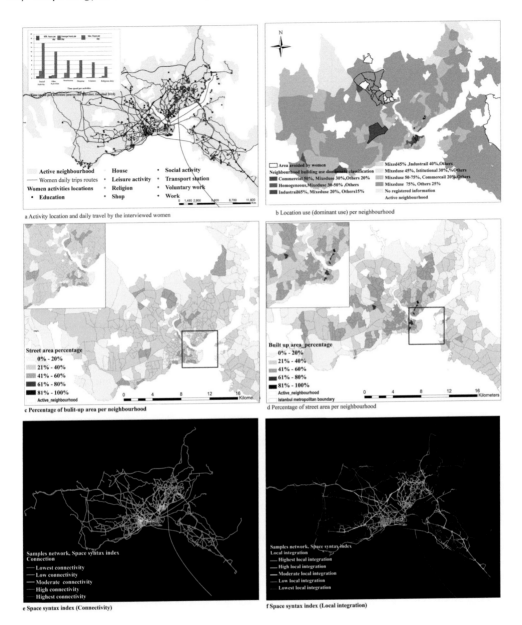

FIGURE 9.5 Indicator maps: (a) activity location and daily travel; (b) location use; (c) percentage of built-up area; (d) percentage of street area; (e) connectivity; (f) local integration.

9.4 RESULTS AND DISCUSSION

9.4.1 SURVEY RESULTS AND ANALYSIS

The questionnaires completed by the women interviewed (street interviews) provided data on their demographics and travel behaviour, which we have summarised in this subsection.

9.4.1.1 Demographics

The sample of 97 women shows diversity in age distribution (Figure 9.6a), marital status (Figure 9.6b) and professional background (Figure 9.6c). Average family income was €450/month (Figure 9.6d).

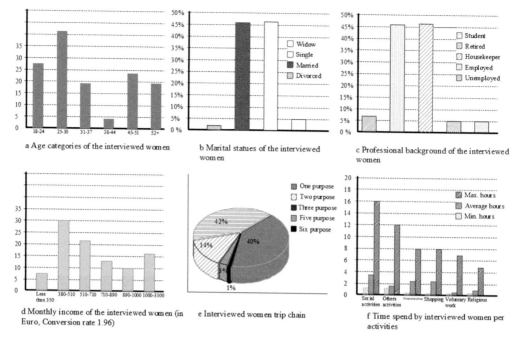

FIGURE 9.6 Demographics of interviewed female travellers and their trip purposes: (a) age categories; (b) marital status; (c) professional background; (d) monthly income; (e) trip chain; (f) time spent per activity.

The interviews were conducted in central Istanbul, with the women interviewed representing a variety of activities and mobility behaviours, e.g. combination of activities (Figure 9.6f).

9.4.1.2 Travel Distance and Purpose

The women interviewed travel on average 11 km per trip. The minimum distance travelled was 0.39 km, while the maximum was 75 km. Urban mobility is highly connected with trip purpose. For the women interviewed in Istanbul, there were six main reasons for undertaking a trip (Figure 9.6e), with social activities being dominant, followed by shopping, work, leisure, study and religious duties.

9.4.1.3 Time Spent per Type of Activity

The women interviewed spend on average two hours per activity; social activities were the most time-consuming (see Figure 9.6f), while religious duties took the least time.

To conclude, there was variation in socio-economic characteristics of the women sampled in terms of age, income, marital status and educational level. The women also displayed active travel patterns and variation in terms of activities, spatial distribution and time spent travelling, reasons for travel and distance travelled.

9.4.2 Sampling Daily Routines and Tracing of Activities

In order to trace women's movements, a participatory mapping technique (i.e. producing a woman's mental map Figure 9.3a), was overlaid (Figure 9.3b) with a time-use survey. GIS-based spatial analysis techniques were employed to trace women's movements.

The results show that it is possible to trace women's movements in relation to the physical structure of the city (streets network, activity locations; see Figure 9.5a, b). The maps in Figure 9.5 show the study area (neighbourhoods), trips made and the spatial distribution of activities.

9.4.3 INDICATOR RESULTS

9.4.3.1 Activity Locations and Accessibility Effect on Travel Patterns

As we have already discussed, a city's morphology determines the location at which activities take place. Likewise, its structural configuration (see Figure 9.7) controls visual permeability, thereby increasing/decreasing the attraction a neighbourhood has for pedestrians. While the proportion of

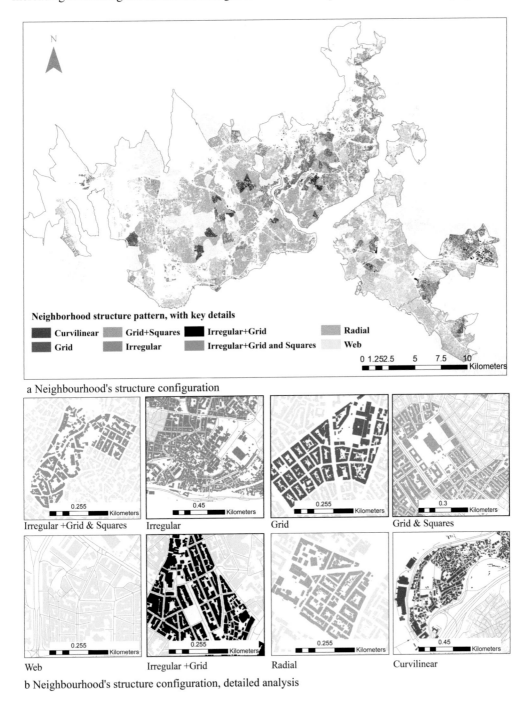

a Neighbourhood's structure configuration

b Neighbourhood's structure configuration, detailed analysis

FIGURE 9.7 Structural configuration of neighbourhoods: (a) overview; (b) detailed analysis.

TABLE 9.3

Neighbourhood Morphology and Attractiveness

Neighbourhood morphology	Attractiveness for people
High % built-up area	Highest attractiveness
Low % built-up area	High attractiveness
High % street area	Low attractiveness
Low % street area	Lowest attractiveness

built-up/street area per neighbourhood (Figure 9.5c, d) has an impact on the number of jobs, opportunities and investments, as people are attracted to well-developed infrastructure and built-up areas of high density, this proportion is also related to street connectivity (Figure 9.5e) and the spatial local integration of activities (Figure 9.5f).

With the connectivity of a neighbourhood in mind and the natural movement theory (Hillier, 1999), an attractive activity works as a magnet, attracting other activities and people to an area. This positively affects its local integration. Table 9.3 provides an overview of neighbourhood attractiveness as related to morphology.

Locations and accessibility of activities hence affect travel patterns of women in Istanbul. An activity's spatial distribution is linearly connected to its accessibility, so a location with a high level

FIGURE 9.8 Physical–spatial indicators overlain with 'areas of fear' as indicated by women travelling on public transport within Istanbul: (a) structure configuration; (b) space syntax; (c) percentage built-up/street; (d) location use.

of accessibility attracts more trips, as does a variety of activities. Figure 9.8d gives the results of SMCE analysis on location use for neighbourhoods, showing variations in the attractiveness index (see also Table 9.4).

9.4.3.2   Socio-Cultural Variances and Urban Structure in Relation to Activity Patterns and Travel Behaviour

Women's travel behaviour in Istanbul is affected by socio-cultural variances and urban structure, particularly in terms of physical spatial configuration (Figure 9.5) and its use (i.e. spatial distribution of activities, location use, proportion of the physical area of buildings to area of streets in neighbourhoods and the spatial arrangement of street networks). Thus, areas with high visual permeability (Figure 9.5d) and a well-developed road infrastructure of regular physical configuration (i.e. grid and/or finely planned physical area; see Figure 9.7) spatially coincide with areas of high concentrations of activities (i.e. mixed use, residential, commercial; Figure 9.5b) and a variety of socio-cultural characteristics.

These results illustrate that women's travel behaviour is affected by socio-cultural variance in terms of activity locations and trip purpose, of which social activities are the most dominant in the latter (see Figure 9.5a).

9.4.3.3   How Can Urban Form Be Defined and Quantified?

Buildings, streets and building use (residential, commercial, etc.) are features of urban form (see *Foundation*, Subsection 9.1.2). These features are defined by physical characteristics, i.e. shape (e.g. linear, curved) and attribute (attractive, connected, integrated), and are measured geometrically (e.g. area or/and volume). Buildings can be defined and quantified by their area, layout and physical arrangement, while for urban form, building use and physical arrangement are the definition of a space (i.e. a curvilinear structure with commercial use). Similarly, streets can be defined by their layout and spatial arrangement and quantified by their number of connections and the spatial relationships between the segments (Figure 9.5e). However, location use is connected with the use of a single building, although at urban scale it can be defined by the number of buildings with the same use to calculate the proportion of use per neighbourhood (Figure 9.5b). Urban form is definable by its geometrical characteristics and can therefore be quantified by the developed attractiveness index (see Figure 9.8).

9.4.3.4   How Do Women's Activities and Cultural Aspects Affect Urban Structure in the City?

Since the interviews were not conducted at random, answering the role of 'cultural aspects' remains uncertain, although women's activities and cultural aspects do have a local spatial connection. As indicated by the highest local integration index (Figure 9.5f), activities are more concentrated in commercial areas (e.g. Istiklal Street) than areas with firm cultural customs (e.g. the dressing style of women interviewed in Beyoğlu *Cihangir* was less conservative than that of women interviewed in Fatih Koca Mustafapaşa).

9.4.3.5   How Does the Street Network Shape Movements in Cities?

'Shaping the movement in the city' is considered to follow two theoretical concepts: *natural movement* and *movement economy* (addressed by Hillier and Iida 2005), where the accessibility and connectivity of streets define an area's attractiveness. Still, they are not the only drivers for attracting visitors. The arrangement of the spatial network (Figure 9.5e), besides influencing connectivity, is also indicative of the attraction of a place, for which *the concept of gravity* defines the spatial distribution of activities. Thus, more movements are attracted to a place when it is highly integrated locally (see Subsection 9.1.6; Figure 9.5f), while departing a place occurs when it is highly integrated globally.

TABLE 9.4

Location Distribution Activities in Respect with Physical Area Indicators

	Total no. of Activity	Leisure Activities	Religious Duties	Household Duties	Education	Shopping	Social Occasion	Transport	Voluntary Work	Work
Lowest attractiveness	13	1	0	3	1	2	6	0	0	0
Low attractiveness	71	30	2	4	0	23	11	0	0	1
Moderate attractiveness	137	23	6	15	4	45	36	1	1	6
High attractiveness	407	54	23	50	8	170	86	1	0	15
Highest attractiveness	161	43	2	24	4	40	21	7	0	20

9.4.3.6 What Barriers Do Women Face on Their Daily Trips?

The barriers women face on their daily trips depend on their perceptions of the surrounding environment. That can be related to either travel services or safety, which are more objective and measurable in terms of describing women's travel in Istanbul, or to the city's urban morphology, i.e. streets, buildings and functions. This is illustrated by the fact that 'areas of fear' do not attract many activities (Figure 9.8) and are not areas of high attractiveness. In short, the observed barriers are relative and are related to the local spatial environment.

9.5 DISCUSSION

Urban mobility is highly linked to urban morphology in Istanbul. The result allowed the mapping of women's 'areas of fear', illustrating the barriers they encounter in their daily activities and movements. In general, areas of fear were identified as having a specific location-use, thus being mono-functional (e.g. industrial, or religious or ethical group residences). These specific location-uses make them less attractive to visit. In addition, the structural configuration (urban pattern) of those areas is irregular, with poor road network connectivity and local integration. These results point to a link between urban morphology and women's perceptions of urban safety (see Table 9.5), where mixed location-use, well-planned structural configuration, high built-up densities, large total area of streets per neighbourhoods and high connectivity make for well-integrated, attractive areas in Istanbul. This relationship was analysed according to the framework we have developed to combine urban design, urban morphology and transport planning, allowing us to link these physical elements with the perceptions of women concerning areas of fear.

9.6 CONCLUSION

In conclusion, this research filled the gaps between urban design and urban transport planning, showing that gender-related perceptions are affected by the configuration of urban space. A theoretical framework was established and tested (for the case of Istanbul) that linked and combined urban design focused on elements of urban morphology, transport planning and behavioural theories (women's perceptions and behaviour) and also included the use of space syntax analysis, PGIS and SMCE. Thus, urban morphology and the mobility of women, and perceptions of urban safety in the city of Istanbul, were linked and mapped. Results showed that travel behaviour is affected by socio-cultural variance among women; social trips are the dominant trip purpose; women's activities and culture are connected to the local spatial characteristics of the place; and high concentration of activities are found in areas that are locally highly integrated. Thus, gender-related (women's) perceptions construct society, while street networks shape the movements of women;

TABLE 9.5
Spatial distribution of trips related to the urban morphology attractive index (in percentage of samples)

	Location Use	Structural Configuration	Physical Area	Space Syntax for City Roads	Space Syntax for Samples Routes
Highest attractiveness	96%	50%	4%	2%	0%
High attractiveness	0%	41%	95%	56%	21%
Moderate attractiveness	3%	6%	1%	41%	55%
Low attractiveness	1%	3%	0%	1%	18%
Lowest attractiveness	0%	0%	0%	0%	5%

place connectivity explains accessibility, and local integration of activities and the spatial distribution of location use explain location attraction. The framework we have developed for this research stresses that for integrated urban design and transport planning, differences in gender-related perceptions (women and men) need to be considered.

REFERENCES

Algers, S., Eliasson, J., & Mattsson, L. (2005). Is it time to use activity-based urban transport models? A discussion of planning needs and modelling possibilities. *The Annals of Regional Science, 39*(4), 767–789. doi:http://dx.doi.org/10.1007/s00168-005-0016-8

Balmer, M., Rieser, M., Meister, K., Charypar, D., Lefebvre, N., Nagel, K., & Axhausen, K. (2009). MATSim-T: Architecture and simulation times. In A. L. C. Bazzan & F. Klügl (Eds.), *Multi-agent systems for traffic and transportation engineering* (pp. 57–78). Hershey, PA, USA: Information Science Reference.

Batty, M. (2004). A new theory of space syntax. *CASA Working Paper, 75*. Retrieved from http://www.bartlett. ucl.ac.uk/casa/publications/working-paper-75

Batty, M. (2007). *Cities and complexity: Understanding cities with cellular automata, agent-based models, and fractals.* Cambridge, MA, USA: MIT Press.

Beall, J. (1996). *Urban governance: Why gender matters. Gender in development monograph* (Vol. 1). New York, USA: United Nations Development Programme.

Bentley, I., Alcock, A., McGlynn, S., Murrain, P., & Smith, G. (1985). *Responsive environments: A manual for designers: The McGraw-Hill Companies.* Oxford, UK: Architectural Press, Elsevier.

Bhat, C., & Koppelman, F. (2003). Activity-based modeling of travel demand. In R. Hall (Ed.), *Handbook of transportation science* (Vol. 56, pp. 39–65). New York, USA: Springer.

Conzen, M. R. G. (1960). Alnwick, Northumberland: A study in town-plan analysis. *Transactions and Papers (Institute of British Geographers), 27*, iii–122. doi:10.2307/621094.

Crabtree, S., & Nsubuga, F. (2012). Women feel less safe than men in many developed countries. *Gallup.* Retrieved from http://news.gallup.com/poll/155402/women-feel-less-safe-men-developed-countries.aspx

de Smith, M. J., Goodchild, M. F., & Longley, P.A. (2009). Geospatial analysis – a comprehensive guide. *Geospatial analysis book online.* Retrieved from http://www.spatialanalysisonline.com/

Feuerschutz, S. (2012). *Gender and urban (in)security in fragile and conflict-affected states.* Retrieved from http://www.nsi-ins.ca/wp-content/uploads/2013/03/Gender-and-Urban-Insecurity-in-FCAS-v111.pdf

FHWA. (1996, October). *Women's travel issues.* Paper presented at the Women's Travel Issues, Second National Conference, Baltimore, USA.

Fox, M. (1995). Transport planning and the human activity approach. *Journal of Transport Geography, 3*(2), 105–116. doi:10.1016/0966–6923(95)00003-L

Frey, H. (1999). *Designing the city: Towards a more sustainable urban form.* New York, USA: Routledge.

Gökariksel, B., & Mitchell, K. (2005). Veiling, secularism, and the neoliberal subject: National narratives and supranational desires in Turkey and France. *Global Networks, 5*(2), 147–165.

Hasson, Y., & Polevoy, M. (2012). *Gender equality initiatives in transportation policy. A review of the literature.* Retrieved from https://il.boell.org/sites/default/files/gender_and_transportation_-_english_1.pdf

Hennig, M. (2011). *Sustainable urban mobility: The example of Istanbul. A short survey* Retrieved from Bonn, Germany: https://sutp.org/files/contents/documents/resources/C_Case-Studies/GIZ_SUTP_CS_ Sustainable-Urban-Mobility-Istanbul_EN.pdf

Hillier, B. (1999). The common language of space: A way of looking at the social, economic and environmental functioning of cities on a common basis. *Journal of Environmental Sciences, 11*(3), 344–349.

Hillier, B., & Iida, S. (2005). Network and psychological effects in urban movement. In A. G. Cohn & D. M. Mark (Eds.), *Spatial information theory* (Vol. 3693, pp. 475–490). Berlin, Heidelberg, Germany: Springer.

Jiang, B. (2015). *Axwoman 6.3: An ArcGIS extension for urban morphological analysis.* University of Gävle, Sweden. http://giscience.hig.se/binjiang/Axwoman/

Johnson, J. H. (1967). *Urban geography: An introductory analysis.* Oxford, UK: Pergamon Press.

Kim, Y. O., & Penn, A. (2004). Linking the spatial syntax of cognitive maps to the spatial syntax of the environment. *Environment and Behavior, 36*(4), 483–504. doi:10.1177/0013916503261384

Küçük, S. (2013). Being a woman in Turkey and in the Middle East. *The Globalist.* Retrieved from http://www. theglobalist.com/being-a-woman-in-turkey-and-in-the-middle-east/.

Kunieda, M., & Gauthier, A. (2007). *Gender and urban transport. Smart and affordable. Sustainable Transport: A Sourcebook for Policy-makers in Developing Cities*. Bonn, Germany: Bundesministerium für wirtschaftliche Zusammenarbeit und Entwicklung (BMZ).

Latham, A., McCormack, D., McNamara, K., & McNeill, D. (2009). *Key concepts in urban geography*. London, UK: Saga Publications Ltd.

Long, Y., Baran, P. K., & Moore, R. (2007, 12–15 June). *The role of space syntax in spatial cognition: Evidence from urban China*. Paper presented at the 6th International Space Syntax Symposium, Istanbul, Turkey.

Lynch, K. (1960). *The image of the city*. Cambridge, MA, USA: MIT Press Ltd.

Mahdzar, S. S. S., & Safari, H. (2014). Legibility as a result of geometry space: Analyzing and comparing hypothetical model and existing space by space syntax. *Life Science Journal, 11*(8), 309–317.

Mauch, M., & Taylor, B. (1997). Gender, race, and travel behavior: Analysis of household-serving travel and commuting in San Francisco Bay Area. *Transportation Research Record: Journal of the Transportation Research Board, 1607*, 147–153. doi:https://doi.org/10.3141/1607-20

Meyer, M. D., & Miller, E. J. (2001). *Urban transportation planning: A decision-oriented approach* (2nd ed.). Boston, USA: McGraw-Hill Higher Education.

Moughtin, C. (2003). *Urban design: Street and square* (3rd ed.). Burlington, MA, USA: Elsevier Science.

Nagel, K., & Flötteröd, G. (2009). *Agent-based traffic assignment: Going from trips to behavioral travelers*. Paper presented at the 12th International Conference on Travel Behaviour Research (IATBR), Jaipur, India.

Nenci, A., & Troffa, R. (2006, 24 Sept.). *Integrating space syntax in wayfinding analysis*. Paper presented at the Space Syntax and Spatial Cognition, Bremen, Germany.

Pacione, M. (2005). *Urban geography: A global perspective* (2nd ed.). London, UK: Routledge.

Phadke, S. (2012). Gendered usage of public space: A case study of Mumbai. In S. Pilot & L. Prabhu (Eds.), *The fear that stalks: Gender-based violence in public spaces* (pp. 51–80). New Delhi, India: University of Chicago Press.

Rakodi, C. (1991). Cities and people: Towards a gender-aware urban planning process? *Public Administration and Development, 11*(6), 541–559. doi:10.1002/pad.4230110603

Sadiqi, F., & Ennaji, M. (2013). *Women in the Middle East and North Africa: Agents of change* (Vol. 2). London: Routledge.

Secor, A. (2002). The veil and urban space in Istanbul: Women's dress, mobility and Islamic knowledge. *Gender, Place & Culture: A Journal of Feminist, 9*(1), 5–22. doi:10.1080/09663690120115010

Southworth, M., & Ben-Joseph, E. (2003). *Streets and the shaping of towns and cities*. Washington, Covelo, London: Island Press.

Tacoli, C., & Satterthwaite, D. (2013). Gender and urban change. *Environment and Urbanization, 25*(1), 3–8. doi:10.1177/0956247813479086

UN Habitat (2008). *Gender Mainstreaming in Local Authorities: Best Practices* (Vol. 16). Nairobi, Kenya: UN-Habitat.

van Nes, A., & Nguyen, T. M. (2009, 8–11 June). *Gender differences in the urban environment: The flâneur and flâneuse of the 21st century*. Paper presented at the 7th International Space Syntax Symposium, Stockholm, Sweden.

Vance, J. E. (1977). *This scene of man: The role and structure of the city in the geography of western civilization*. New York, USA: Joanna Cotler Books.

Wachs, M. (1996, October). *The automobile and gender: An historical perspective*. Paper presented at the Women's Travel Issues, the Second National Conference, Baltimore, USA.

Wesly, S. (1986). Fear of crime and neighborhood change. In A. J. Reiss & M. Tonry (Eds.), *Communities and crime* (pp. 203–229). Chicago, USA: Chicago Press.

10 Children's Perception of Their City Centre

A Qualitative GIS Methodological Investigation in a Dutch City

Haifa AlArasi, Javier Martinez, and Sherif Amer

CONTENTS

10.1   Introduction 163
10.2   Childhood beyond a Dichotomous Framework 164
10.3   Childhood and the City... 164
10.4   Participatory Planning and Children's Insights 165
10.5   Qualities of the Local Living Environment According to Children 166
10.6   Case Study Area 167
10.7   Methods 169
    10.7.1   Qualitative GIS, Mixed-Method Approach 169
    10.7.2   Data Collection 169
    10.7.3   Data Preparation and Analysis 171
10.8   Results... 172
    10.8.1   General Perceptions of Enschede's City Centre 172
    10.8.2   Spatial Perceptions... 173
    10.8.3   Perception, Place and Meaning 175
10.9   Reflection on the qGIS Approach and Results 177
10.10 Conclusion 179
Acknowledgements... 179
References... 179

10.1 INTRODUCTION

Despite the recent sociological change in the concept of 'childhood', which views children as active social beings (Prout, 2005), and increasing international recognition of the importance of creating environments that respond to their needs (UNICEF, 1989,1996), children's experiences are usually invisible in urban discourses, relegated to the margins. Failure to incorporate their experiences in planning policies may contribute to the production of urban environments where children are excluded, become alienated and experience a decrease in their independence of mobility (Christensen and O'Brien, 2003; Gleeson and Sipe, 2006; Kyttä, 2004; Matthews, 1992).

This chapter presents a participatory qualitative approach for capturing children's perceptions of their local living environment in a geographic information system (GIS). We demonstrate this qualitative approach (qGIS) with a case study in the city of Enschede, the Netherlands. This study is the result of close collaboration between us and teachers of the local international school in Enschede. The international school was selected to ensure ease of communication, since the research needed to be conducted in English. In particular, we focused on how children perceive the city centre of

Enschede. Our motivation in choosing the city centre was twofold. First, it would be a location familiar to all the children participating in the research. Second, the city centre has an important role to play as public space, where children are able to experience the social and cultural diversity of the city (Freeman and Tranter, 2011).

10.2 CHILDHOOD BEYOND A DICHOTOMOUS FRAMEWORK

The concept of childhood has not always been static but has rather been contingent on local contexts that vary spatially and temporally. In its current state, it is defined within a dichotomous framework constituting children as the contrasting social other of adults, rendering them from this point of view as dependent, incompetent, passive and in need of protection (Prout, 2005; Stasiulis, 2002). This notion provided the backdrop required to standardise their needs and contributed to their continuous socio–spatial and cultural/political separation from the material practices and spaces of adults (Aitken, 2005; Oswell, 2013). However, one can argue that this concept is ever changing and evolving because of its interconnectedness with the unstable social, economic and cultural conditions of postmodern societies, thus leading to different experiences of childhood in different contexts (Prout, 2005). Scholars have therefore suggested that the absence of a universal experience of childhood challenges the currently constructed and accepted concept of childhood, rendering it outdated and in need of reconceptualization.

At some point in time between Ariès' claim of the modern invention of childhood (Ariès, 1962) and Postman's declaration of the end of it (Postman, 1985), our understanding of childhood started to shift. Acknowledgement of the proliferation of childhood experiences across multiple axes of differences (for example, cultures, classes, ethnicities and genders) led to a socially constructed notion of childhood that was deeply implicated in social, cultural, economic, political and institutional processes, and from this notion a new paradigm of childhood crystallised (Prout, 2005). The paradigm conceives children to be social beings in their own right instead of being measured against the views of adults and adult social structures. It requires us to challenge our perception and encourages us to explore children's own perspectives by acknowledging the limitations of our own cognitive construction of childhood and its associated experiences. (James, 2004; Prout, 2005).

This modern social construct, coupled with an increased collective international awareness of children as human beings with agency, was translated into multiple international policies and pieces of legislation, with *The Convention on the Right of the Child* the most prominent. Established in 1989 and ratified by 140 countries, *The Convention* calls for children (individuals under the age of 18) to be independent subjects with rights to a voice in matters than concern them; see in particular Article 12 and Article 13 in *The Convention* (UNICEF, 1989). The establishment of *The Convention* was not only important in explicitly recognizing the right of meaningful participation for children but also set a new record of viewing children as active citizens and 'full human beings, invested with agency, integrity, and decision-making capacities' (Stasiulis, 2002, p. 509), rather than invisible beings in need of discipline and control by adults – including researchers!

10.3 CHILDHOOD AND THE CITY

The gradual shift in the perception and construct of childhood just described was witnessed in the city as well. Gillespie (2013) notes that children's integration in their city generally, and in the street specifically, was initially unproblematic, as rather a natural extension of their integration in the private realms of their private homes. As a consequence of the reform movements that grew in the 19th century, children's place in the city started to be contested as problematic with the emergence of segregated, private spaces. Gillespie argues – based on 'recovered histories' – that children's segregation (like many other social norms) is a recent, culturally constructed norm (Gillespie, 2013). Of course, the alienation of children was further reinforced in many cities (particularly after the Second World War) with the priority given to the automobile and subsequent zoning bylaws that separated the home from places of schooling, leisure and recreation. Nevertheless, this regulation of public space through surveillance, privatizing and limiting access and, as a consequence, exclusion did not stop children from contesting, occupying and reproducing their local spaces in many ways (Valentine and McKendrick, 1997).

It is therefore easy to recognise that a focus on children's wellbeing in the city accompanied the industrial revolution (and the establishment of the welfare state and its associated reform movements), which emphasised material improvement for all vulnerable groups – including children (Gleeson and Sipe, 2006). In academe, this development translated into a series of studies that drew attention to the experiences and needs of children in their cities (see, for example, works by Ward 1977 and Lynch 1976). According to Gleeson and Sipe (2006), this continued well into the 1980s with explorations of the environmental impact of cities on children's wellbeing. However, the fragmentation of modern institutions that took place in postmodern societies characterised by globalization and individualization resulted in a continuous blurring of the distinction between adulthood and childhood (Lupton, 1999). This blurring was marked at many sites, and as a result, children were successfully negotiating their agency and presenting themselves as competent actors. Most notably in the realm of the home, as demonstrated by the transition to 'negotiated families', but also in the market as consumers 'owning' specific patterns of consumption, and in the social realm as they developed their own subcultures (Atiken, 2005; Jans, 2004; Lupton, 1999;).

As discussed earlier, this prompted the development of policies that recognise the new positioning of children. According to Stasiulis (2002, p. 509), *The Convention on the Rights of the Child* succeeded in setting a new vision of children as active citizens with agency and decision-making capabilities, rather than invisible beings in need of discipline and control by their parents and the state. This view was further established in Chapter 25 of *Agenda 21*, in which the importance of the active involvement of children in promoting economic and social development, as well as their environmental protection, was asserted. Moreover, to ensure the translation of these adopted policies into practical programmes at the local level, UNICEF (1996) established Child-Friendly Cities – an initiative with the sole purpose of helping cities promote children's rights (including the right to participate in decision-making processes).

This resulted in renewed interest in children's geographies, the difference being that studies were based on a new, socially constructed notion of childhood. This meant that researchers became more interested in understanding children's grounded experiences through the utilization of ethnographic methods (Holloway, 2014). These studies indicate that children have excellent knowledge of their living environment and not only are able to identify issues that matter to them but are also capable of suggesting solutions that best address these issues (Laughlin and Johnson, 2011; Lynch and Banerjee, 1977; Travlou et al., 2008; Veitch et al., 2007; Walker et al., 2009).

10.4   PARTICIPATORY PLANNING AND CHILDREN'S INSIGHTS

Although there is no definitive consensus on what constitutes effective participation in planning activities (see Arnstein, 1969; Monno and Khakee, 2012), many advocates see numerous benefits in engaging the public in participatory activities. These benefits include: empowerment of citizens that results in a continuous cycle of engagement (Oldfield, 1990); the potential of the process to be educative, leading to the development of appropriate knowledge and attitudes towards active citizenship (Day, 1997); the affirmation of democracy (Williams, 1976); and the provision of local knowledge necessary for planners to build a complete picture in their planning process (Burke, 1979). Perhaps more importantly for planning scholars, participation affords an opportunity for 'a transformative and deliberate learning and participatory action research', as Forester (1999, p. 6) puts it.

As a result, participatory planning has been growing in recent years in many parts of the world, with a number of successful examples in the U.S., Brazil, Norway and Venezuela, to name but a few (Forester, 1999). Unfortunately, children are not always considered in the participatory process. In fact, they are almost 'invisible' because of the misconstrued belief that children lack the skills needed to effectively participate (Simpson, 1997). Nevertheless, an increasing number of planning scholars have investigated participatory activities with children (see, for example, Cele, 2006; Chawla, 2002; Kyttä, 2004; Lynch and Banerjee, 1977). Categorised by Knowles-Yánez (2005) as being scholarly and non-rights-based approaches to participatory planning, for most of these studies the term *consultation* would be most applicable. According to Driskell (2002), however, even though children were not directly involved in decision-making per se, these studies could be considered as meaningful forms of participation as long as they fulfilled three criteria: that the children understand why they are taking part in the participatory activity; that they have the choice of not participating; and that they are informed about the results.

Numerous examples exist for the use of a variety of participatory methods to engage children in planning processes. Celea (2006), for example, examined children's environments using methods that include focus group discussions, guided walks and studying photographs and drawings made by the (Swedish and English) children in their attempt to understand how they experience and use space in their local neighbourhoods. Kyttä (2004) used open-ended interviews to determine the extent of independent mobility afforded to children in their environment. While Wridt (2010) and Chawla (2002) used participatory mapping, Lynch and Banerjee (1977) used daily recordings written in diaries by the children themselves. More recently, many researchers have attempted to employ technology in an effort to streamline the participation process. Santo, Ferguson and Trippel (2010), for example, trained children to use technology in order to collect data, analyse the information and create maps with computers to register their perceptions of their neighbourhoods. In an earlier study, Berglund and Nordin (2007) developed a method for children to map and record their perceptions in a child-friendly computerised GIS, thus facilitating the process of bringing their insights to the official planning process.

All these examples illustrate the potential qualitative and quantitative methods have to access children's knowledge and engage them in the planning process. The concern then becomes focused on the adaptability of these methods to children's needs and their experiences of their environment, without subjecting them to adult-centric communication and setting bias.

10.5   QUALITIES OF THE LOCAL LIVING ENVIRONMENT ACCORDING TO CHILDREN

One of the prominent studies involving children and adolescents in evaluating their local living environment was done by Lynch and Banerjee (1977) in the 'Growing Up In Cities' (GUIC) project. This project was revived in 1996 by *The Convention on the Rights of the Child* and reaffirmed in *Agenda 21*, which focused on documenting improvements or deterioration of the living environment compared to the situation in the 1970s. The project resulted in the formulation of indicators of social and physical qualities of the local environment (a term that encompasses spaces that children live in, play in or regularly visit) based on the evaluation of the children themselves; qualities can be social or physical, positive or negative (Chawla, 2002). Together, this results in four quadrants – see Figure 10.1 – representing the living environment of the children.

These qualities were the result of the experiences and evaluation of 10–15-year-olds in low- or mixed-income urban centres in several cities in both developed and developing countries. Table 10.1 summarises the qualities and definitions used in Figure 10.1.

Although the framework presented in Figure 10.1 was developed with a focus on residential neighbourhoods, one can argue that such qualities can also be present in a variety of urban settings – including city centres.

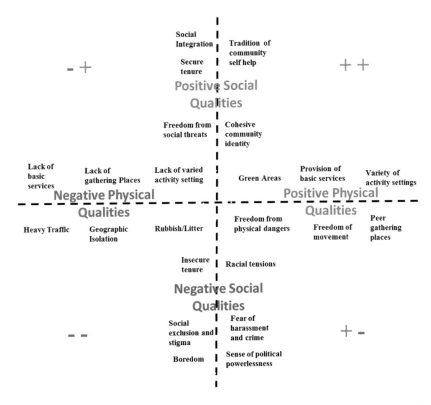

FIGURE 10.1 Qualities of the living environment according to children (10–15-year-olds) (after Chawla, 2002). (Source: Alarasi et al. (2016). Copyright 2016 by Taylor & Francis. Reprinted with permission.)

10.6 CASE STUDY AREA

BOX 10.1   Case Study Area

Enschede is located in the east of the Netherlands, close to the German border, and is the largest city in the province of Overijssel. Enschede covers an area of approximately 143 km². It is a medium-sized city of 158,000 inhabitants (Kennispunt Twente, 2014) and is subdivided into five administrative districts: *Noord, Oost, Zuid, West* and *Centrum*.

The case study area, the city centre of Enschede, is a part of the district *Centrum*, which covers an area of approximately 0.54 km² and has a population of 3,125 inhabitants (CBS, 2016). The city centre is mostly a car-free zone and is pedestrian- and bicycle-friendly. The centre includes an open square, the historic 'Oud Markt', that surrounds an old church.

Several public services border the area, such as bus and train stations and parking facilities for cars and bikes, giving city residents (including children) easy access to and from the centre. A variety of services are concentrated in the city centre, including modern shopping centres, restaurants, bars, hotels, cinemas, a casino and a music centre. It was selected as the case study area, first, because it is a location familiar to all the children participating in the research. Second, it has an important role as a public space where children can experience the social and cultural diversity of the city (Freeman and Tranter, 2011).

Enschede is located in the east of the Netherlands, close to the German border. It is a medium-sized city with around 158,000 inhabitants (Kennispunt Twente, 2014). Since the 1980s, efforts have been made to revitalise Enschede, transforming it into a modern city with a focus on knowledge-intensive industries and institutions of higher education (Enschede-Stad, 2014).

In the mid-1990s, the municipality launched a comprehensive redevelopment plan for the city centre. Motorised traffic was diverted away to create a car-free zone and a large market square was constructed, which now functions as a 'node' for the city centre (Hospers, 2010) and has characteristics of themed public spaces, for example 'funshopping' (Van Melik, 2009). The area around the railway station was earmarked for office use and a new bus station was created. Cultural functions were concentrated in the northern part of the city centre, and offices and apartments were constructed on underutilised locations (Enschede-Stad, 2014).

TABLE 10.1

Summary of Qualities of the Living Environment after Chawla (2002)

Qualities of the Living Environment	Definitions
Physical qualities	*Green areas*: Safe green spaces that are accessible.
	Provision of basic services: Basic services such as food, water, electricity, medical care and sanitation are provided for the children.
	Variety of activity settings: There is a variety of places for children, including places where they can meet friends, talk or play informal games.
	Freedom of movement: Children feel that they can count on adult protection and range safely within their local area.
	Peer gathering places: There are safe and accessible places where children can meet.
	Heavy traffic: Streets and other public places taken over by cars.
	Rubbish/Litter: Children read trash and litter in their environment as signs of neglect for where they live.
	Geographic isolation: The local area is isolated from other communities by a mountain, river or valley.
Social qualities	*Social integration*: Children feel welcome throughout the community and interact with other age groups.
	Tradition of community self-help: Residents are building their community through mutual-aid organizations.
	Cohesive community identity: Children are aware of their community's history and take pride in its accomplishments.
	Secure tenure: Family members have legal rights over the properties they inhabit.
	Racial tension: Children experience stigmatization based on their race or ethnic origin.
	Social exclusion: Children feel unwanted and left out in their local area.
	Fear of harassment and crime: Children avoid specific areas or whole sections of the community due to fear of harassment and violence.
	Boredom: Children express high levels of boredom and alienation because places set aside for them are featureless.
	Sense of political powerlessness: Children and their families feel powerless to improve conditions.

Additional services have been integrated in the fabric of the city centre, such as an underground car and bicycle parking facility, thus giving city residents (including children) easy access to and from the centre. Just outside the car-free zone, there is a hospital, a hotel and a number of small-scale commercial enterprises. The centre itself offers a variety of facilities, among them modern shopping complexes, restaurants, bars, cannabis shops, theatres and a casino. It hosts many activities, including an open market and recreational activities such as festivals and live concerts, making it an ideal space for families to spend their free time. While there are green spaces accessible near the city centre area, the centre itself does not offer any natural green areas, although there are some spaces with greenery and water fountains.

10.7 METHODS

Our study was conducted in the city centre of Enschede, the Netherlands, in close collaboration with its international school (which is located outside the city centre). With the direct cooperation of the school's principal and geography teacher, our study was incorporated into the geography/humanity sessions of the educational program. The study was conducted between September and December 2012 and involved 28 children (19 boys and 9 girls; ages 10–17 years).

Most of the children resided in middle-class neighbourhoods and were from a variety of ethnic backgrounds. Nearly all children had been living in Enschede for a number of years and were well acquainted with the city-centre area.

In line with Hill (2005), ethical considerations were carefully incorporated in the study. First, we obtained informed consent from the children's legal guardians and from the children themselves. Second, contact sessions were organised in an open, non-formal learning space that allowed free arrangement of seating, to ensure very little power distance between the children, the teachers and the researchers involved (us). Third, we made it clear that children could opt out of any session that was of no interest to them. Fourth, we designed the research process in such a way that the one-on-one interviews occurred later on in the research process, to enable the children to feel comfortable communicating with us. Finally, we communicated our preliminary findings and results to the children, parents, teachers and school officials.

10.7.1 QUALITATIVE GIS, MIXED-METHOD APPROACH

To measure the children's perception of socio–physical qualities, we developed a qualitative approach (qGIS) that combines conventional methods (quantitative) of GIS with qualitative methods (Cope and Elwood, 2009). As Figure 10.2 shows, this mixed-method approach enabled a connection to be formed between macro-scale urban phenomenon and the micro-scale of perceptions of the individual child. This approach made it possible to capture children's perceptions associated with different locations.

Dennis (2006) suggests that youth participation using a qualitative GIS approach holds much promise, but given the considerable variety of tools and methods available, it is important to select those that are effective in capturing children's knowledge about their local environment. The different methods used in our study for data collection and analysis (see Figure 10.2) are discussed in the following sections.

10.7.2 DATA COLLECTION

Different data collection methods were used, including participatory mapping, focus group discussions, guided tours and interviews, of which photo-voice recordings were made. First, a participatory mapping exercise was conducted in a 60-minute session with small groups of 4–5 children. The

QGIS Approach

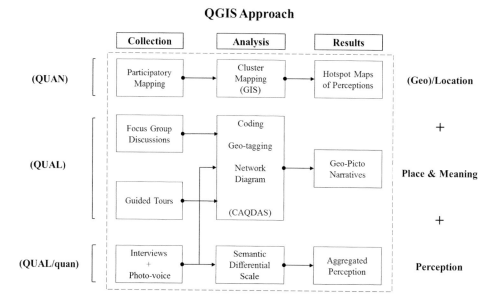

FIGURE 10.2   Mixed-method, qGIS approach. (Source: Alarasi et al. (2016). Copyright 2016 by Taylor & Francis. Reprinted with permission.)

objective of the exercise was for each child to identify particular locations in the city centre that he or she liked or disliked. Google Earth® images of Enschede's city centre were printed on A3 paper, labelled with street names and main features and overlaid with transparent sheets. The children used their school's computers to access Google® maps and its StreetView® function. They were given green and red circular stickers and asked to use these to geocode locations in the city centre that they liked (green) or disliked (red). The children were then asked to associate qualities with the locations identified. They were assisted when needed and supervised by us and their teacher to ensure reliable mapping.

Along the lines of Wridt (2010), at the end of the exercise all transparent sheets were superimposed to give the children immediate insight into how their collective map would look after being entered into the GIS. This also enabled them to see an overview of the locations of interest for the group as a whole.

The second data collection step involved a focus group discussion to obtain better insight into the social and physical qualities of the locations identified. Two focus group discussions were organised, one for the younger children (aged 10–12 years) and one for the older children (aged 13–17 years). This was done to allow later examination of the potential differences in perception between the two age groups. Each session lasted 60 minutes and with the consent of the children was voice- and video-recorded, to enable transcription and further analysis.

The guided tour, adapted from Cele (2006), came as a third step for obtaining a deeper insight into how the children experience the city centre and how they interact with the objects and/or people they encounter. The route was predetermined, based on the locations identified earlier in the mapping exercise. To ensure a degree of flexibility, children were allowed to deviate somewhat from the predetermined route. The children were divided into six groups of 4–5 children, with a researcher or a teacher accompanying each group. Each group was issued a hand-held GPS device to track the walking tour, a camera and a printed map. Within the different groups, each child was asked to take pictures of the positive and negative locations and situations they observed in the city centre.

The fourth and last method used to obtain data consisted of interviews, of which photo-voice recordings were also made. By that time, the children were more familiar with us, which fostered open communication. Conducting the interviews as a final step allowed the children to reflect on and add and/or remove locations on their maps. The interviews were held one-on-one and were recorded for later transcription and analysis. To round off the interview, each child showed the photographs they had taken during the guided tour and explained why he or she took them. To adequately capture the perceptions of the children, a semantic differential scale (Lewis-Beck et al., 2003) was used during the interviews.

After analysing the data collected from the children, additional, semi-structured interviews were conducted with two city planners to find out whether planning professionals see possibilities for incorporating the approaches we have developed and their outcomes into future urban-planning processes.

10.7.3 Data Preparation and Analysis

The maps produced during the participatory mapping exercise were entered into a GIS database to enable spatial analysis. Each child was assigned a unique mapping ID that was linked to personal characteristics such as age and gender. Identified city-centre locations were digitised and a database was created that included attributes such as location ID, child mapping ID and place name. The qualities associated with each mapped location were entered into a separate database accompanied by the location ID and child mapping ID to allow linking of both tables and further analysis of the spatial distribution of the identified qualities. The use of a spatial database enabled the creation of a variety of mapping outputs.

The interviews were transcribed and input in Computer-Assisted Qualitative Data Analysis (CAQDAS) software. The interviews were coded, labelled and sorted according to the themes derived from the main conceptual framework (see Figure 10.1). Google Earth® images of the study area were embedded in the CAQDAS to allow geo-tagging of the codes and linking to relevant textual statements.

The photographs were also geo-tagged and grouped into different categories based on the qualities derived from the children's narratives. These narratives were then coded together with the textual information, using the qualities of the conceptual framework (Figure 10.1) as a main guide.

BOX 10.2   Methods Applied in the Chapter

This chapter illustrates the use of a qualitative GIS approach (qGIS) for measuring children's perception of socio–physical qualities. Our approach combines a mixture of conventional GIS (quantitative) with qualitative methods that include participatory mapping, focus group discussions, guided tours and interviews supported by photo-voice recordings.

The maps resulting from the participatory mapping exercise were incorporated into a GIS database to enable spatial analysis. Each child was given a unique mapping ID linked to personal characteristics such as age and gender. Identified city-centre locations were digitised and a database was created that includes attributes such as location ID, child mapping ID and place name. The qualities associated with each mapped location were entered into a separate database accompanied by the location ID and child mapping ID to allow linking of both tables and further analysis of the spatial distribution of the identified qualities. The use of a spatial database enabled the creation of a variety of mapping outputs. The interviews were transcribed and entered into Computer Assisted Qualitative Data Analysis (CAQDAS) software. The interviews were coded, labelled and sorted according to the themes derived from the main conceptual framework (Figure 10.1). Google Earth® images of the study area were embedded in CAQDAS software to allow the codes to be geo-tagged and linked to relevant textual statements.

Such a qualitative GIS approach can be applied in quality of life studies, participatory mapping, self-enumeration or any other study that elicits people's perceptions.

10.8 RESULTS

Our findings are presented in the three following subsections to illustrate the benefit of the mixed-method qGIS in capturing the children's perceptions. The first subsection (10.8.1) discusses the general perception of the study area and observed similarities and differences between age group and gender. The second subsection (10.8.2) introduces the spatial distribution of the perceptions captured. The third, and last, subsection (10.8.3) discusses the qualities that have emerged from the study. The contextual information collected by the children and us – together with the maps – form what we call *Geo-Picto Narratives*. These give detailed insight into locations in the city centre that were liked and disliked, and the qualities associated with them.

10.8.1 General Perceptions of Enschede's City Centre

The overall perception of the city centre of Enschede was positive (Figure 10.3). The children described it as a safe place where they can enjoy many activities and hang out with their friends and family.

In the instance of the pairs 'Quiet–Noisy' and 'Crowded–Uncrowded', most of the children associated these qualities with a temporal dimension. This is because an open market takes place on the

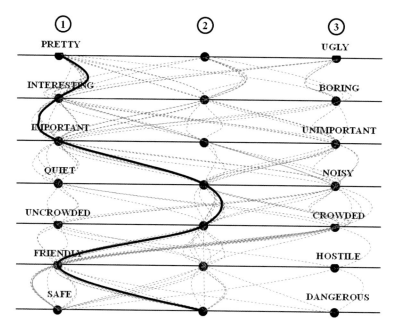

FIGURE 10.3 General children's perception of the city centre. Semantic differential, frequency of response (*N* = 28). (Source: Alarasi et al. (2016). Copyright 2016 by Taylor & Francis. Reprinted with permission.)

square twice a week and attracts many people. Additionally, for the pair 'Safe–Dangerous', most of the children were very clear in identifying areas where they did not feel safe. In some instances, children also associated this feeling with time.

> After 6:00 pm it is dangerous. Before 6:00 pm in winter it is also dangerous, but in summer it is always ok, because if you have light it is ok.
>
> **(Boy, 12 years)**

When looking at variations based on gender, overall the evaluations of boys and girls were both quite similar. The most apparent differences appeared in the qualities 'pretty' and 'friendly'. In both instances, boys scored more positively than girls did. Girls perceived the city centre to be less friendly as they reported feeling uncomfortable when walking through some areas because of unwanted attention there.

In the case of differences between age groups, the older children scored slightly higher on all qualities except for 'Interesting'. This higher score could be expected as older children are better able to negotiate the different spaces and know better what spaces to avoid. For the quality 'quiet', the difference is larger as the younger children appear to be more sensitive to noise, especially on market days and during festivals that take place in the open square. The younger children, however, had higher scores for the quality 'Interesting' as many of them enjoyed the city centre, in particular video-game shops, candy stores, ice cream shops and fast food restaurants.

10.8.2 Spatial Perceptions

The spatial representation of perceptions made it possible to identify positive and negative areas and the convergence or divergence of perceptions between the children. Altogether, the children recorded 235 observations, distributed over 75 locations, in the city-centre area (Figure 10.4). These locations were categorised as being positive, negative or mixed (both positive and negative).

As Figure 10.4 shows, the highest rated positive clusters include locations such as open squares, consumer electronics stores, restaurants and department stores. This is a clear indication that the city centre provides consumer services that children use and appreciate. As pointed out by Chin (1993) and Karsten (2002), these locations are spaces where children can not only socialise with their peers, but also behave and feel as if they are being treated as adults. Some locations were more popular than others because of positive physical qualities (e.g. open plazas with attractive physical features, historical buildings, modern buildings and locations offering variety of activities) and/or positive social qualities (e.g. absence of social threat).

The negative clusters shown in Figure 10.4 include a back street with several 'coffee shops' selling cannabis, the dentist, the hospital, the casino and the bus station. The prime reasons given for perceiving a location negatively were 'Dangerous' and 'Dirty'. 'Dangerous' refers to various forms of perceived threat, such as fear of harassment and crime and heavy traffic. 'Dirty' was associated with features such as litter, 'bad graffiti' and the presence of animals (dogs and birds). Figure 10.4 also shows locations where the children have mixed – both positive and negative – perceptions. These relate in part to negative physical qualities and in part to negative social qualities, depending on the experience of the individual child. For example, although the main square was generally perceived positively, some children perceived it negatively after witnessing a robbery there. The categorization of three types of space (positive, negative, mixed) is in line with the physical–social qualities map originally developed by Chawla (2002), as illustrated in Figure 10.1. Cluster locations that are perceived as fully positive or fully negative can easily be incorporated in this framework, but cluster locations for which perceptions are *mixed* cannot. Nevertheless, the children unanimously agreed on what makes a quality positive or negative. Indeed, several of the qualities included in Chawla's original framework also appeared in our map (Figure 10.5), along with several

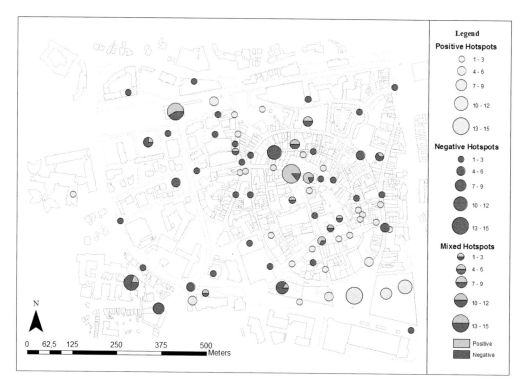

FIGURE 10.4 Spatial perceptions of the city centre based on participants' input. Source: Alarasi et al. (2016). Copyright 2016 by Taylor & Francis. Reprinted with permission.

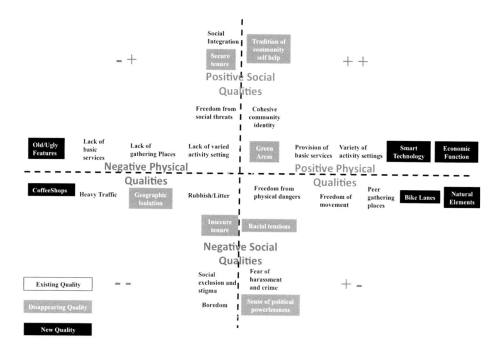

FIGURE 10.5 Emergent qualities based on the registered perceptions of participating children (10–17 years). (Source: Alarasi et al. (2016). Copyright 2016 by Taylor & Francis. Reprinted with permission.)

new qualities that emerged because of the specific context of the Enschede's city centre. Section 10.8.3 provides further insight into this.

10.8.3   Perception, Place and Meaning

Textual analysis of the interviews made it possible to adjust Chawla's original framework presented in Figure 10.1. Several new social and physical qualities became apparent from the analysis, while other qualities that had been observed in previous studies did not emerge from the socio–physical context of this study (see Figure 10.5). This demonstrates in particular that the identified qualities are sensitive to the specific context of the city centre and do not automatically apply to the city of Enschede as a whole.

New qualities that emerged from the textural analysis of the children's interviews are:

- *Natural elements*: A quality similar to 'green areas' but adapted to include other natural features, including water, flowers, sunlight and shade. In their maps, children stated the presence of natural elements as a primary reason for judging those locations positively. In fact, adding more trees was one of the most common improvements mentioned by children when asked what they would like to change in the city centre. The importance of natural elements was also observed during the guided tour, when the children spent a considerable amount of time exploring the various natural elements. Figure 10.6 gives more insight into the perceptions of the children on this quality.
- *Cycle lanes*: The children rated this quality positively, especially when they compared the city centre of Enschede to the cities they lived in before. This quality gives children more freedom of movement as the safety of cycle lanes makes them less dependent on adults to drive them to their various destinations. More than half of the children participating in the study cycled to and from school daily.
- *Smart technologies*: The children explicitly acknowledge technology as a positive quality. For younger boys, smart technologies are associated with playing video and internet games. This was the main reason why a consumer electronics store scored very positively in their maps. The older group of children looked upon smart technologies as a solution for some of the problematic areas they pointed out on their maps.

> I do not like the bus station: there are too many buses and this is bad for the environment – too much CO_2. They should try to look into electric buses; its 2012 after all!
>
> **(Boy, 14)**

  Additionally, most of the children very positively rated locations that offer free Wi-Fi facilities.

- *Economic functions*: This quality frequently occurs in positively mapped locations. It mainly refers to (fast food) restaurants and shops (sweets, video games, consumer electronics). Many of the children pass by the city centre daily, spending a considerable amount of time there. The city centre is therefore not only a place to socialise but also one where they can spend their money, creating a culture of child consumption. This finding is similar to that obtained in Amsterdam by Karsten (2002) and is especially relevant considering that retail and consumption functions played a key role in the redevelopment Enschede's city centre.
- *Coffee shops*: A physical quality related to the specific context of the Netherlands, where cannabis is legally sold. The presence of coffee shops is the one quality that all children evaluated negatively. In fact, the street in which coffee shops were concentrated was the most negative location identified, the most frequently photographed item and a top priority for the children when asked about improvements to the city centre. When children described this quality, they tended to use the words: ugly, dirty and dangerous. Some of the children (especially the younger age group) also perceived the people in these areas as

"Me and my sister keep pressing the button and let the water over flows" (Boy, 12)

1. "I like skipping stones onto the water." (Boy, 12)

2. "I think this place has good air to breath because of the green around it." (Boy, 13)

3. "I love this fountain! I go there and play with my friends, we keep going round and round and round." (Boy, 11)

"It makes you feel more peaceful and the area looks brighter" (Girl, 17)

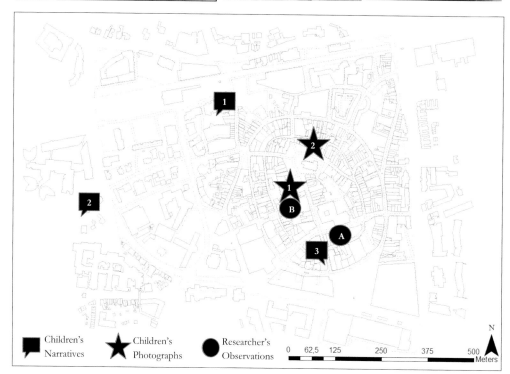

FIGURE 10.6 Natural elements – *Geo-Picto Narrative*. (Source: Alarasi et al. (2016). Copyright 2016 by Taylor & Francis. Reprinted with permission.)

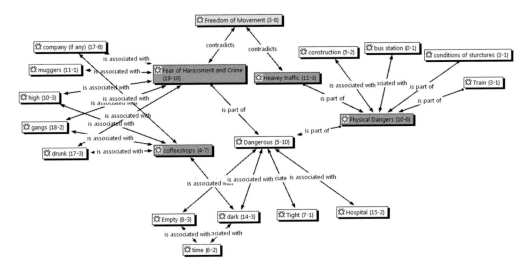

FIGURE 10.7   Relationship between the quality 'coffee shops' and the other physical and social qualities – network diagram. Source: Alarasi et al. (2016). Copyright 2016 by Taylor & Francis. Reprinted with permission.

dangerous. The network diagram in Figure 10.7 shows how this specific quality relates to other physical and social qualities, including fear of harassment and crime, physical danger and freedom of movement.

- *Old/Ugly features:* Children appear to be very sensitive to old features that are not well maintained or features that are 'dull' or 'grey'. Regardless of their age or gender, the children were very sensitive to their surroundings, as they quickly noticed features such as litter, dirty walls, bad smells, poorly maintained buildings and crowded spaces. This is also reported in previous research (Woolley et al., 1999), suggesting that children can observe their environment in considerable detail. Ward (1978) attributes this to the smaller height of children. On the other hand, children reported positively on artistic features. Many pictures were taken of colourful street art, details of buildings and the various statues in the city-centre area.

10.9   REFLECTION ON THE QGIS APPROACH AND RESULTS

This case study illustrates the added value of adopting qualitative GIS when studying children in the context of their neighbourhoods as an essential approach for understanding their experience of place and developing child-centred policies (Christensen and James, 2008). Although a number of studies on children's understanding of maps have been done since the 1970s (see discussion on Matthews, Moore, Spencer and Hart as cited in Holloway and Valentine, 2000, and Aitken, 2005), the conceptualization of these studies were focused on cognition, wayfinding and spatial knowledge acquisition (Aitken, 2005). However, it is clear that adopting a qualitative GIS methodology revolves around visualizations of invisible and unquantifiable experiences and subjects in an effort to position them on a map, which in doing so effectively lends them a sense of legitimacy and authority (Cope and Elwood, 2009).

In fact, the conceptualizing of GIS as a technology, methodology and situated (location and participant dependent) social practice makes the adoption of qualitative GIS for research on children's experiences in their neighbourhood an immediate response to the new sociology of childhood's call for using ethnographic research methodologies that allow children a 'more direct voice and participation in the production of sociological data than is usually possible through experimental or survey styles of research' (Prout, 2005, p. 60). In our specific study, this occurred through incorporating qualitative data and expanding the representational capacities of GIS. And despite inclinations to view similar GIS studies that incorporate these elements as simply adding on qualitative data, it is important to stress the hybrid epistemology and the embedded intersubjectivity, as well as the active reflexivity, of the agent mapper in these studies (Kwan, 2002).

Some may argue that the 'simultaneous' use of multiple research methods could lead to repetitive information. However, the incorporation of multiple forms of data in this research results in a greater understanding of the perceptions and experiences of children. This was also pointed out by Cele (2006) and Jung and Elwood (2010). By overlaying geographical information and exploring relations using querying functions of the textual analysis, we were able to obtain valuable insights into the children's perceptions. This also proved to be insightful for the city planners:

> These methods are very good because by walking around the children can point out exactly what they like and dislike. And from their pictures we can see their point of view […] and the combination between the maps and the picture is really good for us.

Among the results shared with the city planners, the maps ranked highest in importance, in particular when used in combination with the supporting contextual information:

> It is important to know what are the positive areas and why, and also [to know] the negative areas and why. Then you can lay them next to each other. Then you can find the relationships, because when we make our plans we can see and understand what we can do to turn a negative location into a positive one. That is why it is important that we know exactly why it is negative. That is very important for policy-makers. Why! Why! Why! [an area is perceived as negative] Because only then can we see if we can do something about it.

More importantly, the two planners agreed on the importance of including insights of children in urban planning processes. This is important not only for improving current situations but also when considering future plans for a city. Having said that, the planners acknowledged that children are usually forgotten in urban planning processes, as planners generally rely on the viewpoints of adults alone:

> We speak a lot with the parents of children. In the Netherlands, when we plan for a playground, for example, the parents will often make their opinion known, while the children are not in the picture.

The planners also indicated that one of the reasons a policy-maker is more interested in a parent's point of view is because parents have voting power. When asked how this obstacle could be overcome, they pointed out to the importance of studies such as ours. In fact, the collaboration with the international school for this research was seen as 'the power' of the whole study, since a systematic approach can be developed so that the school can schedule this kind of activity into their yearly programme, thus making future collaboration more attainable. At the same time, it must be recognised that conducting participatory sessions in a regular classroom environment may hinder true participation, as children tend to share opinions differently if they perceive something as an assignment rather than a free activity.

Currently, processes of collaboration between municipalities and schools are being introduced in different parts of Australia, New Zealand and the UK (Gleeson and Sipe, 2006); this is also one of the recommendations made by Chawla (2002). Once such collaboration is established, it will become easier, as one of the planners stressed, to replicate and roll out approaches such as this to more schools. This will not only aid in institutionalizing children's participation in existing planning practice but will also help to sensitise – perhaps even 'educate' – planners to more systematically incorporate the perspectives of children in local planning processes. Equally important is that the visual media used in this type of research can be used as a bridge to engage a variety of built-environment professionals. Not only planners, but also architects, (urban) designers and artists could become involved at different stages.

The methodological approach presented here can be applied in different types of living environments, although it might need to be adapted to accommodate the limitations and characteristics of different settings. For example, research done with children from low-income communities who have less access to IT technology may require tools other than Google® maps and, perhaps, even photography. One solution would be to rely on printed maps and drawings made by the children. In cases where the children have difficulties in reading a map, a walking tour in which children take the lead in guiding the researcher to points of interest could also serve as a substitute.

10.10   CONCLUSION

Our study focused on involving children in the town planning processes by capturing their perceptions of their local living environment, in particular those of the city centre of Enschede. Given the specific background of the children we involved in this research, the results reported do not necessarily reflect the general perception of all children on the Enschede's city centre. Based on the framework developed by Chawla (2002), a mixed-method qualitative GIS (qGIS) approach was designed to gain insight into how children perceive the social and physical qualities of the city centre. Finally, the study also illustrates the benefit of a participatory, mixed-method approach that, on the one hand, provides children a platform for expressing their views and, on the other hand, results in a better understanding of these views and perceptions.

ACKNOWLEDGEMENTS

We would like to thank the International School Twente for their generous support and the anonymous referees for their valuable suggestions. Special thanks to Mrs Els Weir and Mr Remko Lulof, who welcomed this research in their classrooms and were actively involved in planning and facilitating it. An exclusive thank you is reserved for all the children who enthusiastically participated in this research.

REFERENCES

Aitken, S. C. (2005). *The geographies of young people: The morally contested spaces of identity.* London: Routledge.

Alarasi, H., Martinez, J., & Amer, S. (2016). Children's perception of their city centre. A qualitative GIS methodological investigation in a Dutch city. *Children's Geographies, 14*(4), 437–452.

Ariès, P. (1962 [1970]). *Centuries of childhood; a social history of family life.* Translated from the French by Robert Baldick. New York, Knopf.

Arnstein, S. R. (1969). A ladder of citizen participation. *Journal of the American Institute of planners, 35*(4), 216–224.

Berglund, U., & Nordin, K. (2007). Using GIS to make young people's voices heard in urban planning. *Built Environment (1978–), 33*(4), 469–481.

Burke, E. M. (1979). *Participatory approach to urban planning.* Human Sciences Press.

CBS (2016) *Kerncijfers wijken en buurten 2016.* https://opendata.cbs.nl/statline/#/CBS/nl/dataset/83487NED/table?ts=1518018275810 (accessed November 2017).

Cele, S. (2006). *Communicating place: Methods for understanding children's experience of place* (Doctoral dissertation, Acta Universitatis Stockholmiensis).

Chawla, L. (Ed.). (2002). *Growing up in an urbanizing world.* Routledge.

Chin, E. (1993). Not of whole cloth made: The consumer environment of children. *Children's Environments, 10*(1), 72–84.

Christensen, P., & James, A. (Eds.). (2008). *Research with children: Perspectives and practices.* London: Falmer Press.

Christensen, P., & O'Brien, M. (Eds.). (2003). *Children in the city: Home neighbourhood and community.* London: Routledge Falmer.

Cope, M., and S. Elwood. (2009). *Qualitative GIS.* London: Sage.

Day, D. (1997). Citizen participation in the planning process: An essentially contested concept? *Journal of Planning Literature*, *11*(3), 421–434.

Dennis Jr, S. F. (2006). Prospects for qualitative GIS at the intersection of youth development and participatory urban planning. *Environment and Planning A*, *38*(11), 2039–2054.

Driskell, D. (2002). *Creating better cities with children and youth: A manual for participation*. London: Earthscan.

Enschede-Stad. (2014). Geschiedenis van enschede. http://www.enschede-stad.nl/geschiedenis.htm (accessed November 3).

Forester, J. (1999). *The deliberative practitioner: Encouraging participatory planning processes*. Cambridge, MA: MIT Press.

Freeman, C., & Tranter, P. J. (2011). *Children and their urban environment: Changing worlds*. London: Earthscan.

Gillespie, J. (2013). Being and becoming: Writing children into planning theory. *Planning Theory*, *12*(1), 64–80.

Gleeson, B., & Sipe, N. (2006). *Creating child friendly cities: Reinstating kids in the city*. London: Routledge .

Hill, M. (2005). Ethical considerations in researching children's experiences. In *Researching children's experience* (pp. 61–86).

Holloway, S. L. (2014). Changing children's geographies. *Children's Geographies*, *12*(4), 377–392.

Holloway, S. L., & Valentine, G. (2000). Children's geographies and the new social studies of childhood. In *Children's geographies: Playing, living, learning* (pp. 1–22). London: Routledge.

Hospers, G. J. (2010). Lynch's the image of the city after 50 years: City marketing lessons from an urban planning classic. *European Planning Studies*, *18*(12), 2073–2081. doi:10.1080/09654313.2010.525369

James, A. (2004). Understanding childhood from an interdisciplinary perspective: Problems and potentials. In Pufall, P. B., & Unsworth, R. P. (Eds.), *Rethinking childhood* (pp. 25–37). New Brunswick, NJ: Rutgers University Press.

Jans, M. (2004). Children as citizens: Towards a contemporary notion of child participation. *Childhood*, *11*(1), 27–44.

Jung, J. K., & Elwood, S. (2010). Extending the qualitative capabilities of GIS: Computer-aided qualitative GIS. *Transactions in GIS*, *14*(1), 63–87.

Karsten, L. (2002). Mapping childhood in Amsterdam: The spatial and social construction of children's domains in the city. *Tijdschrift voor economische en sociale geografie*, *93*(3), 231–241.

Kennispunt Twente. (2014). Enschede in cijfers. Enschede: Kennispunt Twente

Knowles-Yánez, K. L. (2005). Children's participation in planning processes. *Journal of Planning Literature*, *20*(1), 3–14.

Kwan, M. P. (2002). Feminist visualization: Re-envisioning GIS as a method in feminist geographic research. *Annals of the Association of American Geographers*, *92*(4), 645–661.

Kyttä, M. (2004). The extent of children's independent mobility and the number of actualized affordances as criteria for child-friendly environments. *Journal of Environmental Psychology*, *24*(2), 179–198.

Laughlin, D. L., & Johnson, L. C. (2011). Defining and exploring public space: Perspectives of young people from Regent Park, Toronto. *Children's Geographies*, *9*(3–4), 439–456.

Lewis-Beck, M., Bryman, A. E., & Liao, T. F. (2003). *The Sage encyclopedia of social science research methods*. Sage Publications.

Lupton, D. (1999). *Risk*. London: Routledge.

Lynch, K., & Banerjee, T. (1977). Growing up in cities. *New society*, *37*(722), 281–284.

Matthews, M. H. (1992). *Making sense of place: Children's understanding of large-scale environments*. Hemel Hempstead, Hertfordshire: Harvester Wheatsheaf.

Monno, V., & Khakee, A. (2012). Tokenism or political activism? Some reflections on participatory planning. *International Planning Studies*, *17*(1), 85–101.

Oldfield, A. (1990). *Citizenship and community: Civic republicanism and the modern world*. London: Routledge.

Oswell, D. (2013). *The agency of children: From family to global human rights*. Cambridge: Cambridge University Press.

Postman, N. (1985). The disappearance of childhood. *Childhood Education*, *61*(4), 286–293.

Prout, A (2005). *The Future of Childhood: Towards the Interdisciplinary Study of Children*, London: Routledge Falmer.

Santo, C. A., Ferguson, N., & Trippel, A. (2010). Engaging urban youth through technology: The youth neighborhood mapping initiative. *Journal of Planning Education and Research*, *30*(1), 52–65.

Simpson, B. (1997). Towards the participation of children and young people in urban planning and design. *Urban Studies*, 34 (5–6), 907–925.

Stasiulis, D. (2002). The active child citizen: Lessons from Canadian policy and the children's movement. *Citizenship Studies*, 6(4), 507–538.

Travlou, P., Owens, P. E., Thompson, C. W., & Maxwell, L. (2008). Place mapping with teenagers: Locating their territories and documenting their experience of the public realm. *Children's Geographies*, 6(3), 309–326.

UNICEF. (1989). *Conventions on the rights of the child.* http://www.unicef.org/crc/ (accessed September 30, 2016).

UNICEF. (1996). *Child friendly cities.* http://www.childfriendlycities.org (accessed September 30, 2016).

Valentine, G., & McKendrick, J. (1997). Children's outdoor play: Exploring parental concerns about children's safety and the changing nature of childhood. *Geoforum*, 28(2), 219–235.

Van Melik, R. (2009). Visualising the effect of private-sector involvement on redeveloped public spaces in the Netherlands. *Tijdschrift voor economische en sociale geografie*, 100(1), 114–120.

Veitch, J., Salmon, J., & Ball, K. (2007). Children's perceptions of the use of public open spaces for active free-play. *Children's Geographies*, 5(4), 409–422.

Walker, M., Whyatt, D., Pooley, C., Davies, G., Coulton, P., & Bamford, W. (2009). Talk, technologies and teenagers: Understanding the school journey using a mixed-methods approach. *Children's Geographies*, 7(2), 107–122.

Ward, C. (1978). The child in the city. *Society*, 15(4), 84–91.

Williams, S. H. (1976). Citizen participation in city and regional planning: An effective methodology. *Town Planning Review*, 47(4), 349.

Woolley, H., Dunn, J., Spencer, C., Short, T., & Rowley, G. (1999). Children describe their experiences of the city centre: A qualitative study of the fears and concerns which may limit their full participation. *Landscape Research*, 24(3), 287–301.

Wridt, P. (2010). A qualitative GIS approach to mapping urban neighborhoods with children to promote physical activity and child-friendly community planning. *Environment and Planning B: Planning and Design*, 37(1), 129–147.

Measuring the Quality of Streets as
Public Space within a Fragmented
City: The Case of Msasani Bonde la
Mpunga, Dar es Salaam, Tanzania

Bolatito Dayo-Babatunde, Javier Martinez,
Monika Kuffer, and Alphonce Gabriel Kyessi

CONTENTS

11.1 Introduction 184
    11.1.1 Concepts of Urban Fragmentation.. 184
    11.1.2 Manifestation of Urban Fragmentation in African Cities 185
    11.1.3 Streets as Public Space 185
    11.1.4 Measuring Street Qualities Using Quantitative
           and Qualitative Indicators.. 186
11.2 Methodology... 187
    11.2.1 Selection of Streets and Relevant Indicators 188
    11.2.2 Primary and Secondary Data 188
    11.2.3 Data Analysis Methods.. 189
11.3 Results.. 190
    11.3.1 Land Use Characteristics of Msasani Bonde la Mpunga......................... 190
    11.3.2 Profile of the Selected Streets in Msasani Bonde la Mpunga................... 190
    11.3.3 Level of Street Connectivity 191
    11.3.4 Residents' Perceptions of Their Streets 193
           11.3.4.1 Sense of Belonging 194
           11.3.4.2 Social Interaction 195
11.4 Discussion... 197
    11.4.1 Community Integration of Street 1 (Planned/Gated) 197
    11.4.2 Functional Integration of Street 2 (Gated/Unplanned)........................... 197
    11.4.3 Symbolic Integration of Street 3 (Planned/Unplanned) 198
11.5 Conclusions.. 199
References.. 200

11.1 INTRODUCTION

Cities in the Global South have experienced rapid urbanization that has resulted in fragmented spatial patterns of planned and unplanned urban neighbourhoods (Balbo, 1993) – and more recently the spread of gated communities (Asiedu and Arku, 2009). As a result, the quality of life of inhabitants has deteriorated due to a lack of street connectivity and absent social cohesion, not to mention the social exclusion that emanates from such urban fragmentation. For many decades, the street as public space has been identified as a *binding factor* in neighbourhoods, as streets contribute towards social integration of their surrounding urban fragments (Janches, 2011; Loukaitou-Sideris and Banerjee, 1998; Trancik, 1986). Starting in the late nineteenth century, various studies have focused on street designs, street significance, measurement of street qualities and street reclaiming practices that promote pedestrian mobility and restore streets as public spaces (Appleyard and Lintell, 1972; Jacobs, 1961; Mehta, 2013). In a similar vein, UN-Habitat (2013) has recognised streets as drivers of a city's prosperity, affecting the overall quality of life and social wellbeing of people. However, there are only a limited number of studies that focus on the street as public space within African cities; few empirical studies have focused on issues such as housing, land tenure and infrastructure in relation to informal settlements (Kombe, 2005; Nguluma, 2003). Not much is known about the influence of the social dimension of streets as public spaces on the integration of and interaction among different social groups. Such inadequate theoretical and empirical knowledge increases the risk of inappropriate planning interventions (Mrema, 2013).

With the above in mind, with this study we aim to measure the quality of streets as public spaces between various urban fragments – planned, gated and unplanned neighbourhoods – in a fragmented settlement located in Dar es Salaam, Tanzania. In contrast with the UN-Habitat (2013) study, in which only quantitative indicators were measured, our study measures the quality of streets by combining both quantitative (objective) indicators of street connectivity and qualitative (subjective) indicators based on residents' perceptions. Results of this study are relevant for local planning authorities and policy-makers in areas of policy intervention and can contribute to filling the knowledge gap in the literature on African cities and their streets as binding factors.

11.1.1 CONCEPTS OF URBAN FRAGMENTATION

Over the decades, cities, particularly those in the Global South, have continued to exhibit distinct social and morphological characteristics related to a process commonly referred to as *urban fragmentation*. It is therefore pertinent to understand this concept and its implications for social interaction (between residents of different fragments) and processes that affect integration within and between fragments.

Urban fragmentation is a tricky concept: easy to recognise but difficult to define, due to its multi-dimensional nature and its many forms. For example, Harrison (2003) differentiated between institutional fragmentation (related to the decentralization of political structure), social fragmentation (related to increases in gaps between socio–economic classes in society) and spatial fragmentation (understood as the breaking up of urban space into distinct parts). Furthermore, different concepts and terms have been used to refer to the same phenomena, such as the dual city, the disconnected city, the divided city and splintering urbanism (Balbo, 1993; Graham and Marvin, 2001; Madrazo and van Kempen, 2012). The concept of the dual city, for example, reflects social fragmentation by which a city is divided into a society for the rich and another for the poor. The term urban fragmentation also refers to the disintegration of the urban environment into fragments, parts or pieces (Hidding and Teunissen, 2002; Landman, 2011b). Another definition looks at the fragmentation of urban form as a disorderly process of development that leads to splintering of urban space, making the city a mosaic without any distinguishable centrality (Burgress, 2005).

11.1.2 MANIFESTATION OF URBAN FRAGMENTATION IN AFRICAN CITIES

Urban fragmentation of African cities can be traced back to the colonial era. According to Pacione (2009), the processes of urban development reflect a relationship between global economic forces and local cultural contexts, and in this respect African cities exhibit the greatest diversity of urban form. Such diversity developed from an interaction between distinctive traditional forms of dwelling and the influence of the colonial authorities. O'Connor (1983) identifies six types of African cities: indigenous, Islamic, colonial, European, dual and hybrid cities. In the meantime, most African cities have evolved beyond these six types and today show an array of different urban fragments, including urban forms that can be classed as indigenous, planned, unplanned, informal and, more recently, gated, creating a distinctive spatial pattern (Balbo, 1993).

The following three fragments were identified for this study in Dar es Salaam, each based on their peculiar morphological characteristics:

1. *Planned neighbourhoods*: These are referred to as formal developments that are pre-conceived and developed based on urban planning standards and regulations. In these neighbourhoods, efforts are focused on achieving a coherent spatial structure, with streets and building layouts matching the planning scheme. Streets within such neighbourhoods are open and accessible to all. However, local planning authorities in Dar es Salaam face big challenges in managing urban development due to a lack of technical knowledge and expertise to cope with rapid growth (Mrema, 2013; UN-Habitat, 2009). These challenges have resulted in the proliferation of unplanned settlements.
2. *Unplanned neighbourhoods* (or slums): These neighbourhoods mostly exhibit an informal spatial structure in which there is no land tenure. Housing in such neighbourhoods is inadequate and basic infrastructure is lacking, as are regular street patterns (UN-Habitat, 2002). Public spaces are limited, in most cases making streets the only public space available for interaction.
3. *Gated communities*: Gated neighbourhoods are related to a new urban phenomenon, the morphological characteristics of which resemble pre-colonial, walled Islamic African cities. Gated communities are characterised by designated perimeter walls, fences and controlled, gated entrances to prevent access by non-residents (Asiedu and Arku, 2009; Landman, 2011a). Of key interest for this study is that gated communities restrict access to streets, thereby making them no longer public spaces but private ones. Being a 'super block' reduces the connectivity of the entire neighbourhood (Burke and Sebaly, 2001; Victoria Transport Policy Institute, 2014). Thus, as a result of perimeter walls, gated communities are less integrated with their surrounding communities and are in many respects private bodies – with private streets, security services and facilities (e.g. commercial outlets, schools, swimming pools, gymnasiums) (Burke and Sebaly, 2001).

11.1.3 STREETS AS PUBLIC SPACE

Streets are the widest and most accessible public spaces providing opportunities for social activities and interaction (Oranratmanee and Sachakul, 2014). Jacobs (1961, p. 39) stated that 'Streets and their sidewalks, the main public spaces of a city, are its most vital organs'. She furthered buttressed that statement by adding that interesting streets imply an interesting city and, conversely, that if streets are dull, the city will be dull. As the street is a multifunctional space, its role as a public space can be understood from different perspectives. That includes the street as physical space, as a conduit for movement of people and goods, and as a public realm as well as a place.

Apart from playing a major role as an infrastructural link, streets as public spaces play a significant role in the urban structure and life of a city by promoting social–spatial integration that counteracts urban fragmentation (Janches, 2011). The street as a public space within a fragmented city serves as a connector for binding the urban fragments of heterogeneous neighbourhoods, enabling them to become neighbours with others of a different socio-cultural identity. According to Trancik (1986), the linkage function of the street helps to 'glue' the city together and determine the order of buildings and open space for social interaction and exchange. Loukaitou-Sideris and Banerjee (1998) also argue that instead of treating the street as a link only for movement, as in the modernist era, the street should be 'recovered' to play a social role as a connector that 'stitches' together – and often even penetrates – different urban fragments. This physical, spatial perspective on the role of streets as elements of integration in the built environment should be seen as an enabling factor and not as a guarantee of social integration.

However, elements that act as binding factors and facilitate social integration vary among settlements. While studying gated and poor communities in Santiago de Chile, Sabatini and Salcedo (2007) identified three dimensions of social integration. First, *functional integration* can occur through an exchange based on money and power, whereby the poor can participate either as customers or workers, with access to services and facilities. According to Brown (2004), street life encourages activities such as informal street trading, which is prominent in African cities. Second, *symbolic integration* refers to the degree of attachment and compromise an individual feels towards his or her place of residence. Sabatini and Salcedo (2007) argue here that 'sense of belonging' has often been confused in the literature with community integration. However, community integration requires a certain degree of sameness and level of equality, while symbolic integration can also occur under unequal relationships as it relates to the interdependencies and solidarity that may occur, even between groups of different socio-economic classes (e.g. a maid or a caretaker working for a rich family in a gated community). Third, *community integration* goes beyond functional integration and includes mutual recognition and social ties expressed through friendship, family relations and solidarity networks. Awareness of these three dimensions of social integration should assist in identifying how urban fragmentation can enable or hinder certain aspects of social relationships within an urban setting.

11.1.4 Measuring Street Qualities Using Quantitative and Qualitative Indicators

Indicators may be quantitative and/or qualitative variables that help to describe and communicate about a phenomenon (Innes, 1990; Wong, 2006). The main purpose of an indicator is, therefore, to measure features of a phenomenon in an objectively verifiable way and communicate about an assessment of the phenomenon based on those measurements. Quantitative (objective) indicators of urban conditions that impact quality of life relate to observable and measurable conditions, for example the presence of sidewalks. Qualitative (subjective) indicators relate to a perception or level of satisfaction that a person or a community has concerning that domain of urban life (Pacione, 2003; Tesfazghi et al., 2010).

In the literature, several indicators are used to describe the quality of streets. UN-Habitat (2013), for instance, proposed and compiled objective indicators such as street density, intersection density and land area allocated to the street, among others, to measure street connectivity. This study did not, however, include subjective indicators that could measure residents' perception of streets in their neighbourhood. According to Pacione (2003), a combination of the two types of indicators could contribute to the interpretation and better understanding of the phenomena.

11.2 METHODOLOGY

BOX 11.1   Case Study Area

Dar es Salaam is the largest city in and the commercial hub of Tanzania, as well as the largest city in East Africa (UN-Habitat, 2014), covering an area of approximately 1600 km². The city is divided in five administrative districts: Ilala, Kinondoni and Temeke, Kigamboni and Ubungo.

The study area, *Msasani Bonde la Mpunga*, is located in the Kinondoni district and covers an area of 1.17 km²; its population exceeds 13,000 inhabitants. The settlement is one of the fastest-growing areas in the Kinondoni district (Casmiri, 2008) due to its proximity to major institutional, residential and commercial buildings (e.g. the U.S. embassy, a private (referral) hospital, residences of former senior government officials and a major shopping mall). The area was purposely chosen because it contains several types of urban fragments: in the southern part is an unplanned neighbourhood and in the northern part a planned neighbourhood; sandwiched between the two are the gated communities of Highland Villa and Mayfair Apartments (Figure 11.1).

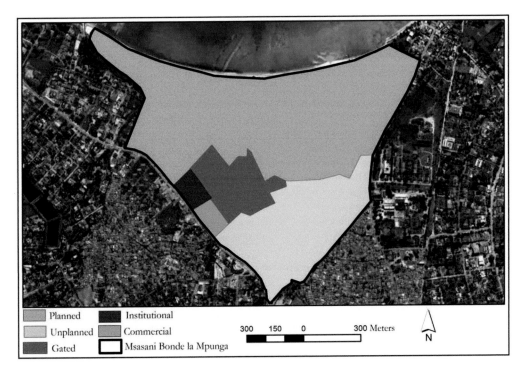

FIGURE 11.1   Existing land use in *Msasani Bonde la Mpunga* in 2014.

11.2.1 Selection of Streets and Relevant Indicators

After a pilot survey was carried out to determine the suitability of streets for our study, we selected three streets based on their location between two different urban fragments and their shared access and street type. In the latter, road hierarchy (e.g. collector or local street) played a significant role in understanding the level of social interaction. To analyse the quality of streets as public space, two main dimensions were used: the street connectivity, and perceptions on street condition and social interaction. Both are important measurements for analysing physical, perceptual and social aspects of street quality. As there are no universal indicators for measuring street quality, the selection of indicators was based on relevant criteria that take into account both quantitative and qualitative indicators (Innes, 1990; Wong, 2006). Table 11.1. summarises the definitions and rationale of selected indicators.

Data were collected in the field during the period September–October 2014, including measurements and observations for the selected indicators. Interviews were conducted to acquire an in-depth understanding of the study area, including the perceptions of residents about the quality of the streets (e.g. safety, cleanness and facilities) and social interactions that occur within and between different neighbourhoods. Due to the lack of a sampling frame, non-probability quota sampling was used, in particular for the unplanned neighbourhood. Table 11.2 gives an overview of types and numbers of persons interviewed.

11.2.2 Primary and Secondary Data

Questionnaire: A survey by questionnaire was conducted among residents and workers to inquire about their perceptions on the quality of the street (e.g. safety, cleanness and

TABLE 11.1
Relevant Indicators Selected to Measure Street Quality

Indicators	Definition and Rationale
Street connectivity	
Proportion of land allocated to street (LAS)	Total land area covered by streets as a percentage of total land area. According to UN-Habitat (2013), sufficient land allocated to streets ensures the integration of neighbourhoods.
Intersection density/ connection	No. of intersections/km². The more connected the network (greater density), the fewer barriers there are for accessing different urban areas. Greater densities also lead to an increase in urban circulation.
Accessibility index	The accessibility index is calculated by dividing the actual travelling distance from the point of origin to a specific destination by the distance of the most direct route (measured as a straight line on the map between the point of origin and destination). An index of 1.0 indicates a person can walk in a straight line to a destination. An average value of 1.5 is considered as a suitability threshold.
Perception on street condition and social interaction	
Safety and security	Experiencing the freedom of dwelling in an area without feeling vulnerable or exposed to harm or danger.
Sense of belonging	Degree of attachment or involvement a person feels towards a place or community in which he/she lives.
Social interaction	Describes the way people talk to and act with those around them. Interaction refers to an exchange between two or more individuals.

Sources: Compiled by the authors from Madanipour, 2010; Mehta, 2013; Montgomery, 1998; UN-Habitat, 2012, 2013.

TABLE 11.2

Overview of Types and Numbers of Persons Interviewed

Interview Form	Interviewees	No. of Interviews
Questionnaire	Residents	60 (20 respondents per three selected streets)
Walking interview	Residents	6
Key informants	Academia (Head of Department Landscape Architecture, Ardhi University)	1
	Government official (Kinondoni municipality/ urban planner)	1

facilities) and their feelings about the neighbourhood and social interaction between and within the different urban fragments. The questionnaire covered a range of issues, including safety and security, sense of belonging and social interaction. The survey was based on a structured questionnaire and scored on a Likert scale of 1–5. The questionnaire was translated to the local Swahili language and run by local fieldwork assistants. The assistants received detailed instructions and training on how to administer the questionnaire. The survey was carried out on both weekdays and during weekends to ensure that the perceptions of residents that may have been working out of the neighbourhood during the day were captured.

Walking interviews: Semi-structured walking interviews[*] were conducted to get an in-depth understanding of the perceived quality of the streets. This style of interview allows ample time for the interviewee to narrate their experiences and for the interviewer to probe for more details. A walking interview is also effective in understanding and capturing the relationship between a perceived quality and the spatial environment (Evans and Jones, 2011) within the context of what people say and where it is said. The interviews and the routes followed were recorded and tracked by GPS.

Secondary data: These data include the land-use map and street network of the study area, which were sourced from Ardhi University (Dar es Salaam, Tanzania) and the University of Twente (Enschede, the Netherlands).

11.2.3 Data Analysis Methods

The quantitative and qualitative data were organised, updated and cleaned before analysis. The available street network dataset of the year 2002 had to be updated in ArcGIS by digitizing new roads and erasing road segments that no longer exist; Google Earth images were used as reference. This enables the calculation of connectivity: the higher the number of intersections, the more connected the street. Data for the indicator land allocated to streets (LAS) are determined by two variables: the total length of the street and its width (averages were 8 m and 3.5 m for planned and unplanned streets, respectively); LAS is calculated as a percentage of the total study area (1.17 km^2).

---

[*] Walking interviews involved walking along with participants (with their permission) and interviewing them in accordance with rules of ethical scientific conduct. All participants were asked for and gave their permission to use the data for academic purposes, provided their personal information remained confidential.

The walking and key informant interviews were translated, transcribed and entered into Atlas.ti, computer-assisted qualitative data analysis (CAQDAS) software. For the text analysis, the interviews were coded, labelled and sorted according to the indicators identified (Table 11.1).

BOX 11.2   Methods Applied in the Chapter

The analysis described in this chapter illustrates the application of quantitative GIS methods in combination with image interpretation, digitization and network analysis. The GIS methods applied include updating the street network by digitization of road segments, spatial analysis of the road network (e.g. street-intersection density) and network analysis, such as travelling time. The GIS methods were complemented with qualitative data gathered from semi-structured walking interviews and text analysis making use of computer-assisted qualitative data analysis (CAQDAS) software. For spatial analysis, ArcGIS was used, but the analysis can be replicated in Q-GIS (e.g. using the GRASS plugin for the network analysis). In addition to primary data, the dataset used comprises secondary spatial data (street network provided by local government) and Google Earth images. The primary data was collected to capture perceptions of street conditions and social interaction; a questionnaire and semi-structured walking interviews were used. The routes followed during the walking interviews were recorded and GPS tracked in order to geo-locate perceptions. The analysis illustrated here for the case of urban fragmentation could also be applied to other urban studies where indicators of street intersections and travelling time are required and the perceptions of residents need to be elicited.

11.3   RESULTS

11.3.1   LAND USE CHARACTERISTICS OF MSASANI BONDE LA MPUNGA

In the Dar es Salaam city master plans of 1968 and 1970, the area of *Msasani Bonde la Mpunga* was designated as hazardous land (Lerise and Malele, 2005). At that time there were only a few (unauthorised/unplanned) houses in the area, which was predominantly used for rice farming. In the mid-1980s there was a rapid increase in residential building there due to liberalization policies and economic reforms that favoured market forces. In 1992 and 2001, the Dar es Salam City Council (DCC) and the Kinondoni municipality, in conjunction with other urban planning agencies, prepared and approved two subdivision schemes (Leader and Lupala, 2009). As a result, amid the existing unplanned/unauthorised land use, plots were formally allocated for the development of private residential estates for gated communities and other land uses. Table 11.3 indicates the main land uses of the study area in 2014.

11.3.2   PROFILE OF THE SELECTED STREETS IN MSASANI BONDE LA MPUNGA

Street 1 (planned/gated): This street marks the boundary of the gated community called Highland Villa (formerly known as the Dar Villa) and provides shared access to a planned non-gated neighbourhood (Figure 11.2a). The street also serves as one of the main entrances to Msasani beach.

Street 2 (gated/unplanned): Locals refer to this street as 'Mayfair Street'. It gives shared access to two different communities (Figure 11.2b): to its right lies an unplanned neighbourhood (*Bonde la Mpunga*); and to its left it forms the boundary of two gated communities

TABLE 11.3
Land-Use Allocation in the Study Area in 2014

Land use	Area (ha)	Proportion (%)
Planned residential	74	63
Unplanned residential	29	25
Gated communities	10	8
Institutional	2	2
Commercial	2	2
Total	**117**	**100**

Sources: Map analysis and field-work inventory in 2014.

(Highland Villa and Mayfair Apartments). The street comprises a clearly delineated segment of 198 m that intersects with an undefined open space connecting several footpaths; these provide access to some parts of the unplanned neighbourhood. The street contains a variety of small commercial activities such as a restaurant, grocery stores and so on. Also located along the street is the Mayfair shopping plaza, although its entrance faces a major road nearby.

Street 3 (planned/unplanned): Known as 'Mehrab Street', this street is named after a foreigner who was the first to build and settle within the planned neighbourhood along the street. Street 3 provides shared access for the planned and unplanned neighbourhoods (Figure 11.2c) in the area. As one travels along the street towards the unplanned neighbourhood, one begins to observe footpaths that are used by some residents of that neighbourhood to access their homes and the neighbourhood in general.

The outcomes resulting from measuring the level of connectivity of the streets in the entire study area and within the individual urban fragments, as well as the outcomes related to the residents' perceptions, are described in the following subsections.

11.3.3 LEVEL OF STREET CONNECTIVITY

Three connectivity indicators were used to measure street connectivity within the road network of *Msasani Bonde la Mpunga*:

- proportion of land allocated to streets (LAS)
- street density and
- intersection density.

In total, 9.5% of land within the study area is allocated to streets (LAS), resulting in a street density of 12 km/km². The LAS is relatively lower and the street density is slightly higher when compared with the average reported by UN-Habitat (2013) for the entire city of Dar es Salaam. Intersection density is an important element in determining street connectivity, in which the higher the number of intersections, the more connected and walkable the street. Figure 11.3 depicts the street connectivity of the study area. To calculate the connectivity, network nodes were reclassified into two groups, the real nodes representing intersections and dangle nodes representing dead ends. For the entire study area, the intersection density was 50 km⁻², although the planned neighbourhood accounted for 76% of the connections. Some problems in the level of connections can be attributed to barriers posed by the gated communities and irregular, short street patterns in the unplanned

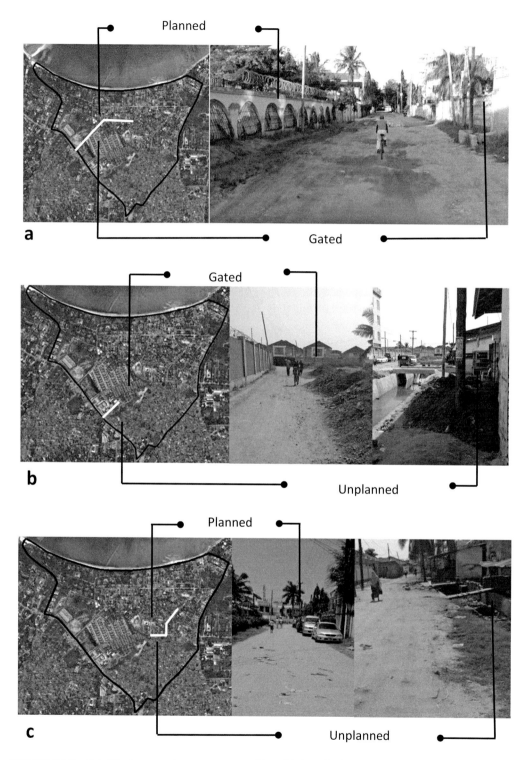

FIGURE 11.2 Satellite imagery (street represented in white) and photos of (a) Street 1, (b) Street 2 and (c) Street 3.

FIGURE 11.3 Map of street connectivity for the entire study area.

neighbourhood, where dead ends predominate (Figure 11.3). This contributes to the low level of social integration of the settlement.

The siting of the gated community contributes further to the disconnection of the streets in terms of accessibility to facilities, as Figures 11.4 and 11.5 show; these figures show the travelling distance from a point of origin (residential) to the west of the gated community to a destination (shopping plaza) within the same neighbourhood. The location of the gated community brought about an increase of the previous travelling distance of 714 m (Figure 11.4) to the current travelling distance of 2155 m (Figure 11.5). The direct distance (straight line) in both cases is 563 m. Consequently, the accessibility index (actual distance/direct distance) increased from the previous 1.3 to a current 3.8. Thus, before the development of the gated community, a resident living at that point of origin could easily walk to the destination, while under the current level of connectivity the route is almost four times longer than the direct route (2155 m versus 563 m). Such differences in travelling distance might affect choices made by residents as to how to get to their destination, thereby reducing opportunities for social encounters.

11.3.4 RESIDENTS' PERCEPTIONS OF THEIR STREETS

Residents' perceived quality of their streets and their usage was measured using qualitative indicators that include safety and security, comfort, sense of belonging and social interaction (based on the questionnaires and walking interviews). Perception was measured on a Likert scale of 5, ranging from strongly agree to strongly disagree. The pattern of perception varies across neighbourhoods, as well as for socio-economic status and background.

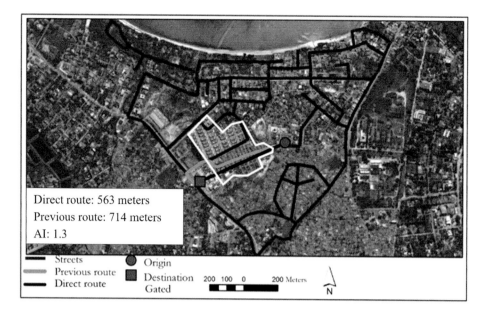

FIGURE 11.4 Example of accessibility index prior to siting of the gated area. Origin=residential; destination=shopping plaza. The white line represents the boundary of the gated area.

11.3.4.1 Sense of Belonging

Figure 11.6 depicts residents' sense of belonging. Respondents were asked if they feel at home, i.e. whether they identify with and experience a sense of attachment to the neighbourhood. Findings from the questionnaires and walking interviews reveal that the residents are generally proud of their neighbourhood, albeit in varying degrees; more than half of all respondents feel at home in their neighbourhood. A small number of respondents were not satisfied with their neighbourhood: for instance, in Street 1 (planned/gated), one respondent did not feel at home.

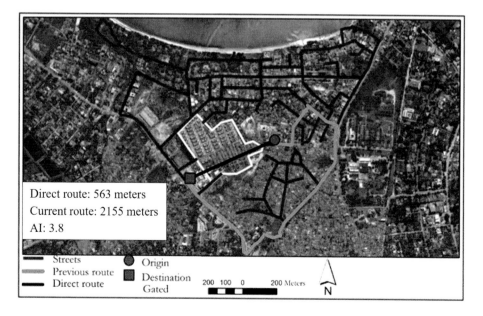

FIGURE 11.5 Street connectivity for the entire study area after siting of the gated area. Origin=residential; destination=shopping plaza. The white line represents the boundary of the gated area.

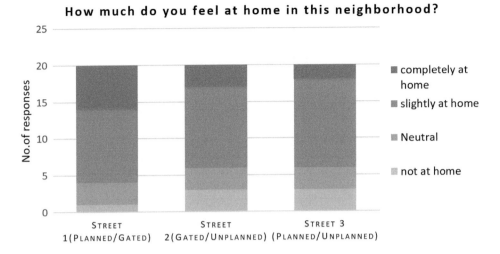

FIGURE 11.6   The sense of belonging of inhabitants of each fragment.

The sense of belonging was also captured by asking the residents during the walking interview which place they like most in their neighbourhood. The beach and the shopping plaza were the most frequently mentioned locations. In general, the results reveal a sense of belonging and attachment of the residents to the street environment.

11.3.4.2   Social Interaction

The questionnaire revealed higher levels of social interaction *within* neighbourhoods and, conversely, lower levels *between* neighbourhoods. Figure 11.7 shows an overwhelming level of social interaction within their neighbourhood. On Street 3 (planned/unplanned), for instance, 14 out of 20 respondents interacted with most people. However, as indicated in Figure 11.8, a difference in interaction can be noticed between neighbourhoods. The social interaction between the fragments bordering Street 2 (gated/unplanned) is much lower: only six people interact with just a few others, while more than half the respondents interact with no one. To explain the level of interaction, one interviewee stated:

> This area used to be friendly until the foreigner [residents of the gated community] came and create a gap between the haves and the have nots. They fenced themselves inside. They do not encourage friendship.

(Interviewee resided in an unplanned neighbourhood)

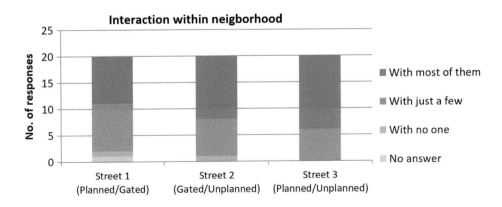

FIGURE 11.7   Perceptions of social interaction within neighbourhoods.

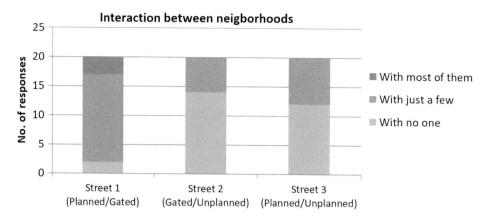

FIGURE 11.8   Perceptions of interaction between neighbourhoods.

Figure 11.8 furthermore reveals that interaction between the planned and unplanned neighbour-hoods on Street 3 is slightly higher, with eight respondents interacting with just a few people and 12 with no one. However, the survey suggests a higher level of social interaction between residents in the neighbourhood of Street 1 (gated/planned). One reason for this might be related to the (to some extent) similar socio-economic profiles of the neighbourhoods.

Figure 11.9 shows the type of relationships and interactions that exist between the different urban fragments, revealing a mixture of perceptions. More than half of the respondents on Street 2 (gated/unplanned) indicate no relationship between themselves and the residents of the neighbour-hood across the street. Business/professional relationships were the next most common relation-ship referred to. An interviewee from the gated community confirmed that residents of the gated community employ residents from the unplanned neighbourhoods for menial tasks, such as child-minding (nannies) or gardening.

One resident from the gated neighbourhood responded:

Hmmm, I don't really know many people outside this villa, but some of my neighbours employ people living in the unplanned area.

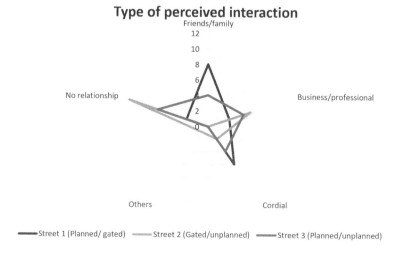

FIGURE 11.9   Perceptions of types of interaction.

Another interviewee from the unplanned neighbourhood said the interaction was

only on business, because there are many activities going on and so there are enough customers; and even from Mayfair apartments [referring to the gated community]) they come and
buy things from me, even though they come only occasionally.

Figure 11.9 shows that relationships between residents of Street 1 (planned/gated) are more
cordial and family–friend related than those of Street 2 (gated/unplanned) and Street 3 (planned/
unplanned), for which the levels of interaction are minimal and weak, as expressed by some
respondents.

11.4 DISCUSSION

The discussion about binding factors is based on our findings from the walking interviews, observations and questionnaire-based survey. The categories discussed here follow the three dimensions of
social integration identified by Sabatini and Salcedo (2007).

11.4.1 COMMUNITY INTEGRATION OF STREET 1 (PLANNED/GATED)

Our study reveals social and business ties. The results show that two-thirds of the respondents
interact with at least a few other people, while more than half the residents' interaction is based on
friendships, family ties or business relationships. This situation can be related to the community integration describe by Sabatini and Salcedo (2007), in which such relationships are expressed in terms
of friendships, family ties and business associations that recognise others as equals. Respondents
from both the planned and the gated communities feel a sense of equality with one another. The
results also suggest that the socio-economic similarities observed between the two neighbourhoods
might contribute to community integration, as interpersonal communication is easier.

Furthermore, residents of the planned neighbourhood did not perceive the fencing of the gated
community as a problem. In fact, most of the private buildings observed within the planned neighbourhood are also (individually) fenced off for security.

Interviewees from the gated community preferred to interact with people around the swimming
pool, the gymnasium and other public spaces provided within Highland Villa (gated community)
rather than outside the villa, where the street quality is perceived as being poor.

In short, community integration exists between the neighbourhoods along Street 1. However,
social interaction rarely occurs on the street: this takes place in other, specific spaces such as the
beach and the shopping plaza. Social interaction occurred beyond the street, and this interaction,
together with the participants socio-economic similarities, might be responsible for the community
integration observed. Reasons suggested for the low level of interaction on the street are related to
the perceived poor quality of the street in facilitating such interaction, as well as the motorised lifestyle of most residents.

11.4.2 FUNCTIONAL INTEGRATION OF STREET 2 (GATED/UNPLANNED)

Functional integration driven by economic factors is most conspicuous on Street 2, where
there is interaction is based on power and money, as discussed by Sabatini and Salcedo (2007).
Economic activities that encourage social interaction were most prominent because of the
high volume of transactions – both formal and informal – that are related to local businesses,
e.g. the restaurant, pharmacy and shopping plaza. According to some interviewees, residents
within the unplanned neighbourhood perceive a sense of integration in terms of patronage of
their businesses by residents of the gated community, especially for those businesses along the
street. Although there are negative perceptions of the gated community, its arrival has increased

economic opportunity. As one interviewee from one of the unplanned neighbourhoods explains during a walking interview:

> I think it's all because of the influx of foreigners that even renting a home is also expensive. There are positive and negative impacts of their coming and that is why the activity I tried to engage in is business, because I know it will be good. In the end there are some local people who have benefited by doing business.

Others also commented on the type of interaction, which is specifically an exchange of money in the form of buying and selling goods. Furthermore, there are employment opportunities especially for menial jobs, such as child-minding (nannies) or cleaning houses. This functional integration allows for interaction between two extremely different social classes, as commented by a key academic informant during an in-depth interview:

> At the shopping area, there is now a lot of socializing between the two social classes. The shops are owned by the locals (lower class), who know what the rich people (upper class) want and they provide it in their shops. Therefore – economically – they are co-existing.

Nevertheless, residents of the gated community and the unplanned neighbourhood still express resentment towards each other. The network views generated during the text analysis (Figures 11.10 and 11.11) indicate interviewees' perceptions of the gated community and unplanned neighbourhoods. They show that residents from the unplanned neighbourhood associate the gated community with rich people and a place of economic opportunity, while gated-community residents associate the unplanned area with illegal activities and see it as being unsafe.

11.4.3 Symbolic Integration of Street 3 (Planned/Unplanned)

The dimension of integration most noted in this study is that of symbolic integration, which occurs under conditions of unequal relationships and, to some extent, functional integration. Residents of the unplanned neighbourhood commented that they have casual relationships and, at some levels,

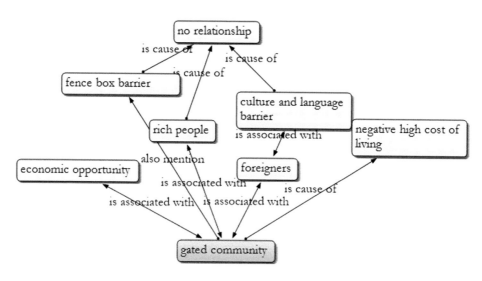

FIGURE 11.10 Network view of perceptions of the gated community by residents of the unplanned neighbourhood.

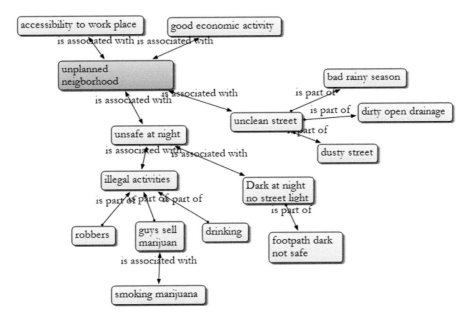

FIGURE 11.11 Network view of perceptions of the unplanned neighbourhood by residents of the gated community.

friendships with residents of the planned neighbourhood. One interviewee from the unplanned neighbourhood stated:

> Yes, I do have a relationship with them – a friendly one – since I have many friends [there] and most of the time we meet here and chat.

Indeed, the questionnaire-based survey showed that some respondents from the unplanned neighbourhood have friends within the planned neighbourhood, and vice versa. Though the number is relatively low, it is remarkable considering the heterogeneity of the two neighbourhoods.

This level of symbolic integration might be due to the degree of connectivity between the neighbourhoods, as there are no barriers to the use of the streets as public space, in contrast with the situation for Street 2 (gated/unplanned). Consequently, the residents of the unplanned neighbourhood felt a sense of belonging and attachment to the place and do not feel stigmatised.

11.5 CONCLUSIONS

Our aim with this study was to contribute to the development of a more holistic planning policies for African cities, embracing both the physical and social functions of the street. Our study shows that the street as a public space may function as a binding factor, but only between certain types of urban fragments (e.g. gated and unplanned) and for specific types of integration (functional integration). Future street design and intervention programmes for improving city planning need to focus on the integration of neighbourhoods (in particular by avoiding the fencing off of neighbourhoods) and allowing a more flexible variety of accommodation for all users and uses. This relates in particular to the quality of the street itself, since residents from gated and planned communities avoid socializing in the street due to a variety of infrastructural, aesthetic and safety issues.

Planning authorities in cities of the Global South need, therefore, to consider how to improve the quality of the streets, reduce physical fragmentation and observe the needs of residents in different urban fragments. For instance, planning schemes and subdivisions should focus on designs that integrate rather than fragment neighbourhoods, for example by reducing the number of 'super blocks' and the amount of fencing intended for gated communities,* and increasing the ratio of complementary, mixed uses. Doing so would create an environment that improves connectivity and increases on-street activities, thus facilitating integration of fragmented neighbourhoods. Such an approach would allow standards and functions to co-exist and consequently improve the quality of life and wellbeing of all inhabitants.

REFERENCES

Appleyard, D., & Lintell, M. (1972). The Environmental Quality of City Streets: The Residents' Viewpoint. *Journal of the American Institute of Planners*, *38*(2), 84–101. https://doi.org/10.1080/01944367208 977410

Asiedu, A. B., & Arku, G. (2009). The Rise of Gated Housing Estates in Ghana: Empirical Insights from Three Communities in Metropolitan Accra. *Journal of Housing and the Built Environment*, *24*(3), 227–247. https://doi.org/10.1007/s10901-009-9146-0

Balbo, M. (1993). Urban Planning and the Fragmented City of Developing Countries. *Third World Planning Review*, *15*, 23–35.

Brown, A. (2004) *Claiming rights to the street: the role of public space and diversity in governance of the street economy*. N-AERUS Conference Barcelona, 15th-16th September.

Burgess, R. (2005). Technological Determinism and Urban Fragmentation: A Critical Analysis. *9th International Conference of the ALFA-IBIS Network on Urban Peripheries, Pontificia Universidad Catolica de Chile, Santiago de Chile, July 11th–13th 2005*.

Burke, M., & Sebaly, C. (2001). Locking in the Pedestrian? The Privatised Streets of Gated Communities. *World Transport Policy & Practice*, *7*(4), 67–74.

Casmiri, D. (2008). *Vulnerability of Dar es Salaam City to Impacts of Climate Change*. EMPS. Dar es Salaam.

Evans, J., & Jones, P. (2011). The Walking Interview: Methodology, Mobility and Place. *Applied Geography*, *31*(2), 849–858. https://doi.org/10.1016/j.apgeog.2010.09.005

Graham, S., & Marvin, S. (2001). *Splintering Urbanism: Networked Infrastructures, Technological Mobilities and the Urban Condition*. London: Routledge.

Harrison, P. (2003). Fragmentation and Globalisation as the New Meta-Narrative. In Harrison, P., Huchzermeyer, M., & Mayekiso, M. (Eds.), *Confronting Fragmentation: Housing and Urban Development in a Democratising Society* (p. 15). Cape Town: University of Cape Town University Press.

Hidding, M. C., & Teunissen, A. T. J. (2002). Beyond Fragmentation: New Concepts for Urban–Rural Development. *Landscape and Urban Planning*, *58*(2–4), 297–308. https://doi.org/10.1016/S0169-2046 (01)00228-6

Innes, J. E. (1990). *Knowledge and Public Policy: The Search for Meaningful Indicators*. New Brunswick: Transaction Publishers.

Jacobs, J. (1961). *The Death and Life of Great American Cities*. London: Penguin Books.

Janches, F. (2011). *Significance of Public Space in the Fragmented City Designing Strategies for Urban Opportunities in Informal Settlements of Buenos Aires City* (Working Paper No. 13/2011).

Kombe, W. (2005). Land Use Dynamics in Peri-urban Areas and Their Implications on the Urban Growth and Form: The Case of Dar es Salaam, Tanzania. *Habitat International*, *29*(1), 113–135. https://doi.org/10.1 016/S0197-3975(03)00076-6

Landman, K. (2011a). *Gated Communities in South Africa: Comparison of Four Case Studies in Gauteng*.

Landman, K. (2011b). Urban Fragmentation: Different Views on Its Causes and Consequences. In Geyer, H. S. (Ed.), *International Handbook of Urban Policy* (pp. 39–60). Cheltenham: Edward Elgar Publishing.

Leader, R. K., & Lupala, J. (2009). *Mainstreaming Disaster Risk Reduction in Urban Planning Practice in Tanzania*.

Lerise, F., & Malele, B. (2005). *Community initiatives in managing urbanization and risk accumulation processes: Lessons from Dar es Salaam, Tanzania – Risk accumulation in the development of Msasani Bonde la Mpunga* Dar Es Salaam: AURAN.

---

* This suggestion should be complemented by government policies to increase safety and security, since studies have shown that the major reason people live in gated communities is that these offer security against crime (Asiedu and Arku, 2009; Landman, 2011a).

Loukaitou-Sideris, A., & Banerjee, T. (1998). *Urban Design Downtown: Poetics and Politics of Form.* Berkeley, CA: University of California Press.

Madanipour, A. (2010). *Whose Public Space?: International Case Studies in Urban Design and Development.* Hoboken: Taylor and Francis.

Madrazo, B., & Van Kempen, R. (2012). Explaining Divided Cities in China. *Geoforum, 43*(1), 158–168. https://doi.org/10.1016/j.geoforum.2011.07.004

Mehta, V. (2013). *The Street: A Quintessential Social Public Space.* Hoboken: Taylor and Francis.

Montgomery, J. (1998). Making a City: Urbanity, Vitality and Urban Design. *Journal of Urban Design, 3*(1), 93–116. https://doi.org/10.1080/13574809808724418

Mrema, L. K. (2013). Creation and Control of Public Open Spaces: Case of Msasani Makangira Informal Settlement, Tanzania. *Online Journal of Social Science Research, 2*(7): 200–213.

Nguluma, H. (2003). *Housing Themselves: Transformations, Modernisation and Spatial Qualities in Informal Settlements in Dar es Salaam, Tanzania.* PhD Thesis, Stockholm: The Royal Institute of Technology.

O Connor, A. (1983). *The African City.* London: Routledge.

Oranratmanee, R., & Sachakul, V. (2014). Streets as Public Spaces in Southeast Asia: Case Studies of Thai Pedestrian Streets. *Journal of Urban Design, 19*(2), 211–229. https://doi.org/10.1080/13574809.2013.870465

Pacione, M. (2003). Quality-Of-Life Research in Urban Geography. *Urban Geography, 24*(4), 314–339. https://doi.org/10.2747/0272-3638.24.4.314

Pacione, M. (2009). *Urban Geography: A Global Perspective* (3rd ed.). London: Routledge Taylor & Francis Group.

Sabatini, F., & Salcedo, R. (2007). Gated Communities and the Poor in Santiago, Chile: Functional and Symbolic Integration in a Context of Aggressive Capitalist Colonization of Lower Class Areas. *Housing Policy Debate, 18*(3), 577–606. https://doi.org/10.1080/10511482.2007.9521612

Tesfazghi, E. S., Martinez, J. A., & Verplanke, J. J. (2010). Variability of Quality of Life at Small Scales: Addis Ababa, Kirkos Sub-City. *Social Indicators Research, 98*(1), 73–88. https://doi.org/10.1007/s11205-009-9518-6

Trancik, R. (1986). *Finding Lost Space: The Theories of Urban Design.* New York: Van Nostrand Reinhold.

UN-Habitat. (2002). Expert Group Meeting on Urban Indicators Secure Tenure, Slums and Global Sample of Cities, October, 33. Retrieved from http://www.citiesalliance.org/sites/citiesalliance.org/files/expert-group-meeting-urban-indicators%5B1%5D.pdf

UN-Habitat. (2009). *Tanzania: Dar es Salaam City Profile.* Nairobi, Kenya.

UN-Habitat. (2012). *Streets as Tools for Urban Transformation in Slums: A Street-Led Approach to Citywide Slum Upgrading.* Nairobi, Kenya.

UN-Habitat. (2013). *Streets as Public Spaces and Drivers of Urban Prosperity.* Nairobi, Kenya: UN-Habitat.

UN-Habitat. (2014). *The State of African Cities 2014. Re-imagining sustainable urban transitions. United Nations Human Settlements Programme, Nairobi.* Nairobi: UN-HABITAT.

Victoria Transport Policy Institute. (2014). Online TDM Encyclopedia. Retrieved February 7, 2018, from http://www.vtpi.org/tdm/tdm12.htm

Wong, C. (2006). *Indicators for Urban and Regional Planning. The Interplay of Policy and Methods.* London and New York: Routledge Taylor and Francis group.

Part II

The 'Compact-Competitive City' and the 'Resilient City'

12 Modelling Urban Growth in the Kathmandu Valley, Nepal

Sunita Duwal, Sherif Amer, and Monika Kuffer

CONTENTS

12.1 Introduction 205
12.2 Materials and Methods 206
    12.2.1 Study Area..206
    12.2.2 Data Sources..207
    12.2.3 Identifying Urban Growth Patterns 208
    12.2.4 Spatial Logistic Regression (SLR) Model 210
    12.2.5 Preparation of Input Variables for SLR 211
12.3 Results... 211
    12.3.1 Spatio-Temporal Pattern of Urban Growth... 211
    12.3.2 Spatial Pattern of Infill and Expansion... 211
    12.3.3 Driving Factors of Urban Growth 213
        12.3.3.1 Drivers of Overall Urban Growth... 213
        12.3.3.2 Drivers of Infill Growth.. 213
        12.3.3.3 Drivers of Expansion 215
    12.3.4 Model Evaluation... 216
        12.3.4.1 Model Evaluation for Overall Urban Growth 216
        12.3.4.2 Model Evaluation for Infill Growth 218
        12.3.4.3 Model Evaluation for Expansion...220
    12.3.5 Validation of the Prediction Model 220
12.4 Discussion and Conclusion 221
References...222

12.1 INTRODUCTION

Urbanization has in recent decades been the subject of a great deal of scientific attention due to its dynamic nature and the profound changes that cities have experienced over time. These changes are linked to population growth and human activities (Alqurashi et al., 2016) resulting in the loss and fragmentation of agricultural and natural land that at regional scales support biodiversity (Naghibi et al., 2016; Vaz and Arsanjani, 2015). Moreover, the spatial heterogeneity of factors such as topography, accessibility, market conditions and population all influence urban morphology, leading to different typologies of urban growth (Chen et al., 2016). A considerable body of literature has evolved to explain these typologies. Liu et al. (2010) and Shi et al. (2012), for example, have identified three types of urban growth: infilling (occurring on vacant land within cities that has so far not been built upon (Xu et al., 2007)); edge expansion (usually occurring at the urban fringe, spreading outwards from existing urban areas); and outlying expansion (new urban growth that has no direct spatial connection with any existing urban fabric (Berling-Wolff and Wu, 2004)). Our study focuses on overall urban growth, infill growth and expansion (including edge and outlying expansion).

Since the 1950s, urban planners, geographers and ecologists (Berling-Wolff and Wu, 2004) have developed a number of methodological approaches for simulating and forecasting future urban growth, among them cellular automata (CA), spatial statistical models, agent-based models (ABM), artificial neural networks (ANN), fractal-based models and decision-tree modelling (Triantakonstantis and Mountrakis, 2012). Although all these approaches have the common goals of modelling and projecting future urban growth, they differ widely in terms of underlying methodologies, assumptions, resolution and scale.

The extensive body of research shows that statistical urban growth models are commonly used, as they model the relationships between land use changes and the drivers underlying those changes. They are effective in identifying factors affecting urban growth, as well as allowing projections of future growth to be made (Zeng et al., 2008). They provide, moreover, quantitative information about the magnitude of the contribution made by such factors (Hu and Lo, 2007). Conveniently, computational requirements for this type of model are not as intensive as for CA models and requirements for input data are relatively easy to fulfil, especially if data is scarce (Dubovyk et al., 2011) or if only an initial analysis of structural relationships between identified determinants and urban growth is required.

In recent decades, Earth Observation and GIS techniques have been used in the study of urban development in the Kathmandu valley, in particular for monitoring its growth. Thapa and Murayama (2011) and Haack and Rafter (2006), for instance, used CA modelling and spatial analysis, respectively, to model urban growth in the Kathmandu valley. Despite its suitability for modelling urban growth and its driving factors, a logistic regression model for modelling urban growth has never been attempted for the Kathmandu valley. Nor have previous studies attempted to model different types of urban growth occurring in the valley – yet, this would provide important information for city planners, decision makers and resource managers to regulate urban expansion in the future. Therefore, to contribute to improving the understanding of dynamic factors that underlie different types of urban growth and their drivers, this study uses a spatial logistic regression (SLR) model. Furthermore, our study aims to shed light on the potential and limitations of SLR models in predicting overall urban growth in the future – as well as growth specifically due to infilling and expansion.

12.2 MATERIALS AND METHODS

12.2.1 STUDY AREA

The Kathmandu valley, with an average altitude of 1,300 m above mean sea level, lies in the central foothills of Nepal. It consists of three districts – Kathmandu, Latitpur, and Bhaktapur – which together cover an area of approximately 696 km². The only river draining the valley is the Bagmati River, which serves as a major source of drinking water and irrigation throughout the year. The valley contains one of the country's most populated urban areas, accommodating about 10% of Nepal's total population, although it accounts for less than 1% of the country's area (CBS, 2011). The valley is the centre of Nepal's economic activity and as a result has attracted huge numbers of migrants from different parts of the country. This has led to a rapid increase in demand for housing and, consequently, a substantial increase in building construction. As a result, between 2010 and 2017, a sizeable amount of predominantly rural and agricultural land has been subject to urban development.

The extent of the study area matches that of two previous studies, one by Haack and Rafter (2006) and one by Thapa and Murayama (2011): it is delineated by the watershed boundary of the valley. The study area consists of 18 urban and three rural municipalities (Figure 12.1).

BOX 12.1   Case Study Area

Located in the central hills of Nepal, **Kathmandu Valley** is one of the country's most populous metropolitan regions. It consists of three districts – Kathmandu, Latitpur and Bhaktapur – and covers an area of approximately 696 km². The only river draining the valley is the Bagmati River, which serves as a major source of drinking water and irrigation throughout the year.

Containing the capital city of Nepal, Kathmandu valley has faced urbanization pressure since the 1980s. As the centre of Nepal's economic activity, it accommodates a population of about 2.5 million people, which grows annually at a rate of about 4.3% (CBS, 2011). As a result, the city is facing considerable unplanned sprawl and haphazard growth, particularly in peri-urban areas. It is likely that such uncontrolled growth will eventually threaten agricultural and forest areas around the city. The case study described here aims to improve understanding of the dynamic processes that underlie different types of urban growth and their drivers.

BOX 12.2   Methods Applied in the Chapter

The chapter illustrates the application of GIS and remote sensing methods together with spatial logistic regression (SLR) as a means for identifying the primary drivers of urban growth in Kathmandu valley and forecasting future patterns of urban growth. Such patterns were identified by processing four multi-temporal Landsat datasets for the years 1989, 1999, 2010 and 2017 using maximum-likelihood supervised classification. To examine the influence of different factors on urban growth patterns, three binary SLR models were constructed for overall urban growth, infill growth and expansion. The models were then evaluated on the basis of percentage of correct prediction (PCP). Erdas Imagine™ 11, ArcGIS™ 10, Change analyst software, Map comparison Kit 3 and IBM SPSS™ statistics 20 were used in the study. The datasets used for analysis comprise spatial (remotely sensed data and vector datasets) and non-spatial data (population and land value) collected through various government organizations and websites. Primary data for building an understanding of the key driving factors of urban growth were obtained from interviews with academicians and planners in the government and private sectors who deal with urban planning issues. Field data was collected using a handheld GPS unit to verify the training samples for the accuracy of the land-cover classification.

12.2.2   Data Sources

Datasets required for the study were obtained from various sources (Table 12.1 and Table 12.2). Landsat imagery of moderate resolution was acquired from the US Geological Survey (USGS, 2017) and projected to WGS 84, UTM Zone 45. High resolution imageries of 1992, 2001 and

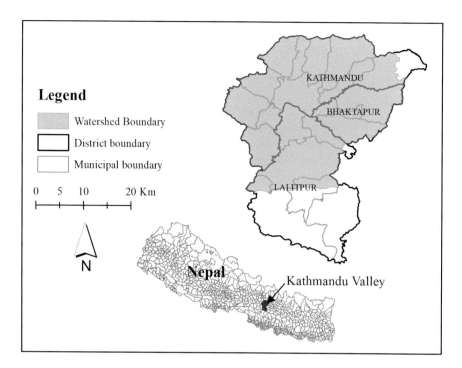

FIGURE 12.1   Study area (watershed boundary).

2010 were acquired from the Kathmandu Valley Town Development Committee (KVTDC). Vector datasets were obtained from the Department of Survey (DOS) and KVTDC for the years 1995 and 2008, respectively. Statistical data comprise population data for the years 2001 and 2011 at ward level. These data were obtained from the Central Bureau of Statistics (CBS) of Nepal. Primary data were collected during five interviews with local experts to understand the key driving factors of urban growth in the Kathmandu valley. Field data collection combined with Google Earth was done to verify training samples and to collect reference points for assessing the accuracy of land-cover classification (generating two independent sets). For this purpose, a recent Google Earth image, a WorldView image for 2010, an ortho-rectified aerial image for the year 1992 and an Ikonos image from 2001 were used.

12.2.3   Identifying Urban Growth Patterns

Four multi-temporal Landsat datasets (Landsat TM and ETM+) were processed to identify changes in land-cover patterns in the valley for the years 1989, 1999, 2010 and 2017. These images were classified separately using maximum-likelihood supervised classification. Four general land-cover classes – urban, water, forest and arable – were identified and used to analyse changes in land-cover patterns, as indicated in Table 12.3.

After the post-classification refinements,* the land-cover maps were assessed to determine their accuracy. Furthermore, to obtain infill and expansion for 1989–1999 and 1999–2010, a continuous

---

* Post-classification refinement is an approach to improve the accuracy of classified results. It follows two steps: the first step uses a majority filter of 3 × 3 window size to eliminate the noise of misclassified pixels; the second step overlays the river networks and waterbodies with the filtered image to retain the waterbodies in land-cover maps.

TABLE 12.1
Remotely Sensed Data

Dataset	Acquisition Year/Date	Scale/ Resolution	Source	Projection	Purpose
Landsat TM	1989/10/31	30 m	USGS	UTM zone 45	Landsat images were used
Landsat ETM+	1999/11/04	30 m	USGS	(WGS 84)	for preparing land-cover
Landsat ETM+	2010/03/07	30 m	USGS		maps.
Landsat OLI_TIRS	2017/05/05	30 m	USGS		
Ortho-rectified aerial image	1992	1 m	KVTDC	Everest 1830 Modified UTM	High resolution images were used for verification
Ikonos image	2001	1 m	KVTDC	UTM zone 45 (WGS84)	of training sample sets.
WorldView image	2010	0.5 m	KVTDC	UTM zone 45 (WGS84)	
Google Earth image	2017	0.5 m	Google Earth	UTM zone 45 (WGS84)	

TABLE 12.2
Vector Data Set

Data	Year	Scale	Source	Description	Purpose
Administrative boundary	1995	1:25,000	Survey department	Valley, District, Municipality, Ward, VDC	Delineating extent of study area.
Road network	1995	1:25,000	Survey department	Road centrelines categorised as national highway, ring road, feeder road, urban road	Road maps were used as driving factors of urban growth by using proximity analysis.
	2008	1:10,000	KVTDC	and district road	
Rivers	1995	1:25,000	Survey department	River centrelines	River maps were used as driving factors of urban growth by using proximity analysis.
Public services	1995	1:25,000	Survey department	Facilities such as schools, hospitals, post offices, petrol	Facilities maps were used as driving factor of urban growth by using proximity analysis.
	2008	1:10,000	KVTDC	pumps, temples, etc.	
Slope	1995	1:25,000	Survey department	Digital elevation map of cell size 20 m	Slope maps were used as driving factor of urban growth.
Designated areas	1995	1:25,000	Survey department	Conservation areas categorised as wild life reserve and national parks	Conservation areas were used as driving factors of urban growth.
Land use	1995	1:25,000	Survey department	Land-use map with categories: airport, waterbodies, ponds, built areas, etc.	Land-use maps were used as a reference for preparing land-cover categories.

TABLE 12.3
Land-Cover Classification

Land-Cover Classes	Description
Urban	Urban fabric: industrial, commercial, transportation and other built areas.
Water	Surface water, e.g. rivers, streams, lakes, ponds and so on.
Forest	Tree cover, e.g. deciduous forests, evergreen forests, mixed forests and so on.
Arable	Irrigated and non-irrigated agricultural land, including bare areas.

polygon was used to delineate the 1989 and the 1999 built-up areas. Infill growth for 1999, 2010 and 2017 was defined as the built area within 200 m of the 1989, 1999 and 2010 polygons (Masser et al., 2005; Dijkstra and Poelman, 2014).

12.2.4 SPATIAL LOGISTIC REGRESSION (SLR) MODEL

SLR has been widely applied to examine the influence of various factors on the pattern of changes in land cover or land use. The model computes the probability of change in land cover from non-built areas into built areas and estimates the empirical relationship between such urban growth and the driving factors considered (Huang and Sin, 2010). The dependent variable (urban growth) in a binary SLR model is dichotomous and the independent variables (driving factors) are predictors of the dependent variable. Independent variables can be measured on a nominal, ordinal, interval or ratio scale. The general form of logistic regression is expressed as:

$$P\left(y = 1 \middle| X\right) = \frac{e^{\left(b_0 + \sum b_i x_i\right)}}{1 + e^{\left(b_0 + \sum b_i x_i\right)}}$$

where:

x_i $(i = 1,2,\ldots\ldots,n)$ are driving factors or explanatory variables
y is the dependent variable, equal to 1 if urban growth has occurred in a cell and equal to 0 if urban growth has not occurred
b_0 is an intercept of the model
b_i $(i=1,2,\ldots\ldots,n)$ are regression coefficients or model parameters which are estimated
P is the probability of transition of a cell from non-urban to urban (Cheng and Masser, 2003; Verburg et al., 2002)

Regression coefficients b_i indicate the contribution of each of the driving factors to probability value P. A positive sign indicates that the particular factor will contribute to an increasing probability of change, while a negative sign indicates that the particular factor will contribute to a decreasing probability of change (Cheng and Masser, 2003).

Before starting logistic regression analysis, all independent variables were checked for the presence of multicollinearity considering the Variance Inflation Factor (VIF), an index that measures the variance of an estimated regression coefficient (O'brien, 2007). All collinear variables were eliminated in a stepwise manner from the analysis until all the variables had a VIF value lower than 10 (Field, 2013).

The SLR models were computed using only the selected variables from this process. To determine whether a variable is a significant predictor of the model, the T-Wald statistic was used (Field, 2013), while the goodness of fit of the model was determined with the Chi-square test. The statistically significant variables were selected by backward elimination. This process was applied because it has an advantage over forward elimination for joint predictive capability of the variables:

backward elimination begins with all variables in the model, whereas forward elimination fails to identify this effect. To check for the presence of spatial autocorrelation in the model, Moran's I index was calculated on the map of model residuals.

12.2.5 Preparation of Input Variables for SLR

The input factor maps were prepared for 1999 and 2010 using ArcGIS 10.1. The 1999 factor maps (Figure 12.2) were used to calibrate the model and 2010 factor maps were used for prediction of urban growth in 2017. Because of lack of data for 1989, SLR was only done for 1999 and 2010. All input data were created in raster of 30 m resolution (to match the resolution of Landsat imageries). The factor map data are either dichotomous or continuous. The continuous data were normalised for the range 0–1. Normalization was done by minimum–maximum linear transformation of the input raster to achieve a similar data range. Use of this method is important, as in logistic regression all continuous variables should have the same scale to enable comparison of the value within the same range (Huang et al., 2009; Priyanto, 2010).

The evaluation of the model was performed for the period 1999–2010; validation of model performance was performed for the period 2010–2017. First, the land-cover status of sample cells from 1999 and 2010 was determined. Then for each cell in the study area, the probability of change was computed using the estimated parameters from the model. The model was then evaluated and validated on the basis of the percentage (%) of correct predictions (PCP) and Kappa statistics. Table 12.4 shows the list and short description of the variables prepared for the logistic-regression model.

12.3 RESULTS

12.3.1 Spatio-Temporal Pattern of Urban Growth

The results of the classified land-cover maps (Figure 12.3a) show that there has been notable change in patterns of agricultural and built land use between 1989 and 2017. During the period 1989–1999, around 15.8 km² of agricultural land, mainly near major roads and existing urban areas, was lost (Table 12.5). In this period a huge amount of this agricultural land was converted into urban land, with the pattern of urbanization following road networks and existing built-up areas. For the period 1999–2010, a different trend of land-cover transformation can be observed. Not only did the urbanization continue near major roads and existing built areas, but urban land also expanded at the expense of forest areas. Around 39.5 km² of forest was lost, largely due to conversion for urban use in this period (Table 12.5). During the period 2010–2017, there has been a dramatic increase in urban area around previously existing built areas and the city periphery, once again at the expense of agricultural land and forest areas. In those seven years, the urban area increased by approximately 37.7 km².

The overall accuracy of the classified images for the years 1989, 1999, 2010 and 2017 is 94%, 89%, 91% and 92%, respectively. These data were classified separately using a supervised classification technique (using maximum-likelihood supervised classification) and post-classification change detection for extracting the temporal dynamics.

12.3.2 Spatial Pattern of Infill and Expansion

Figure 12.3b shows the spatial pattern of infilling and expansion of the valley during 1989–1999, 1999–2010 and 2010–2017. During 1989–1999 infill development was relatively low. Only approximately 5.7 km² of vacant land inside the city was subject to infill, amounting to only 35% of total urban growth (see Table 12.6). Conversely, 65% of urban growth occurred on the fringes of the city as expansion, mainly along the four major transportation corridors and inner ring road around the city periphery and arterial roads in western, northern and southern directions. During

FIGURE 12.2   1999 independent variables used for SLR model.

1999–2010, large areas along the major transportation routes and around existing urban peripheries were converted to urban use. Infill development increased to 14.2 km², i.e. about 40% of total urban growth, while some 60% of land around the city periphery was urbanised through expansion. During 2010–2017, both infill and expansion became more prominent with a conversion of 15.1 km² of vacant land inside existing built-up areas as infill and 22.6 km² as expansion.

TABLE 12.4

List of Variables and Descriptions Prepared for the Logistic-Regression Model

Factor	Variable in SLR Models	Description	Nature of Variable
Dependent variable urban growth	y	1 = urban growth 0 = no urban growth	dichotomous
Independent variables			
Biophysical characteristics	Slope	Degree of slope	continuous
Designated area	Forest	1 = forested; 0 = not forested	dichotomous
Population density	Population density	Population density (person/km²)	continuous
Proximate causes	Distance to major road	Distance to major roads (m)	continuous
	Distance to minor road	Distance to minor roads (m)	continuous
	Distance to pipeline	Distance to water supply pipelines (m)	continuous
	Distance to CBD	Distance to CBD (m)	continuous
	Distance to industries	Distance to industrial areas (m)	continuous
	Distance to river	Distance to major rivers (m)	continuous
	Distance to urban centre	Distance to urban centres (m)	continuous
	Distance to health facility	Distance to health facilities (m)	continuous
	Distance to educational facility	Distance to educational facilities (m)	continuous
Existing urban cluster	Proportion of urban (%)	Proportion of urban area in the surrounding area (a rectangular neighbourhood of 7×7)	continuous
	Distance to urban cluster	Distance to existing urban clusters (m)	continuous
Economic factor	Land value	Value of land in each ward	continuous

12.3.3 Driving Factors of Urban Growth

Altogether, 15 factors were hypothesised as determinants of (total) urban growth for 1999–2010. Multicollinearity diagnosis indicated that distance to industries have a high collinearity with distance to urban centres. Therefore, it was removed from the analysis and the final model was constructed with only 14 variables. Three SLR models were built, one each for overall urban growth, infill growth and expansion. For each model, the number of significant driving factors, percentage of correct prediction (PCP) and Moran's *I* were analysed (Table 12.7).

12.3.3.1 Drivers of Overall Urban Growth

The estimated parameters for the model of overall urban growth show that urban growth from 1999 to 2010 was mainly driven by distance to pipelines (coefficient value = −10.26), followed by distance to urban clusters (coefficient value = −6.47) and distance to minor roads (coefficient value = −5.11); see Table 12.8. This indicates that areas with better access to infrastructure – along the transportation network, near existing built-up areas – have a high probability of becoming urban. The strong negative correlation indicates that the closer to infrastructure, the higher the probability of urban growth.

The final model for overall urban growth was obtained on the fifth backward step; factors such as distance to the urban centre, distance to health facilities and distance to rivers and forest were eliminated since their T-Wald statistics were greater than the assigned confidence level (0.05). The model was significant with a Chi-square value of 7465.285 and *p*-value less than 0.05.

12.3.3.2 Drivers of Infill Growth

A second SLR model specifically characterised infill development. The dependent variable was specified as 1 for cells in which infill growth occurred and 0 for cells in which infill growth had not occurred.

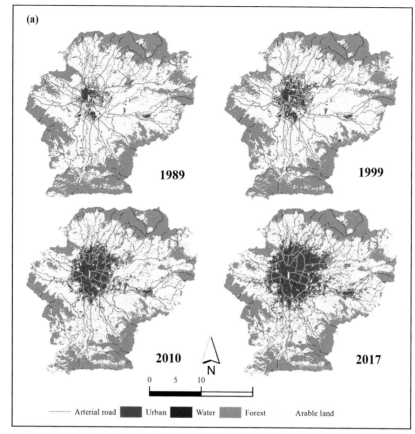

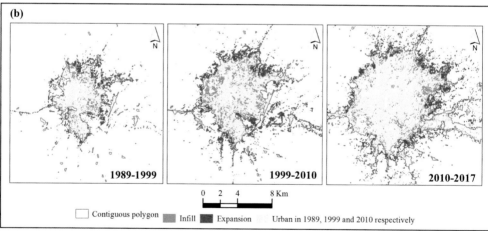

FIGURE 12.3 Image classification results: (a) land-cover classification maps of Kathmandu valley; (b) spatial pattern of infill development and expansion between 1989 and 2017.

The estimated parameters of the model show that the distance to pipelines plays a significant role in the transformation of vacant land for urban use inside the existing urban area (coefficient value = −11.83); see Table 12.9. Similarly, distance to minor roads has also a strong negative correlation (coefficient value = −4.18), indicating that this has a strong influence on the probability of infill growth. This is evident from the existing trend of urban development around road networks, which

TABLE 12.5
Land-Cover Distribution

Land Cover Classes	Area (km²)								Change in Area (km²)		
	1989	%	1999	%	2010	%	2017	%	1989–1999	1999–2010	2010–2017
Urban	19.8	3	35.8	5	73.3	10	111	16	16.1	37.5	37.7
Water	12.2	2	12.4	2	12.3	2	12.3	2	0.2	−0.2	0
Forest	259.3	37	258.9	37	219.4	32	217.1	31	−0.4	−39.5	−2.3
Arable land	404.7	58	388.9	56	391.2	56	355.9	51	−15.8	4	−35.3
Total	696.2	100	696.2	100	696.2	100	696.2	100			

TABLE 12.6
Urban Growth by Type

Type of Growth	Area (km²)					
	1989–1999	%	1999–2010	%	2010–2017	%
Expansion	10.9	65	23.3	62	22.6	60
Infill	5.7	35	14.2	38	15.1	40
Total	16.6	100	37.5	100	37.7	100

TABLE 12.7
Moran's Index

Type of Model	Sampling Size	Moran's I	Z-Score
Overall urban growth model	7×7	0.002	1.74
Infill growth model	3×3	−0.002	−0.79
Expansion model	7×7	0.00016	1.92

A = 0.05; Critical z-score = ±1.96.

is mainly due to ease of access to transportation. Distance to the urban centre has a positive correlation (coefficient value = 2.9), which shows that the probability of urban growth is lower if a cell is closer to the urban centre (as most of the central areas have already been built upon).

The SLR model for infilling was obtained on the fifth backward step after eliminating factors such as forest, land value, distance to educational facilities and distance to the CBD. The model was significant with a Chi-square value of 503.98 and p-value less than 0.05.

12.3.3.3 Drivers of Expansion

A third SLR model specifically explains urban expansion. The dependent variable was defined as 1 for cells undergoing such growth and as 0 for cells not undergoing urban expansion.

The model shows that the major driving factors of urban expansion are distance to pipelines (coefficient value = −11.38); see Table 12.10. This means that the closer a cell is to water supply facilities, the greater the probability of urban expansion. Other major driving forces of urban expansion are distance to minor roads (coefficient value = −9.28), followed by distance to an urban cluster (coefficient value = −7.01) and distance to a health facility (coefficient value = −3.01).

TABLE 12.8

Model Parameters for Overall Urban Growth, 1999–2010

Variables	b (Coefficient)	S.E.	Wald (z-value)	T-Wald Test (p-value)	Odds Ratio (O.R.)	95% C.I. for O.R. Lower	Upper
Constant	3.404						
Population density	2.271	0.650	12.201	0.000	9.690	2.709	34.656
Proportion of urban	1.363	0.615	4.911	0.027	3.908	1.171	13.050
Land value	0.999	0.400	6.242	0.012	2.716	1.240	5.946
Distance to major road	−1.102	0.313	12.360	0.000	0.332	0.180	0.614
Slope	−1.319	0.647	4.163	0.041	0.267	0.075	0.949
Distance to educational facilities	−2.476	0.670	13.655	0.000	0.084	0.023	0.313
Distance to CBD	−2.753	0.487	31.985	0.000	0.064	0.025	0.165
Distance to minor road	−5.113	1.003	25.977	0.000	0.006	0.001	0.043
Distance to urban cluster	−6.472	0.561	133.113	0.000	0.002	0.001	0.005
Distance to pipeline	−10.262	1.223	70.400	0.000	0.000	0.000	0.000

An SLR model for expansion was obtained on the eighth backward step. The model was significant with a Chi-square value of 1136.46 and p-value less than 0.05.

12.3.4   MODEL EVALUATION

12.3.4.1   Model Evaluation for Overall Urban Growth

The performance of the urban growth model was evaluated by comparing the probability map of urban growth in 2010 with the actual map of urban growth in that year. The PCP value indicates that

TABLE 12.9

Model Parameters for Infill Growth, 1999–2010

Variables	Coefficient (b)	S.E.	Wald (z-value)	T-Wald (p-value)	Odds Ratio (O.R.)	95% C.I. for O.R. Lower	Upper
Distance to urban centre	2.984	0.751	15.782	0.000	19.757	4.534	86.096
Proportion of urban area	2.234	0.272	67.327	0.000	9.335	5.475	15.916
Distance to major road	0.649	0.331	3.856	0.050	1.914	1.001	3.661
Population density	0.000	0.000	136.176	0.000	1.000	1.000	1.000
Distance to urban cluster	−0.831	0.249	11.157	0.001	0.435	0.267	0.709
Distance to river	−1.734	0.493	12.386	0.000	0.177	0.067	0.464
Distance to health facility	−1.813	0.607	8.918	0.003	0.163	0.050	0.536
Slope	−2.410	0.991	5.921	0.015	0.090	0.013	0.626
Distance to minor road	−4.180	1.377	9.207	0.002	0.015	0.001	0.228
Distance to pipeline	−11.832	2.743	18.601	0.000	0.000	0.000	0.002
Constant	0.681	0.285	5.692	0.017	1.976		

TABLE 12.10

Model Parameters for Expansion, 1999–2010

Variables	Coefficient (b)	S.E.	Wald (z-value)	T-Wald (p-value)	Odds Ratio (O.R)	95% C.I. for O.R. Lower	Upper
Constant	3.972	0.669	35.226	0.000	53.072		
Proportion of urban area	2.979	1.124	7.020	0.008	19.666	2.17	178.13
Land value	1.657	0.672	6.080	0.014	5.242	1.40	19.56
Distance to CBD	−2.721	0.814	11.190	0.001	0.066	0.013	0.324
Distance to health facility	−3.018	0.908	11.044	0.001	0.049	0.008	0.290
Distance to urban cluster	−7.011	0.993	49.884	0.000	0.001	0.000	0.006
Distance to minor road	−9.286	1.865	24.782	0.000	0.000	0.000	0.004
Distance to pipeline	−11.384	2.261	25.347	0.000	0.000	0.000	0.001

the overall urban growth model predicts 96.3% of y occurrences correctly (Table 12.11). However, the correct prediction of the urban area is lower, equal to 72.5%. This suggests that, on the one hand, not all the driving forces are represented in the SLR model, while, on the other hand, the phenomenon of urban growth cannot be forecast with complete accuracy. Nevertheless, when compared with results reported elsewhere in the literature, a PCP greater than 70% appears to be acceptable for urban growth models (Li and Yeh, 2001; Huang et al., 2009). Furthermore, the overall Kappa value of 0.783 means there is 78.3% agreement between observed and predicted urban growth, which can be categorised as an excellent level of agreement (Pontius 2000).

Figure 12.4a compares urban growth simulated by the model for 2010 with actual urban growth in 2010. Although most of the urban and non-urban areas are correctly predicted, the model does not estimate urban expansion with complete accuracy. Figure 12.4b shows that the model over-predicts infill growth in areas located within the contiguous built-up area. In other words, according to the model, vacant pockets inside the existing urban area have a higher probability of being converted into urban land than is in reality the case. Conversely, according to the model, areas outside of and disconnected from the built-up area have a lower probability of being converted into urban land than is actually the case.

To estimate its real performance, the overall growth model was again evaluated, but only considering the cells that changed between 1999 and 2010. In this case, now only 49.1% of actual change is explained by the model (see Table 12.12). This is not surprising, however, as urban growth is a more complex phenomenon than models can capture, particularly in rather data-poor environments.

TABLE 12.11

Evaluation of Model Results for Overall Urban Growth, 1999–2010

Observed (Pixels)		Predicted (Pixels) Non-urban (0)	%	Urban (1)	%	Total	%
	Non-urban (0)	672,437	99.1	6,344	0.9	678,781	100
	Urban (1)	21,724	27.5	57,405	72.5	79,129	100
	Total	694,161		63,749		757,910	

Correct predictions: 729,842

Incorrect predictions: 28,068

Percentage of correct predictions (PCP): 96.3%

PCP for urban area only: 72.5%

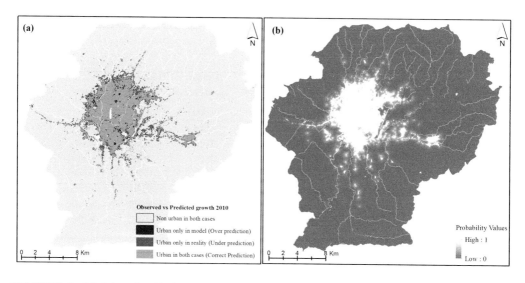

FIGURE 12.4 Model results: (a) observed and predicted urban growth for 2010; (b) probability map for 2010.

Nevertheless, almost 50% of the growth can be explained by the factors used in the model, which provides an overall idea of the main driving forces of urban development. While this level of accuracy may look low given it only accounts for changing cells, it actually implies that the model is good for predicting the overall urban growth. Indeed, one should realise that this is possible, in part, because Kathmandu has grown in a relatively organic manner. In this sense, while infill is less than expansion, the results suggest a large proportion of the expansion has followed the systematic patterns that govern the location of overall urban activities and is not as scattered as the descriptive results of the growth typology would suggest.

12.3.4.2 Model Evaluation for Infill Growth

Figure 12.5a shows that although the model correctly predicts a large amount of infill development (overall PCP of 76% shown in Table 12.13), it still over-estimates a considerable amount of infill growth. The over-prediction mainly occurs in locations that already have a large proportion of built-up land and are located close to a ring road. This is probably due to restricted areas not being included as a factor in the regression model: vacant land in the urban fabric is often vacant for a reason – one not usually reflected in the determinants of urban growth. The model assumes vacant pockets (at central locations) have a high potential for transformation into built-up land, particularly if there is a high proportion of urban land nearby; it may not take into account spatially random causes that disqualify the land from being included in the supply for potential urban use (such as, for example, land speculation or land set aside for public parks).

TABLE 12.12
Model Evaluation for Changed Pixels

		Predicted (Pixels)		
		Non-Urban (0)	**Urban (1)**	**Total Changed Cells**
Observed (Pixels)	**Urban (1)**	21,161	20,451	41,612
	%	50.9%	49.1%	100%
	Correct predictions: 20,451			
	Incorrect prediction: 21,161			
	Percentage of correct prediction (PCP): 49.1%			

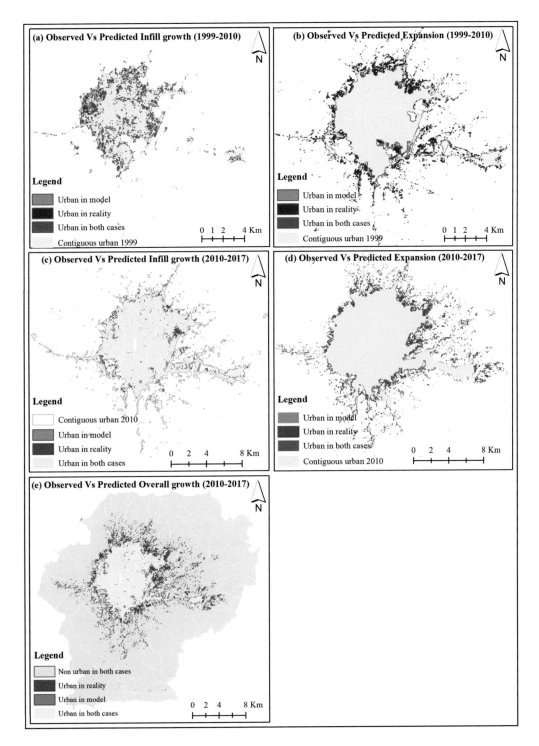

FIGURE 12.5 Model results: (a) observed and predicted infill growth, 1999–2010; (b) observed and predicted urban expansion, 1999–2010; (c) observed and predicted infill growth, 2010–2017; (d) observed and predicted urban expansion, 2010–2017; (e) observed and predicted overall growth, 2010–2017.

TABLE 12.13

Evaluation of Model Results for Infill Growth, 1999–2010

		Predicted (Pixels)					
		Non-Urban (0)	%	Urban (1)	%	Total	%
Observed (Pixels)	Non-urban (0)	536	10	4,657	90	5,193	100
	Urban (1)	405	3	15,219	97	15,219	100
	Total	941		19,876		20,412	

Correct predictions: 15,755

Incorrect prediction: 5,062

Percentage of correct prediction (PCP): 76%

PCP for built-up area only: 97%

Percentage of over-prediction: 90%

12.3.4.3　Model Evaluation for Expansion

The overall accuracy of the expansion model was 97%, with a PCP of only 22% for built-up areas (see Table 12.14 and Figure 12.5b). In other words, 78% of urban cells were under-predicted (i.e. predicted to be non-urban). This is the result of the much larger amount of non-urban cells as compared to expansion cells, i.e. the study area is predominantly rural. Most of the non-urban cells remained non-urban during the study period. Moreover, there was a rapid growth of housing areas, colonies and other land development sites, which have been mushrooming in Kathmandu valley. Such sudden growth and the inherent randomness of any process of urban expansion make growth difficult to capture in a model. Nevertheless, in spite of the relative simplicity of the model and its low spatial resolution, it is able to successfully identify the causes of these changes and their overall pattern – if not their specific location.

12.3.5　VALIDATION OF THE PREDICTION MODEL

All urban growth models were validated to determine the quality of their predictions in relation to the actual urban-growth map of 2017. To do this, the model was calibrated for 1999–2010 input factors and then used to predict 2017 urban growth based on input factors for 2010–2017. The overall accuracies of the overall urban growth, infill and expansion models are 94.7%, 89.3% and 95.6%, respectively. Unfortunately, although the PCPs are relatively higher for the overall urban growth and

TABLE 12.14

Model Evaluation for Urban Expansion, 1999–2010

		Predicted (Pixels)					
		Non-Urban (0)	%	Urban (1)	%	Total	%
Observed (Pixels)	Non-urban (0)	669,650	99%	3,702	1%	673,352	100
	Urban (1)	18,722	78%	5,384	22%	24,106	100
	Total	688,372		9,086		697,458	

Correct predictions: 675,034

Incorrect prediction: 22,424

Percentage of correct prediction (PCP): 97%

PCP for built-up area only: 22%

TABLE 12.15

Model Validation for Overall Urban Growth, Infill Growth and Expansion, 1999–2010

Prediction	Overall Urban Growth	Infill	Expansion
Non-urban to non-urban	625,328	6,367	618,961
Non-urban to urban	10,907	6,294	4,613
Urban to non-urban	29,506	5,299	24,207
Urban to urban	93,820	90,738	3,082
Correct prediction	719,148	97,105	622,043
Incorrect prediction	40,413	11,593	28,820
Percentage of correct prediction (PCP)	94.7%	89.3%	95.6%
PCP for urban cell	76.1%	94.5%	11.3%

expansion, the accuracy of correct prediction of urban areas was lower: 76.1% and 11.3%, respectively (see Table 12.15). This suggests that the models are not able to capture urban growth precisely. Furthermore, the spatial distribution maps of predicted and actual built areas show some disagreement. Most of the expansion is under-estimated by the model, while infill growth is over-estimated. Figures 12.5c, 12.5d and 12.5e compare observed and predicted urban growth for infill, expansion and overall urban growth during the period 2010–2017.

12.4   DISCUSSION AND CONCLUSION

Our findings show that three of the four land-cover classes, i.e. urban, arable land and forest areas, have been subject to pronounced change during the entire study period. Classified digital imagery, together with other spatial data, demonstrate the models' ability to provide information on the rate, direction and location of land-cover change as a result of urbanization. Some degree of uncertainty exists, however, as a result of class inconsistency, which has not been examined in our research.

Results from all three SLR models shows that urban growth of Kathmandu valley is associated with access to pipelines (i.e. infrastructure access), distance to minor roads and distance to urban clusters. The results suggest that the Kathmandu city tends to grow in areas adjacent to those that are already agglomerated. Furthermore, the evaluation of the models indicates that the models are good at predicting overall urban growth. This was confirmed by comparing our results against those found in similar studies, such as those of Cheng and Masser (2003) and Huang et al. (2009). Evaluation of results from the infill growth and expansion models shows, however, that SLR modelling is better at predicting infill growth than that of urban expansion.

The underlying problem in this study might be a problem of endogeneity,[*] which is extremely difficult to deal with. This type of problem is usually common in studies that use a conditional stepwise technique for including or excluding variables (George 2005) – with the result that significant predictors may be excluded.

The main conclusion from this study is that although SLR-based urban-growth models have some limitations, they are suitable for exploring the structural effect of determinants on the landscape and for identifying their relative importance. Despite the simplicity of our SLR models, we have learnt several valuable lessons from this study. Firstly, we must be cautious about the endogeneity problem

---

[*] Endogeneity is a term used in econometric analysis in which the value of one independent variable is dependent on the value of other predictor variables, so a significant correlation can exist between the independent and dependent variables, which results in biased estimators (George 2005).

when using a SLR model to study urban growth, since it can lead to the use of biased estimators. Secondly, SLR models produce a probability map that indicates the possible locations of future urban growth but does not indicate when it is going to happen. In this respect, further research needs to be done to find a self-modifying approach that can update model variables automatically. Thirdly, there is an inherent randomness to any process of urban expansion that cannot be fully captured by a model. Use of data of lower spatial resolution and fewer land-use categories negatively affect the reliability of model outcomes. Conversely, data of higher spatial resolution and finer-grained land-use categories could improve accuracy.

REFERENCES

Alqurashi, Abdullah F., Lalit Kumar, and Khalid A. Al-Ghamdi. 2016. "Spatiotemporal modeling of urban growth predictions based on driving force factors in five Saudi Arabian cities." *ISPRS International Journal of Geo-Information* 5 (8):139.

Berling-Wolff, Sheryl, and Jianguo Wu. 2004. "Modeling urban landscape dynamics: A case study in Phoenix, USA." *Urban Ecosystems* 7 (3):215–240.

CBS. 2011. "National population census." Accessed 23 August. http://census.gov.np/index.php?option=com_content&view=article&id=22&Itemid=22

Chen, Jianglong, Jinlong Gao, and Feng Yuan. 2016. "Growth type and functional trajectories: An empirical study of urban expansion in Nanjing, China." *PloS One* 11 (2):e0148389.

Cheng, J., and I. Masser. 2003. "Urban growth pattern modeling: A case study of Wuhan city, PR China." *Landscape and Urban Planning* 62 (4):199–217.

Dijkstra, L., and H. Poelman. 2014. "A harmonised definition of cities and rural areas: The new degree of urbanisation." *Regional Working Paper 1/2014*: European Commission. Accessed 7 May 2018. http://ec.europa.eu/regional_policy/sources/docgener/work/2014_01_new_urban.pdf.

Dubovyk, Olena, Richard Sliuzas, and Johannes Flacke. 2011. "Spatio-temporal modelling of informal settlement development in Sancaktepe district, Istanbul, Turkey." *ISPRS Journal of Photogrammetry and Remote Sensing* 66 (2):235–246.

Field, Andy. 2013. *Discovering Statistics Using IBM SPSS Statistics*. Sage.

George, Avery. 2005. "Endogeneity in logistic regression models." *Emerging Infectious Disease Journal* 11 (3):503. doi:10.3201/eid1103.040462.

Haack, Barry N., and Ann Rafter. 2006. "Urban growth analysis and modeling in the Kathmandu Valley, Nepal." *Habitat International* 30 (4):1056–1065.

Hu, Zhiyong, and C. P. Lo. 2007. "Modeling urban growth in Atlanta using logistic regression." *Computers, Environment and Urban Systems* 31 (6):667–688.

Huang, B., and H. L. Sin. 2010. "Uncovering the space–time patterns of change with the use of change analyst–case study of Hong Kong." *Advances in Earth Observation of Global Change*:255.

Huang, Bo, Li Zhang, and Bo Wu. 2009. "Spatiotemporal analysis of rural–urban land conversion." *International Journal of Geographical Information Science* 23 (3):379–398.

Li, Xia, and Anthony Gar-On Yeh. 2001. "Calibration of cellular automata by using neural networks for the simulation of complex urban systems." *Environment and Planning A* 33 (8):1445–1462.

Liu, Xiaoping, Xia Li, Yimin Chen, Zhangzhi Tan, Shaoying Li, and Bin Ai. 2010. "A new landscape index for quantifying urban expansion using multi-temporal remotely sensed data." *Landscape Ecology* 25 (5):671–682.

Masser, Ian, Mrs Bernadette Williams, and R. Williams. 2005. *Learning from Other Countries: The Cross-National Dimension in Urban Policy Making*. Routledge.

Naghibi, Fereydoun, Mahmoud Reza Delavar, and Bryan Pijanowski. 2016. "Urban growth modeling using cellular automata with multi-temporal remote sensing images calibrated by the artificial bee colony optimization algorithm." *Sensors* 16 (12):2122.

O'brien, R. M. 2007. "A caution regarding rules of thumb for variance inflation factors." *Quality & Quantity* 41 (5):673–690.

Pontius, R. G. 2000. "Quantification error versus location error in comparison of categorical maps." *Photogrammetric Engineering and Remote Sensing* 66 (8):1011–1016.

Priyanto, A. Tunggal. 2010. "The impact of human activities on coastal zones and strategies towards sustainable development: A case study in Pekalongan, Indonesia." University of Twente, Faculty of Geo-information and Earth Observation ITC.

Shi, Yaqi, Xiang Sun, Xiaodong Zhu, Yangfan Li, and Liyong Mei. 2012. "Characterizing growth types and analyzing growth density distribution in response to urban growth patterns in peri-urban areas of Lianyungang City." *Landscape and Urban Planning* 105 (4):425–433.

Thapa, R. B., and Y. Murayama. 2011. "Urban growth modeling of Kathmandu metropolitan region, Nepal." *Computers, Environment and Urban Systems* 35 (1):25–34.

Triantakonstantis, Dimitrios, and Giorgos Mountrakis. 2012. "Urban growth prediction: A review of computational models and human perceptions." *Journal of Geographic Information System* 4 (06):555.

USGS. 2017. "U.S. Geological Survey." Accessed 10 July. https://www.usgs.gov/.

Vaz, Eric, and Jamal Jokar Arsanjani. 2015. "Predicting urban growth of the greater Toronto area-coupling a Markov cellular automata with document meta-analysis." *Journal of Environmental Informatics* 25 (2):71–80.

Verburg, Peter H., Welmoed Soepboer, A. Veldkamp, Ramil Limpiada, Victoria Espaldon, and Sharifah S. A. Mastura. 2002. "Modeling the spatial dynamics of regional land use: The CLUE-S model." *Environmental Management* 30 (3):391–405.

Xu, C., M. Liu, C. Zhang, S. An, W. Yu, and J. M. Chen. 2007. "The spatiotemporal dynamics of rapid urban growth in the Nanjing metropolitan region of China." *Landscape Ecology* 22 (6):925–937.

Zeng, Y. N., G. P. Wu, F. B. Zhan, and H. H. Zhang. 2008. "Modeling spatial land use pattern using autologistic regression." *The International Archives of the Photogrammetry, Remote Sensing and Spatial Information Sciences* 37 (B2):115–118.

13 Stakeholder-Based Assessment
Multiple Criteria Analysis for Designing Cycle Routes for Different Target Populations

Amy Butler, Mark Brussel, Martin van Maarseveen and Glen Koorey

CONTENTS

13.1 Introduction 225
    13.1.1 Research Scope...225
    13.1.2 Study Area 226
13.2 Methodology 227
    13.2.1 Traditional MCA Methods..228
        13.2.1.1 Step 1: Define Constraint Criteria and Remove Areas Violating Them.........229
        13.2.1.2 Step 2: Define Compensatory Criteria and Their Performance Measures......229
        13.2.1.3 Step 3: Performance Measurements to Standardised Criteria-Performance Scores 230
    13.2.2 Modifying MCA for Cycle Route Design...230
        13.2.2.1 Step 4: Target Populations Participating as Stakeholders.........................230
        13.2.2.2 Step 5: Stakeholder Preference Sets to Aggregated Group Ranks 232
        13.2.2.3 Step 6: Sensitivity Analysis for Group Weights.......................233
13.3 Application of MCA for Cycle Route Design in Christchurch............................233
    13.3.1 Ranks Converted to Main Criteria Weights 237
    13.3.2 Ranks Converted to Sub-Criteria Weights.......................................238
13.4 Improvements for the Future 241
13.5 Conclusion 242
References..242

13.1 INTRODUCTION

13.1.1 RESEARCH SCOPE

Potential cyclists, beginner cyclists, utility cyclists, recreationalists and trail riders – similar categories have been defined as target populations by cycle programmes worldwide. This study refers to 'target populations' as those groups of people that have been defined by local, regional or national policy for specific cycling programme interventions. Europe's PRESTO Programme (Urbanczyk, 2010) explains the concept of target populations as part of their suggested marketing strategy for segmentation and targeting of different cycling groups. Partners of the PRESTO Programme believe a systematic application of infrastructural design and marketing will provide socially positive behavioural change through cycling programmes. This idea is not just applicable in Europe. As Damant-Sirois et al. (2014) discovered in Montreal, 'Cyclists react heterogeneously to interventions and infrastructure. Building a network adapted to different cyclist types and emphasizing its convenience, flexibility and speed could be an effective strategy for increasing cycling-mode share and frequency among the various groups'. Thus, a methodology that incorporates pre-construction assessment must be presented that accounts for the presence of these different groups. There are a number of existing assessment systems, such as the bicycle level of service (BLOS) assessment (Landis et al., 2003) and traditional transport multiple criteria analyses (MCAs) (Thomopoulos and Grant-Muller, 2012; Yang and Regan, 2012). Yet these assessments are not well developed for cycling programmes that need a systematic process to support decision-making.

Different countries are likely to define different target populations based on local travel behaviour (Kroesen and Handy, 2013) and the objectives of regional transport bodies (Thomopoulos and Grant-Muller, 2012). Similarly, route design criteria affecting a target population's safety and perception of the cycling environment will differ according to the city's situation. The objective of the research we describe in this chapter was, therefore, to create a methodology that can account for these sorts of variations. The methodology must be flexible enough for local engineers and designers to choose their own design-criteria hierarchy. It must also provide detailed segment and junction information when high resolution results are needed, but it must also be scalable and allow comparison with assessments of any of the city's other routes.

With these concerns in mind, in this chapter we present a methodology based on multiple criteria analysis (MCA) for the design of cycle routes that take into account any given city's target populations and their preferences. After the study area is introduced, the methods section covers six steps from defining criteria for stakeholder participation, performance measurements, standardised criteria performance scores, aggregated group ranks and a sensitivity analysis. The analysis section displays this methodology in practice at the study area in Christchurch, New Zealand. The chapter concludes with improvements for future work.

13.1.2   STUDY AREA

BOX 13.1   Case Study Area

Christchurch is a major city on New Zealand's south island and has a population of 360,000 people. It is rebuilding its infrastructure after the 2010 and 2011 earthquakes that destroyed major parts of the city; this includes the development of an extensive cycleway programme. The building of cycling infrastructure is seen as a way to promote cycling, in particular for major groups such as school-going children and people who currently commute to work by car. The development of such public infrastructure is accompanied by all kinds of spatial decision-making problems such as which locations and communities to serve, which routes to develop, which designs to apply in which environments and so on. These problems are inherently complex due to physical limitations, finite resources, involvement of a number of parties and their mixed interests. For this reason, decision-makers use policy-driven and objective-based criteria to evaluate options, such as route alternatives, to help them compare and prioritise projects that are most suitable to their needs.

In 2014, the cycle programme manager for the city of Christchurch (New Zealand) requested that research be done on the use of multiple criteria analysis (MCA) for cycle route design. The city of

Christchurch (population 360,000) suffered a series of earthquakes in 2010 and 2011 that destroyed large parts of its infrastructure, which it is still in the process of rebuilding. The city has placed importance on improving its cycle network, with a total of NZD 156 million (approx. €90 million) to be allocated over a period of seven years, NZD 65 million of which was to be spent over a period three years on its 13 main routes (Christchurch City Council, 2016). The cycle network is meant to safely connect the city centre of Christchurch with major suburbs and activity centres. In this manner, the council intends to facilitate commuting by bicycle, as well encouraging cycling to school and for leisure activities and shopping trips.

Christchurch and many other cities in New Zealand want to encourage more people to take up cycling, mainly through educational programmes and infrastructural investment (Canterbury Regional Transport Committee, 2012). In Christchurch, about 7% of commuting trips and 3% of all trips are made by cycle (Butler, 2015). Unfortunately, there is no national framework for legally regulating efforts for the planning, design and implementation of bicycle facilities, although the New Zealand government recently injected NZD 330 million into a three-year Urban Cycleway Program (NZ Transport Agency, 2016). The quality of regional cycling projects is dependent upon the experience and judgement of locally available experts; national planning and design guidance is just now being prepared or updated.

The goals and road designs of these regional cycling projects are based on universally accepted supply-side criteria, yet provision of these infrastructure standards may not be enough to significantly increase a city's cycling modal share. As the current New Zealand national cycle network and route planning guide states, 'A perennial problem in cycle route network planning is the reliance on bright ideas and pet projects that may not have been critically evaluated for usefulness and value for money'. Similar to any other publicly funded infrastructural project, cycling routes should undergo assessment and review before being finalised (Land Transport Safety Authority, 2004). The Cycleway Program Manager and the lead Senior Traffic Engineer for Christchurch suggested that MCA be used to assess a section of the Norwest Arc, an 8 km planned orbital bicycle route that city designers had previously identified in an ad-hoc manner. A study area was chosen (see Figure 13.1) along this planned route to include two simple route options for assessment.

13.2 METHODOLOGY

BOX 13.2   Methods Applied in the Chapter

The methodology we developed in this case study is based on multi criteria analysis (MCA) techniques that have been modified to the problem of infrastructure appraisal and route selection. Based on focus group discussions, criteria were identified that were important for route quality. These main criteria were subdivided into 17 sub-criteria. Both main and sub-criteria were weighted on the basis of ranks that had been derived for the three main stakeholder groups identified: parents of school-going children; commuter cyclists; and potential cyclists. As a result, scores for two simple route options were calculated. The GIS-based approach allows for the analysis and visualization of individual route segments and junctions. For each route, this is broken down into their sub-criteria scores, overall route (total) scores and how these routes scores might be weighted to reflect a particular stakeholder perspective.

This section first describes the traditional MCA approach and then shows how it can be modified and improved for application to cycle route design. Figure 13.2 summarises the six steps used in this MCA study.

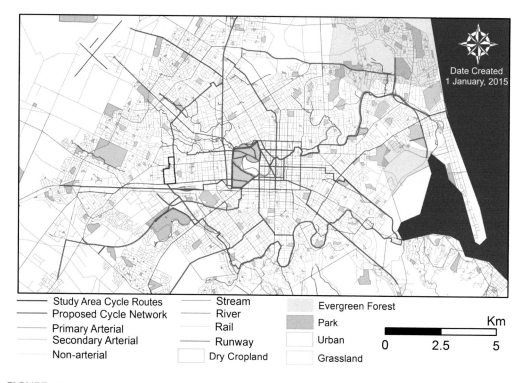

FIGURE 13.1   Map of study area in Christchurch and its surrounding areas.

13.2.1   TRADITIONAL MCA METHODS

MCA's strength lies in its vast base of different industry users and the variety of applications developed by these users. The largest benefit of MCA is its ability to provide a structured government decision-making process in the face of conflicting criteria and stakeholder priorities. Value-focused and not alternative-focused, MCAs allow flexibility so criteria can be removed or altered and

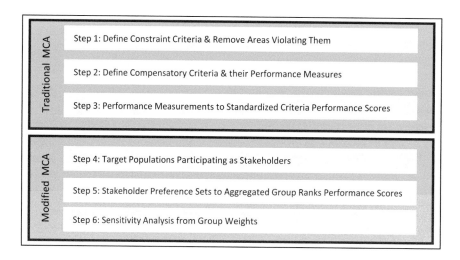

FIGURE 13.2   Steps in traditional and modified multiple criteria analysis (MCA).

preference sets can be assessed (Sharifi et al., 2006). Then, through performance measurement, standardization and weighting (of the multiple criteria according to their relative importance in a particular person's preference set), a variety of options can be analysed and compared to find which is the most suitable for each stakeholder group and their related policy visions or managerial objectives (Keshkamat et al., 2009; Sharifi, 2004).

13.2.1.1 Step 1: Define Constraint Criteria and Remove Areas Violating Them

Constraints are criteria or variables that are non-compensatory, or under conditions of strict dominance, and have the potential to cancel out the usefulness of the other criteria being assessed (Hajkowicz and Higgins, 2008; Pomerol and Barba-Romero, 2000), and as such can be included in thresholds via value functions but cannot be ranked. All domination options should be spatially excluded from a rank-based decision set before MCA is applied.

In the case of cycle route design, constraint criteria are especially important if running an MCA over a large area that has many roadway options. This is because constraints cannot be scaled or compared against other design criteria, and areas where they prevail must be removed from the spatial route options. Designers and engineers of cycle infrastructure must decide what the constraints of their city would be. If local designers or engineers believe any section of roadway is too hazardous or too expensive to provide bicycle-friendly infrastructure, then it is under the influence of an unavoidable and prohibitive constraint. Once defined, constraint criteria will eliminate some of the possible planned-route locations. Depending on local concerns, there can be any number of these constraint criteria. Throughout the rest of this chapter, these will be referred to as 'constraints', and any further mention of 'criteria' will solely refer to those criteria that are compensatory.

13.2.1.2 Step 2: Define Compensatory Criteria and Their Performance Measures

Unlike constraints, compensatory criteria are not prohibitive and have (to some extent) advantages and disadvantages. Compensatory criteria can be scored and ranked. These compensatory criteria are a standard of judgment or rule on the basis of which alternative decisions can be evaluated and ordered according to their desirability (Malczewski, 2006). Once compensatory criteria are defined, then a performance measure must be defined for each. Performance measures are indicators, a decision-option's raw score against a criterion (Hajkowicz and Higgins, 2008). Performance measures generally have units of measure.

For cycle route design, compensatory criteria can be based on international best practices and recent discoveries in the literature, but ultimately those local designers and engineers who will be using the output information must, in consultation with stakeholders, decide which criteria are most deterministic for the bicycle-friendliness of their city. Compensatory design criteria should then be given meaningful performance measures that can deal with the variation present within the city landscape. Examples of criteria might be 'slope' or 'outdoor attractiveness', with potential performance measures being 'average gradient per kilometre' or 'percentage vegetation land cover adjacent to road'. This value-based MCA allows different criteria and different performance measures to be input.

The process is the same regardless of which criteria or performance measures are used. After determination of the criteria scores, a total route suitability score can be calculated. This total score can be calculated in one of two ways: with, or without, compensating for the length of the segment. With compensation, each performance measure's scores are normalised on a scale from 0–1 and assigned to the segment or junction. The total route score is then a summation of all segment and junction scores divided by the sum of the number of segments and junctions. Without compensation, the scores of segments are normalised based on their length, then normalised on a scale from 0–1.

13.2.1.3   Step 3: Performance Measurements to Standardised Criteria-Performance Scores

To be comparable to each other, performance measures must be standardised into unitless scales. Linear maximum standardization is favoured among participatory suitability studies because it is easy to understand for stakeholders and it does not cause undue exaggeration between small measurement differences. Such small differences may be of only minor importance and may even be the result of measurement or estimation error. Equations 1a and 1b are common in linear maximum standardization practice in value-based MCAs, as demonstrated by Geneletti (2010) and others. This criteria standardization will result with criterion scores from ranging from 0–1.

$$\text{Standardized Cost Subcriterion Performance Score} = 1 - \left(\frac{\text{actual score}}{\text{maximum score}} \right) \quad (13.1a)$$

$$\text{Standardized Benefit Subcriterion Performance Score} = \left(\frac{\text{actual score}}{\text{maximum score}} \right) \quad (13.1b)$$

To be implemented in detailed cycle route design, this third step assumes the data for each data set of performance measures are available, clean and ready to use, and that all criteria performance measures have given their unstandardised score for each segment and junction. It also assumes each performance measure's score is either beneficial (positive) or detrimental (negative) to the overall bicycle-friendliness of the road segment and junction. Once these assumptions are fulfilled, they must be transformed into unitless scales with either Equation 13.1a or Equation 13.1b. To determine a total route suitability score, this standardization must be done for each segment and junction's performance measures. This preservation of detailed information is unused by most value-based MCAs, but it will allow useful information to be given to the cycle route designers, as can be seen in Figures 13.5 through 13.7. Information at the segment and junction level is also the key for any researchers who in the future would like to run MCA cycle route design assessments on an entire network.

13.2.2   Modifying MCA for Cycle Route Design

Above, Section 13.2.1 describes the beginning three steps traditionally followed in MCA. In this section we modify this traditional method in a further three steps (Steps 4–6) to make the MCA suitable for use in cycling infrastructure design. To our knowledge, these steps have never been used in MCA or by the cycling infrastructure design community.

13.2.2.1   Step 4: Target Populations Participating as Stakeholders

This MCA for cycle route design requires target populations to be defined by local cycle programme managers. These will probably be based on past cycling surveys, traffic counts and travel behaviour studies. However, policy-defined target populations are not always based on groups with homogeneous travel behaviour. Known heterogeneous populations should be handled with current best statistical practices.

Defining target populations enables focused research and engagement. Representative samples of each target population can be invited to participate in the design process as stakeholders. This stakeholder participation could take the form of survey campaigns, focus groups, web-based discussion and forums. The purpose of this participation is to: (1) present each participant with the design criteria set(s) and receive in return their ranking, stating their personal preferences; and (2) receive participants' feedback on past cycle route designs and concerns about upcoming plans. This additional information can also be a mechanism for target populations to state their post-construction satisfaction about certain facility designs. The method presented here (see Figure 13.3) suggests that setting a schedule to periodically seek people's comments on past projects (successes and failures)

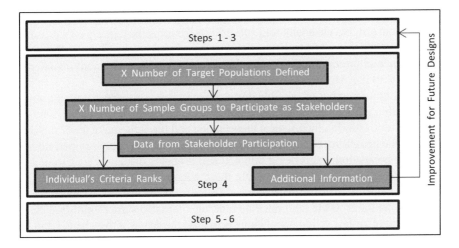

FIGURE 13.3   Target populations participating as stakeholders.

can help to regularly update the pre-construction design process and hopefully improve future cycle route designs.

Criteria can be presented to stakeholders for ranking in two main ways: hierarchically and non-hierarchically. There is a trade-off between a large number of criteria and more manageable assessments with a small number of criteria. An MCA with more design criteria allows for more specific answers, but it also allows greater variation between each two individuals' stated preference sets. If a large number of design criteria were chosen, then a hierarchical presentation should be chosen. Presentation of criteria matters to the reliability of the ranking results. With poorly presented criteria it will be hard to find useful results as every participant could give a drastically different ranking set. See the example in Figure 13.4, which has 12 arbitrary cycle-route criteria {a…l}.

Giving a participant/stakeholder a non-hierarchical set of 12 design criteria would allow the participant a total of 12! (i.e. 47,900,160) possible ranking alternatives (see Figure 13.4). This can

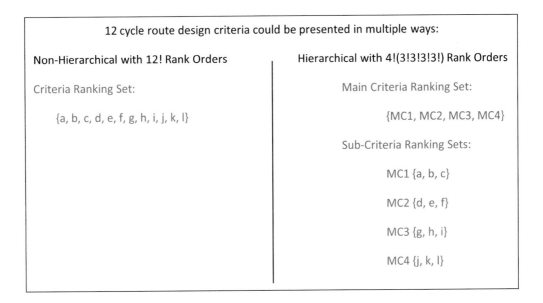

FIGURE 13.4   Cycle route design criteria organised in non-hierarchical and hierarchical ranking sets.

be dramatically reduced when the rankings are presented hierarchically. Splitting these criteria into three groups of main criteria, each with four sub-criteria, reduces the participant's possible ranking alternatives to 3!(4!4!4!) (i.e. 82,944). Even better, if these criteria are divided into four groups of main criteria, each with three sub-criteria, then each participant's possible ranking alternatives are reduced to 4!(3!3!3!3!) (i.e. 31,104, as shown in Figure 13.4).

There are an infinite number of possible design criteria sets, so there is not an optimal hierarchy. The presentation must be left to the judgment of the designers and engineers who choose the criteria. Theoretically, criteria's hierarchical presentation and the subsequent fewer ranking alternatives will reduce the possibility for variability of the answers. This is highly important considering each target cyclist population that is being sampled will probably display at least some heterogeneity of personal preferences. To properly sample these populations would already require a fairly large sample size and various recruitment sources. Thankfully, the necessary sample size for robust results is likely to be lower as the total possible ranking alternatives are diminished. Sample sizes for the participation of target cycling populations is beyond the scope of this chapter, but it certainly would be an interesting topic for future research.

13.2.2.2   Step 5: Stakeholder Preference Sets to Aggregated Group Ranks

Most traditional MCAs do not use group preferences; instead, each person's preference ranking is turned into their personal weighting scheme. These traditional MCAs are impossible when including the public in the design process. See Equation 2, modified from the MCA work of Mendoza and Martins (2006), who talk about the three methods for heterogeneous group opinions (fuzzy situations). If a city's cycle programme wants to avoid extremely pessimistic and extremely optimistic data transformations, the compromising midpoint should be chosen. Furthermore, when results show skewed distributions of criteria, then caution would dictate use of the median. Median criteria ranking sets of the three target cyclist groups can then be transformed into weights with the following equation:

$$W_i = \frac{R_i}{\sum R_i} \tag{13.2}$$

where:

W_i = the weight assigned to the criterion i;
R_i = aggregated target cyclist group rank to the criterion i.

Once the weights are aggregated for each target cyclist group, they can be multiplied by their respective standardised criteria scores. MCA's traditional weighted summation equation (Geneletti, 2010; Hajkowicz and Higgins, 2008; Pomerol and Barba-Romero, 2000) can be modified, as shown in Equation 3, to combine the scores into a single-route suitability score. This adapted version allows for the same sub-criteria to appear many times within the route (e.g. an individual visibility score for each junction along the route) without averaging. This is important. In order to prioritise design interventions, any route options must be addressed not only by their total suitability scores but also by the detailed performance of the road junctions and segments forming them. A bicycle route assessment providing only one final score is of very little use to designers and engineers. This approach, however, allows each segment and junction to maintain its broken-down scores before being included in the actual route sum. This actual route sum is then divided by the total possible route sum.

$$S = \frac{\text{actual route sum} \sum_{i=1}^{n} W_i X_i}{\text{total max route sum} \sum_{i=1}^{n} W_i \max X_i} \quad (i = 1, 2 \ldots n) \tag{13.3}$$

where:

$$\sum_{i=1}^{n} W_i = 10 < W_i \le 1$$

S = total route suitability score;
n = number of criteria;
W_i = weight assigned to the criterion i;
X_i = normalised score of criterion i.

13.2.2.3 Step 6: Sensitivity Analysis for Group Weights

A traditional MCA sensitivity analysis as seen in Geneletti (2010) changes weights at equal intervals to see if there is a reversal point in which an option scores as the 'best'. This would typically indicate how robust the scores were when under the influence of possible preferences of different decision-makers.

For the design of cycle routes, a similar but different sensitivity analysis could reference how each of the changing weights from each participating target cycling group changed the scores. This could be done for each junction and segment but is also summarised by the total route suitability score. If there is a change in the most suitable route option for different participating target cycling groups, then this signals a reversal point. Reversal points in cycle route design are interesting because they may highlight a route's relative weakness for a particular target population. If design interventions were made for the weakest point of this route, then the reversal point may disappear and the route's interventions could be assumed to be more robust for more target populations. This could benefit the design process of a cycle route that was currently facing public opposition.

13.3 APPLICATION OF MCA FOR CYCLE ROUTE DESIGN IN CHRISTCHURCH

Steps 1 and 2 are context dependent and the final design criteria must be accepted by the route designers who will be using the analysis results. Christchurch officials decided not to have constraint criteria create 'black spots' or 'no-go zones' for planned cycle routes. They wanted to emphasise the use of compensatory design criteria.

In the case of Christchurch's urban cycle routes, 49 design criteria were considered. These were narrowed down to 17 after being reviewed against both the cycling research literature and the city's needs. Table 13.1 below shows the seven main criteria and 17 sub-criteria and their performance measures in the hierarchy that was approved by city authorities. The hierarchy did not affect the equal-weight standardised criteria scores shown in Figures 13.5 through 13.7 (which are based solely on roadway performance measures); it only affected the scores weighted by the target cycling participants, as shown in Figures 13.8 and 13.9.

Christchurch's chosen criteria took into account both infrastructure supply and social demands. The city is situated in a temperate zone, predominantly on a flat sprawling plain, with the sea to its east and hills to its south. If the study area had been situated in a city with complex terrain or extreme environmental conditions, then the design criteria would have probably included characteristics such as steep slopes, road areas prone to ice accumulation etc. The methods we describe in this chapter are able to deal with whatever design criteria a local city would like to choose.

Some data were not available and were unable to be measured during our fieldwork due to lack of time and equipment. These proxies are listed in Table 13.1. Table 13.2 shows the data sets and how they were obtained. Most city representatives were associated with the CCC (Christchurch City Council), or the UC (University of Canterbury).

Once data is collected, Step 3 transforms the criteria performance scores into route suitability scores. The performance measures were computed for their respective road segments and junctions in ArcGIS ™ attribute tables. The raw data shows there are variations present in the micro-environments

TABLE 13.1

Chosen Criteria Hierarchy and Performance Measures (Acoustic Engineering Services, 2009; Christchurch City Council, 2014; Landis et al., 2003; Landis et al., 1997)

Main Criteria	Sub-Criteria (Segment or Junction) Data for Test Area)	Performance Measure Computed As
Comfort	S1_Non-slip surface (segment surface material chip size)	Chip size as proxy for macro-texture skid resistance
	S2_Roughness (average per road segment)	Link NAASRA Average = ((sum (tilt counts/20 metres)) / number of NAASRA measures per link)
Junction safety	J1_Visibility (junction average metres to potential obstruction)	Average visibility = ((sum of distances to surrounding properties) / number of surrounding properties)
	J2_Speed & Volume (junction speed as km/h & volume as 24-hour, four-day average ADT)	Speed $\times$ volume
	J3_Facility Capability (junction average reserve width)	Average Reserve Width = (Sum of roadway reserve widths) / number of roads at junction
Road capacity	S3_Effective width (segment width relative to 24-hour, four-day average ADT)	Wv = Effective width as a function of traffic volume Wt = Total pavement width of shoulder and outside lane Wv = Wt if ADT > 4000 vehicles/day Wv = Wt (2-.00025$\times$ADT) if ADT $\leq$ 4000 veh/day and if the carriageway is unstriped and undivided Adopted from: (Landis, Vattikuti, & Brannick, 1997)
	S4_Traffic Composition (Segment % non-light vehicles)	% medium and heavy vehicles (categorized by weight and specified by RAMM definitions)
Directness & efficiency	S5_detour factor (DF segment $\times$ DF route)	Segment detour score = (link length / optimal link length) $\times$ (route length / optimal route length)
	J_4_Right-hand turns (junction turn count)	Sum turn counts for both directions
	J5_Delay (seconds average per junction)	Average delay = ((sum of the junction's delays along the route directions) / number of directional delays)
Connectivity & Transit Cohesion	S6_Connectivity (segment length)	Measured from cyclable cross-street to cyclable cross-street (unnamed residential and commercial cul-de-sacs)
	S7_Bus stops (# within 100 m network distance of segment ends	Count of bus stops within 100 m of road segment
Attractiveness	S8_Art/Parks/Public Areas (segment % frontage)	% public frontage = metres of public frontage along route link / total metres of route link
	J6_Noise & pollution (junction estimated noise as dBA Leq/day & volume of vehicles which expose cyclists to more PM10 estimated as vehicles/day) N	Intensity of noise & pollution emitting vehicles = (24-hour dBA Leq within 10 m) $\times$ ((24-hour ADT) $\times$ (% heavy emitting vehicles)) Adopted from: the (Acoustic Engineering Services, 2009) report completed for Christchurch City Council
	S9_Street lighting	Link lighting = no. of street lights along link / ((Total Carriageway Width) $\times$ (Route Link Length))
Trip generators & attractors	S10_Population adjacent to segment	Population adjacent to link = (No. dwellings adjacent to link) $\times$ (average household size) / (Route link length) Adopted from: Christchurch City Council (2014a), which reported an average 2013 household size of 2.5 people per dwelling, and the bicycle Latent Demand Score (Landis et al., 1997).
	S11_Destinations adjacent to segment	Destination adjacent to link = number of non-residential destinations with direct access to link / (Route Link Length) Adopted from: the bicycle Latent Demand Score (Landis et al., 1997), which uses attractions such as employments, shopping centres, parks, and schools.

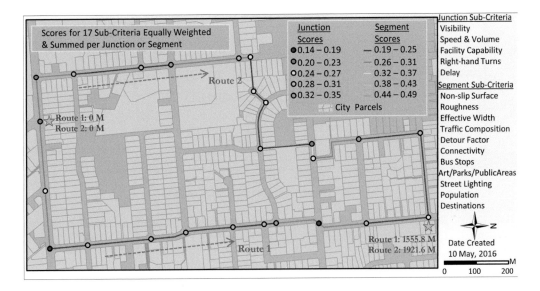

FIGURE 13.5 Map of route segment and junction scores when equally weighted.

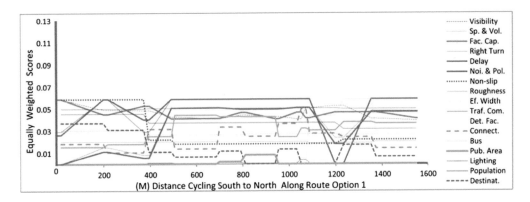

FIGURE 13.6 Route 1 sub-criteria segment and junction scores when equally weighted.

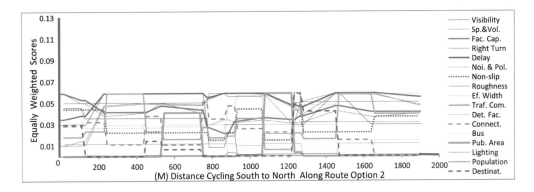

FIGURE 13.7 Route 2 sub-criteria segment and junction scores when equally weighted.

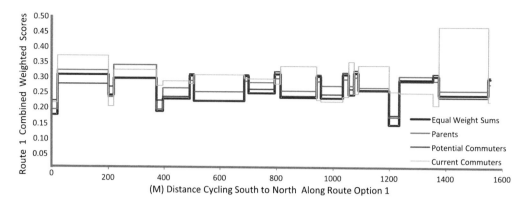

FIGURE 13.8 Route 1 summed sub-criteria segment and junction scores when weighted by target populations.

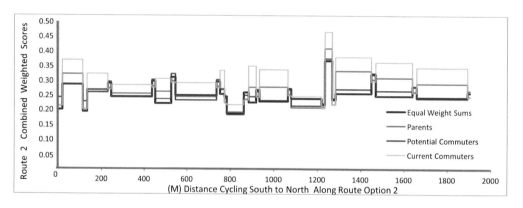

FIGURE 13.9 Route 2 summed sub-criteria segment and junction scores when weighted by target populations.

TABLE 13.2
Data Sets Used for the Study Area

Data Set	Pertinent Information	Information Obtained From (Local Organization)
July 2014 Bicycle Survey	Cycling perceptions & frequencies of > 1500 Christchurch residents	Karyn Teather (CCC Asset & Network Planning & UC Alumni)
Road Asset and Maintenance Management (RAMM)	Chip size, NAASRA roughness, ADT, traffic composition, reserve width & carriageway width	Binaya Sharma (CCC Asset & Network Planning, City Infrastructure Division) & updated via Counts website http://www.ccc.govt.nz/cityleisure/projectstoimprovechristchurch/transport/trafficcount/index.aspx
Cadastral parcels	Land use, frontage, dwelling units, commercial tenant	Josh Neville (UC Alumni) & updated via fieldwork
Road centre lines	Block length & road name	Aimee Martin (UC Alumni)
Roads miscellaneous	Speeds, facility photos, right-hand turn counts, directional delay	Manually recorded during fieldwork, samples of directional delay were timed during 8–9 am peak morning traffic for 20-minute intervals at each junction that would require a right turn
Bus	Bus stops, routes & shelters	Shannon Boorer (Environment Canterbury)

of segments and junctions. For instance, National Association of Australian State Road Authorities (NAASRA) roughness is different for each metre along the route, and in some places it is much worse than others. Despite these existing real-world variations, the performance measures require some level of aggregation to be practical. Segments and junctions are the building blocks of the network and thereby suited for scaling up to the city network. Thus, they were selected for this study.

Step 4 involves getting target populations to participate as stakeholders. Due to Christchurch's goal of increasing cycling's modal share, our research classified three target populations: current cycling commuters, potential cycling commuters and parents with children aged 10–17 years. Accordingly, three small sample groups were created. The participation in surveys and focus groups together totalled 66 individuals ($n=66$). The results produced included each person's preference ranking, which was then aggregated into a group criteria weighting scheme for their target population. These sample groups were not assumed to be entirely representative of the target populations of Christchurch. Rather, they served as an example of how personal preferences can be turned into weighting schemes for each target population. The 17 criteria were presented to them hierarchically with the seven main criteria as shown in Table 13.1.

It is then possible to visualise the individual sub-criteria scoring graphs and how they change with distance at different segments and junctions along each route. Figure 13.5 shows the summed version of these equal weight scores, displaying how many of the segments within the study area score only moderately in terms of bicycle-friendliness. This is because low scoring criteria such as lighting and adjacent non-residential destinations, as well as parks, displayed art and public areas (shown in the graphs as Pub. Area), were displayed as being equally important to effective width, detour factor and other sub-criteria. It was expected that these aforementioned criteria would score low, as this study area is a fairly typical Christchurch residential neighbourhood that borders large industrial and commercial districts.

These standardised segment and junction scores were then weighted to show the preferences of people who had participated as stakeholders. Remember, these are small samples, not significant representations of the true preferences of the target cyclist populations in Christchurch. The participants first ranked the main criteria and then the sub-criteria. Participants' results were then aggregated into the median for their target cyclist group. These tables show the rounded weights, but the procedure used fractions with non-rounded weights summing to the normalised '1'. This satisfies the major assumption of traditional MCA weighted summation.

13.3.1 Ranks Converted to Main Criteria Weights

Tables 13.3 through 13.5 present the main criteria weights (MCW) of the three cyclist groups targeted as derived from their rankings (Table 13.6).

TABLE 13.3
Main Criteria Weights of 18 Potential Cycling Commuters

Main Criteria Rank Set (Highest Ranking Criteria Listed First)	Median Stakeholder Rank	SDSS Weight (Rank/Rank Sum)
Junction safety	6	0.211
Directness & efficiency	4	0.140
Connectivity & transit Cohesion	4	0.140
Attractiveness	4	0.140
Trip generators & attractors	4	0.140
Capacity	3.5	0.123
Comfort	3	0.105
Sum	28.5	1.000

TABLE 13.4

Main Criteria Weights of 32 Current Cycling Commuters

Main Criteria Rank Set (Highest Ranking Criteria Listed First)	Median Stakeholder Rank	SDSS Weight (Rank/Rank Sum)
Junction safety	5.5	0.200
Directness & efficiency	4	0.145
Connectivity & transit cohesion	4	0.145
Capacity	4	0.145
Attractiveness	3.5	0.127
Trip generators & attractors	3.5	0.127
Comfort	3	0.109
Sum	27.5	1.000

TABLE 13.5

Main Criteria Weights of 16 Parents with Children Aged 10–17 Years

Main Criteria Rank Set (Highest Ranking Criteria Listed First)	Median Stakeholder Rank	SDSS Weight (Rank/Rank Sum)
Junction safety	7	0.226
Capacity	6	0.194
Trip generators & attractors	5	0.161
Directness & efficiency	4	0.129
Comfort	3	0.097
Attractiveness	3	0.097
Connectivity & transit cohesion	3	0.097
Sum	31	1.000

13.3.2 Ranks Converted to Sub-Criteria Weights

Having established relative weightings attached to different criteria, these criteria could then be scored for the two route options trialled for the Norwest Arc. Figure 13.5 shows the route segment and junction scores for the two options (start and end locations indicated by stars). Figures 13.6 through 13.9 show the segment scores for the criteria and targeted cyclist groups.

Both route options scored 6–7% higher for the current commuters than for potential commuters and parents of children 10–17 years. As we have noted previously, this is owing to the type of roads in this study area and how they score better with the combination of criteria preferred by the current commuters. These results support the theory that not all roads are equally suitable for groups with different levels of confidence and abilities (CROW, 2007). If these trends manifested themselves with representative sampled target populations, then it might require special interventions to accommodate these different groups.

As we expected, there was a reversal point (see Table 13.7) after the weights were significantly altered from equal criteria weighting. This indicates that the route chosen would have to be improved at its worst scoring junctions and segments prior to becoming significantly more suitable than the other route option.

This big shift in overall suitability was produced by the weights acting as linear transformations of the original performance values. In other words, the cyclist preferences and weighting schemes

TABLE 13.6
Sub-Criteria Weights of 32 Current Cycling Commuters

Main Criteria	Median Stakeholder Rank	SDSS Weight (Rank/Rank Sum)	Sub-Criteria Rank Sets	Median Stakeholder Rank	SDSS Weight (Rank/Rank Sum) × mc Rank
Junction safety	6	0.211	Visibility	2	0.084
			Volume & speed	1	0.042
			Facility capability	2	0.084
			sum	5	0.211
Directness & efficiency	4	0.140	Detour factor	1	0.028
			Right turns	2	0.056
			delay	2	0.056
			sum	5	0.140
Connectivity & transit cohesion	4	0.140	Connectivity	2	0.093
			Bus stops	1	0.047
			Sum	3	0.140
Attractiveness	4	0.140	Public place	2	0.047
			Noise & pollution	2	0.047
			Street lights	2	0.047
			Sum	6	0.140
Trip generators & attractors	4	0.140	Population	1.5	0.070
			Destinations	1.5	0.070
			Sum	3	0.140
Capacity	3.5	0.123	Effective width	2	0.082
			Traffic composition	1	0.041
			Sum	3	0.123
Comfort	3	0.105	Roughness	2	0.070
			Non-slip	1	0.035
			Sum	3	0.105
Sum	28.5	0.999	N/A	N/A	N/A

TABLE 13.7
Total Route Suitability Scores

Weighting Scheme	Total Route Suitability Scores	
	Route 1	Route 2
Equal weights	0.13841	0.13844
Current commuter*	0.65093	0.65097
Potential commuter	0.59580	0.58407
Parents of children aged 10–17	0.59417	0.58805

* Indicates a reversal point (when the 'best' scoring route changes).

change, but the original road scores remain the same. Changing weights leaves the potential for a reversal point (where the 'best' option changes) to be caused by a target population's different preferences for one route over another (the routes having different road types, transecting different neighbourhoods, different densities of attractors and generators etc.). Weights change the total route suitability score and let it range from bad to good on the bicycle-friendliness scale of 0–1, with 1 being most friendly for that target population. Note also that the total suitability scores for

Routes 1 and 2 show very little difference when compared by the same target population. This is because both routes were similar in street design. Both route options score reasonably well for current cycling commuters, but less so for potential cycling commuters and parents of children aged 10–17 years. For these last two groups, both Routes 1 and 2 would require significant design improvements to be considered bicycle-friendly routes.

Target populations of any city likely have shifting preferences and experiences as time passes and cycling facilities improve. This is why regular feedback from these target populations should be gathered periodically to improve the design process. As shown in Table 13.8, additional information can be gained from these stakeholder sample groups: many times, the participants in our study mentioned past facilities as examples of either good or poor design in meeting their preferences and needs. This additional information highlighted where Christchurch's comprehensive cycling program could improve its efforts. Maps were presented to the participants, and by the end of the stakeholder engagement process, 66 people had marked intersections and segments that were on their way to work or school with ideas and suggestions for facility improvements.

TABLE 13.8
Comment Summary of Christchurch Cycling Focus Group

Comment Category	Comment Sub-Category	No. Times Mentioned in Focus Groups	%
Behaviour	Cyclist behaviour	7	4.5
	Driver behaviour	18	11.5
	Media/public perception/initiatives	11	7.1
	Pedestrian behaviour	2	1.3
Connectivity	Lack of options	4	2.6
Good facilities	Cycle lane separation	5	3.2
	Intersections	5	3.2
	Parked cars	4	2.6
Maintenance	Broken glass	2	1.3
	Roadworks	4	2.6
Navigation	General road segment Difficulty/danger	6	3.8
	Lane change difficulty	3	1.9
	Left-turn difficulty/danger	3	1.9
	Right-turn difficulty/danger	13	8.3
	Roundabout difficulty/danger	6	3.8
	Through intersection difficulty/danger	10	6.4
Obstruction/visibility	Parked cars	7	4.5
Poor facilities	Cycle paths designed around car parks/ bus stops	3	1.9
	Disjoint segment cycle lanes	2	1.3
	Major cycle paths too narrow	4	2.6
	No cycling facilities	8	5.1
	Shared cycle lane/footpath	6	3.8
	Transfer between segment cycle facilities & junctions with no facilities	4	2.6
	Unclear design	9	5.8
Traffic related	Bus conflict	4	2.6
	Congestion blocks junction cycle lane	2	1.3
	Road is too busy	2	1.3
	Truck conflict	2	1.3
Total		156	100.0

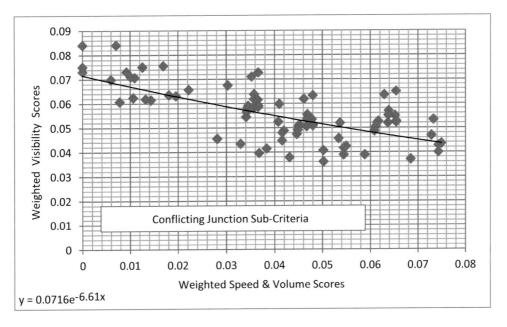

FIGURE 13.10 One of Christchurch supply-side criteria conflicts.

Knowing which designs and past projects the target populations perceive as successes or failures is important, as future projects will probably have to assign priorities among design ideals. This is especially important when target populations state their safety is being compromised by a past design (which may have got a high score in other, conflicting criteria).

Two desirable, bicycle-friendly criteria may work against each other, so that an increase of one leads to a decrease in the other. The results from our study also contain this general conflict, as illustrated in Figure 13.10. Even when different weighting schemes are applied, as shown in the figure, there remains a non-linear, moderately strong negative correlation between the study area's increasing speed and volume in relation to junction visibility scores. Visibility, volume and speed are some of the most important variables for reducing collision severity and fatality rates (Ehrgott et al., 2012; Environment Canterbury, 2005), and these are commonly considered in facility engineering designs in New Zealand (Land Transport Safety Authority, 2004). Designing for a single cycle route will bring with it both types of criteria. Mitigating the effects of the inherent compromises this demands is the difficult job facing facility designers and engineers. A standardised bicycle route assessment would provide a way of structuring the complications and prioritizations involved with these compromises.

13.4   IMPROVEMENTS FOR THE FUTURE

For stakeholder participation to become representative of Christchurch's target cycling populations, involvement would need to be implemented on a larger scale than that of our study. This could be undertaken through surveys or public-opinion websites with a larger sample of participants for each target population. Without anticipating sample bias, results from any assessment (even assessments with fewer criteria or other MCA techniques) may give misleading conclusions.

Though scalability was not within the scope of this study, the equations we have presented were modified from their original sources so as to be scalable in theory. If future research demonstrates that these types of MCA equations for cycle route design are indeed scalable to whole networks, then it would dramatically change how the industry assesses infrastructural interventions for target populations. Network assessment could identify vulnerabilities for a city's target cycling populations due to network fragmentation and/or other design failures. This would be useful for countries like the U.S.A, where many city-wide cycleway programmes have left networks fragmented by well-designed, yet disconnected individual projects (Schoner and Levinson, 2014). Future studies could look at how the needs of multiple target populations could be streamlined across a cycle network and whether multiple cycle networks are needed within any given city.

13.5 CONCLUSION

A real cycle route is not simply an aggregated score but rather the sum of its many diverse parts, and its bicycle-friendliness can change over space and time. Consequently, cycle route designers are better equipped if they have access to quantitative spatial assessments that: (1) give detail at junction and segment levels; (2) can assess the preferences of any given target group for any of the criteria involved and the overall design of the planned cycle route; and (3) can improve future designs by triggering reactions to past designs. Systematically integrating target populations into the process of cycle route design can strengthen the justification for city-wide cycle programmes and encourage public support for any individual construction project.

Future studies could explore the dynamics of implementing standardised bicycle-route assessment procedures in different situations and different city environments. It would especially help policy-makers to better understand how stakeholder participation can be applied to cycle route design on a city-wide scale. Standards need to be better defined in order for quantitative bicycle-route assessments to operate efficiently within city managements. Although assessment results support the monitoring and processing of detailed data, ultimately reaching strategic transport targets requires laws and policies for a strong and comprehensive foundation. Without some kind of regional or nation-wide assessment, bicycle routes will still be designed using criteria of a priori importance, but the quality of the work will continue to be dependent on locally available experts and is likely to vary from project to project.

REFERENCES

Butler, A. (2015). *Multiple Criteria Bicycle Route Assessment Integrating Demand, Supply & Stakeholder Perceptions for a Spatial Decision Support System in Christchurch, New Zealand.* Retrieved from https://webapps.itc.utwente.nl/librarywww/papers_2015/msc/upm/butler.pdf

Canterbury Regional Transport Committee. (2012). *Canterbury Regional Land Transport Strategy 2012–2042.* Christchurch, New Zealand. Retrieved from http://ecan.govt.nz/our-responsibilities/regional-land-transport/pages/regional-land-transport-review.aspx

Christchurch City Council. (2016). *Major Cycle Route Network Updates: Christchurch City Council.* Retrieved April 8, 2016, from http://www.ccc.govt.nz/transport/cycling/cycleways/latest-cycleways-news/

CROW. (2007). *CROW Design Manual for Bicycle Traffic.* CROW. Retrieved from http://www.crow.nl/shop/productDetail.aspx?id=889&category=90

Damant-Sirois, G., Grimsrud, M., & El-Geneidy, A. M. (2014). What's your type: A multidimensional cyclist typology. *Transportation, 41*(6), 1153–1169. https://doi.org/10.1007/s11116-014-9523-8

Ehrgott, M., Wang, J., Raith, A., & Van Houtte, C. (2012). A bi-objective cyclist route choice model. *Transportation Research Part A: Policy and Practice, 46*(4), 652–663. https://doi.org/10.1016/j.tra.2011.11.015

Environment Canterbury. (2005). *Cycling in Canterbury: Strategy for the Development of a Regional Network of Cycle Routes.* Christchurch, New Zealand. Retrieved from http://ecan.govt.nz/publications/Plans/CinC.pdf

Geneletti, D. (2010). Combining stakeholder analysis and spatial multicriteria evaluation to select and rank inert landfill sites. *Waste Management (New York, N.Y.), 30*(2), 328–337. https://doi.org/10.1016/j.wasman.2009.09.039

Hajkowicz, S., & Higgins, A. (2008). A comparison of multiple criteria analysis techniques for water resource management. *European Journal of Operational Research, 184*(1), 255–265. https://doi.org/10.1016/j.ejor.2006.10.045

Keshkamat, S., Looijen, J., & Zuidgeest, M. (2009). The formulation and evaluation of transport route planning alternatives: a spatial decision support system for the Via Baltica project, Poland. *Journal of Transport Geography*, *17*(1), 54–64. https://doi.org/10.1016/j.jtrangeo.2008.04.010

Kroesen, M., & Handy, S. (2013). The relation between bicycle commuting and non-work cycling: results from a mobility panel. *Transportation*, *41*(3), 507–527. https://doi.org/10.1007/s11116-013-9491-4

Land Transport Safety Authority. (2004). *Cycle Network and Route Planning Guide*. Retrieved from http://www.nzta.govt.nz/resources/cycle-network-and-route-planning/

Landis, B. W. Vattikuti, V.R., Brannick, M.T. (1997). Real-Time Human Perceptions: Toward a Bicycle Level of Service. Transportation Research Record, (1578), pp. 119–131.

Landis, B., Vattikuti, V., Ottenberg, R., Petritsche, T., Guttenplan, M., & Crider, L. (2003). Intersection level of service for the bicycle through movement. *Transportation Research Record: Journal of the Transportation Research Board*, *1828*, 101–106. Retrieved from http://trb.metapress.com/content/r6t63r2g723661u8/

Malczewski, J. (2006). GIS-based multicriteria decision analysis: A survey of the literature. *International Journal of Geographical Information Science*, *20*(7), 703–726. https://doi.org/10.1080/13658810600661508

Mendoza, G., & Martins, H. (2006). Multi-criteria decision analysis in natural resource management: A critical review of methods and new modelling paradigms. *Forest Ecology and Management*, *230*(1–3), 1–22. https://doi.org/10.1016/j.foreco.2006.03.023

NZ Transport Agency. (2016). *Urban Cycleways Programme*. Retrieved May 11, 2016, from http://www.nzta.govt.nz/walking-cycling-and-public-transport/cycling/for-people-involved-in-cycling-programmes-and-projects/urban-cycleways-programme/

Pomerol, J.-C., & Barba-Romero, S. (2000). *Multicriterion Decision in Management: Principles and Practice*. Springer Science & Business Media. Retrieved from http://books.google.com/books?hl=en&lr=&id=mNOKayvMqH4C&pgis=1

Schoner, J. E., & Levinson, D. M. (2014). The missing link: Bicycle infrastructure networks and ridership in 74 US cities. *Transportation*, *41*(6), 1187–1204. https://doi.org/10.1007/s11116-014-9538-1

Sharifi, M. (2004). Site selection for waste disposal through spatial multiple criteria decision analysis. *Journal of Telecommunications and Information Technology*, *3*, 28–38. Retrieved from http://yadda.icm.edu.pl/baztech/element/bwmeta1.element.baztech-article-BAT3-0012-0021

Sharifi, M., Boerboom, L., Shamsudin, K., & Veeramuthu, L. (2006). Spatial multiple criteria decision analysis in integrated planning for public transport and land use development study in Klang Valley, Malaysia. *ISPRS Technical Commission II Symposium*, 85–91. Retrieved from http://www.isprs.org/proceedings/XXXVI/part2/pdf/sharifi.pdf

Thomopoulos, N., & Grant-Muller, S. (2012). Incorporating equity as part of the wider impacts in transport infrastructure assessment: an application of the SUMINI approach. *Transportation*, *40*(2), 315–345. https://doi.org/10.1007/s11116-012-9418-5

Urbanczyk, R. (2010). *PRESTO Cycling Policy Guide Promotion of Cycling*. Retrieved from http://www.rupprecht-consult.eu/uploads/tx_rupprecht/PRESTO_Cycling_Policy_Guide_Promotion.pdf

Yang, C. H., & Regan, A. C. (2012). A multi-criteria decision support methodology for implementing truck operation strategies. *Transportation*, *40*(3), 713–728. https://doi.org/10.1007/s11116-012-9432-7

14 Post-Resettlement Socio-Economic Dynamics
The Case of Ahmedabad, India

Rushikesh Kotadiya, Monika Kuffer,
Richard Sliuzas, and Sejal Patel

CONTENTS

14.1 Introduction 245
14.2 Displacement and Resettlement in Ahmedabad 246
14.3 Method 247
14.4 Analysis of Changes in Risks of Impoverishment 249
    14.4.1 Landlessness 250
    14.4.2 Housing and Basic Services (Water and Sanitation) 252
    14.4.3 Joblessness 253
    14.4.4 Access to Education and Health Facilities 255
    14.4.5 Health Risks (Morbidity, Mortality and Food Security) 255
    14.4.6 Marginalization 257
    14.4.7 Social Disarticulation 260
    14.4.8 The Dynamics of Impoverishment 261
14.5 Discussion 261
14.6 Conclusions 262
References 264

14.1 INTRODUCTION

A substantial proportion of the urban population in Indian cities resides in slums, many of which are well located for access to livelihood opportunities. These slum settlements are often located on (previously vacant) government land. If these areas are not subject to severe environmental constraints or hazards (UN-Habitat, 2015), some city authorities consider such settlements to be situated on prime land for infrastructural improvements or other urban development projects. Slum dwellers, therefore, face displacement through forced land acquisition for the provision of urban infrastructure and development projects (Cernea, 1997a; Patel and Mandhyan, 2014; UN-Habitat, 2015). Due to their location and conditions of tenure, slum dwellers are often vulnerable to forced evictions despite their rich socio-cultural character, socio-economic networks with their surroundings and compact population density (Hooper and Ortolano, 2012; UN-Habitat, 2003).

Development-induced displacement and resettlement (DIDR) mostly affects the weaker sections of the society – the urban poor. Commonly, forced eviction causes loss of assets, loss of livelihoods and increased inequality. Involuntary DIDR affects the physical and mental health of those resettled, often resulting in increased poverty (Cernea, 1997b). In addition, displaced people are often poorly compensated for their loss of assets and have no access to proper assistance to restructure their livelihoods. Without an adequate rehabilitation policy and practical support, DIDR increases the risk of impoverishment (Cernea, 2008; Patel, Sliuzas and Mathur, 2015).

In general, urban development projects aim to improve the regional economy and individual livelihoods. Yet, such projects often have a negative impact on some part of the population. Programmes for developing, for example, water supply (dams, reservoirs, irrigation canals, riverfront and lakefront development), transportation (roads, highways, canals), mining, power plants and parks and forest reserves often result in development-induced displacement, requiring many people to resettle and rebuild their lives elsewhere (Cernea, 1997b; Jackson and Sleigh, 2000; Robinson, 2003).

BOX 14.1   Case Study Area

Ahmedabad is the fifth largest city of India, with a population of 5.6 million (Census of India, 2011). The city can be divided into three large sub-areas: the historic walled city; the industrial area to the east of the walled city, mixed with areas of poor-quality housing and some recent middle-class housing and commercial development; and the wealthier western side of the city, where the rich and the middle class have access to better services, public spaces, schools and universities and other institutions and amenities. The case study covers the entire city and focuses on a number of resettlement areas under the Basic Services for Urban Poor (BSUP) component of the Jawaharlal Nehru National Urban Renewal Mission (JNNURM). Most resettlement areas have been sited in the east, where land values and the levels of services and amenities are lower. BSUP sites were set up to provide security of tenure and affordable housing, along with access to basic services such as water supply and sanitation (MHUPA, 2009). The total project cost is being shared between the central government, the state government, Ahmedabad Municipal Corporation, Ahmedabad Urban Development Authority and the beneficiary (resettled slum household). Only slum dwellers who could prove their 'eligibility' were compensated with a small apartment on a resettlement site. Nevertheless, relocated households faced many risks of impoverishment (e.g. loss of livelihood due to increased distance to job opportunities) (Patel et al., 2015).

14.2   DISPLACEMENT AND RESETTLEMENT IN AHMEDABAD

Ahmedabad Municipal Corporation (AMC) has a vision of creating a world-class city (Desai, 2014). To achieve their vision, AMC has been undertaking many large-scale urban development projects, such as the Sabarmati Riverfront Development (SRFD), Kakariya Lakefront Development and the Bus Rapid Transit System (BRTS). However, the availability and cost of land are major bottlenecks to these projects. AMC has, therefore, been targeting some sites occupied by the urban poor for redevelopment, resulting in large-scale displacement of slum dwellers. Between 2003 and 2011, AMC demolished around 29,000 houses in 67 slum areas and relocated the slum dwellers to Basic Services for Urban Poor (BSUP) sites on the outskirts of the city (Patel et al., 2015).

Our research investigates how displacement and resettlement to these BSUP sites have changed the lives of the affected households. It builds upon an earlier study of DIDR impact that was based on a 2011 survey of 396 affected households (Patel et al., 2015). We revisited a large number of the same households in 2015, and our current analysis focuses on the changes that took place in the period 2011–2015. Cernea's (1997b) impoverishment risk and reconstruction (IRR) model provided the basis for our analysis. This model assesses socio-economic conditions after displacement with specific contextual indicators (see Table 14.1); it has been widely used for DIDR studies (e.g., Alexandrescu, 2013; Heggelund, 2006; Kaida and Miah, 2015; Muggah, 2000; Quetulio-Navarra et al., 2014). Cernea (1997a) argues that displacement disrupts lives and social welfare, and that livelihoods should be a central part of reducing risk of impoverishment in any resettlement program. The impact of resettlement is assessed using eight impoverishment risk parameters (see Figure 14.1).

The purpose of our study was to revisit resettled households that had been identified by Patel et al. (2015) in 2011, detect changes in their lives that had taken place between 2011 and 2015 and, consequently, gain insight into the dynamics of resettlement risks and their impacts. To date, only a few studies have analysed the dynamics of impoverishment caused by resettlement (e.g. Perlman, 2010). It is important to understand the roles of non-government organizations (NGOs), resident welfare associations (RWAs) and local authorities in the reconstruction of the lives of these resettled urban poor. Within this context, we posed three main questions to gain an understanding of the changes in the lives of resettled households in the period 2011–2015:

- What are the level and dynamics of impoverishment in resettlement areas?
- What factors are responsible for changes in the impoverishment of resettled households?
- Have the linkages between resettlement areas and surrounding neighbourhoods improved between 2011 and 2015?

14.3   METHOD

To investigate these questions, we revisited the households that had been surveyed in 2011 by Patel et al. (2015). In 2011, indicators were used to assess the direct effects of displacement and the resettlement process on households' lives. Although we used a similar set of indicators (see Table 14.1) for our study, some were adapted to address subsequent changes in households' lives between 2011 and 2015.

BOX 14.2   Methods Applied in the Chapter

This chapter presents an analysis of impoverishment caused by resettling slum dwellers on resettlement sites that are commonly in remote and poorly serviced locations. The analysis focuses on the socio-economic aspects of impoverishment and the contribution of spatial aspects to impoverishment. This includes an analysis of access to livelihood, public services, amenities and the linkages of resettlements sites with their surrounding neighbourhoods. The accessibility analysis was done using the network analyst of ArcGIS™ (Version 10.2); the neighbourhood analysis also used ArcGIS. The data set employed was extracted from the household survey (locational information), land use, public services and amenities data obtained during fieldwork. The accessibility results were influenced by the adequacy of the network model and the precision of the spatial locations obtained from interviewees. The neighbourhood analysis was influenced by the nature of the spatial data employed, e.g. content, spatial detail and temporal consistency. The spatial analysis provided an aggregate picture of the surveyed households.

TABLE 14.1

Indicators for Assessing Resettled Households' Lives, 2011 and 2015

Form of Impoverishment	Indicators in 2011 (Patel et al., 2015)	Indicators in 2015
Landlessness	Loss of land – interpreted as distant relocation leading to increased distances to livelihoods, social amenities, marketplaces, etc., as well as loss of access to opportunities	Land – interpreted as distant relocation leading to long distances to travel to livelihoods, social amenities, marketplaces, etc., as well as access to opportunities
Joblessness	Sustained loss of job Loss of working days Increased distance to work and associated transport costs Increased monthly expenditure as percentage of income Increased debt	Employment No. of working days Distance to work and associated transport costs Increased or decreased monthly expenditure as percentage of income Debt
Homelessness	Loss of assets Cost of transporting assets Cost of reconstruction on new plot Loss of group cultural spaces, resulting in a sense of placelessness	Presence of group cultural spaces, resulting in a sense of community
Marginalization	Lower socio-economic status in new location Coercive displacement Deliberate subjugation of vulnerable groups, i.e. women and minorities Hostility from host community Loss of standing in community	Socio-economic status in new location after resettlement Coercive resettlement Deliberate subjugation of vulnerable groups, i.e. women, children and minorities Hostility from host community
Health risks (combination of two categories food security, morbidity and mortality)	Lack of access to safe drinking water and sewerage Inadequate solid-waste management Decreased access to fair-price shops (public shops that provide subsidised goods to poor) Decreased access to primary and higher-order health facilities	Access to safe drinking water and sewerage Solid-waste management Access to fair-price shops (public shops that provide subsidised goods to poor) Access to primary and higher-order health facilities
Loss of access to common facilities	Decreased access to education and health services School dropout ratio and loss of days school attended Increased distance, travel costs and monthly costs for education vis-à-vis income Decreased access to hospitals in emergencies	Access to education and health services School dropout ratio and loss of days school attended Distance to school, travel costs and monthly costs for education vis-à-vis income Access to hospitals in emergencies
Social disarticulation	Fragmentation of social units Dissatisfaction and discord with new community Loss of community institutions Application for transfer to reunite with kin Alienation and abuse of common resources	Community restructuring Dissatisfaction and discord with new community Role of community institutions Maintaining community assets Application for transfer to reunite with kin Alienation and abuse of common resources

Displaced households were relocated to various BSUP resettlement sites in 2011, irrespective of their livelihoods. As a result, resettled households had to rebuild their lives according to the locational characteristics of their resettlement area. To understand the reasons behind changes in their lives between 2011 to 2015, indicators such as distance to work, number of days for employment throughout the year, access to social amenities and access to health and education were examined

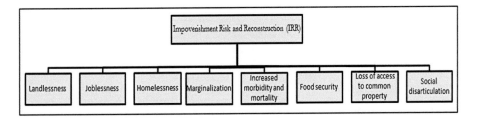

FIGURE 14.1 Eight parameters of impoverishment risk and reconstruction (IRR) (Cernea, 1997b).

in 2015. Furthermore, there was a need to assess indicators that gave an understanding of aspects of social cohesion: e.g. availability of cultural space, sense of community, role of community institutions and characteristics of women, children and minorities. In 2011, some indicators focused on the direct effects of displacement, such as loss of assets, cost of transporting assets and cost of house reconstruction on the new plot; in 2015, these were no longer relevant for assessing changes in a resettled household's lives.

Between 2003 and 2010, some 29,000 households were displaced from slums to resettlement areas. And although 3,275 households of these were resettled to the seven BSUP sites shown in Figure 14.2, only 2,717 of these residential units were occupied in 2011 (see Table 14.2). Some of the remaining displaced households were relocated to an interim site, but others, who could not prove their eligibility for BSUP housing, stayed at the demolished sites in makeshift shelters for as long as possible (Patel et al., 2015).

An objective of the fieldwork was to trace and revisit the resettled households at various BSUP sites that had been interviewed in 2011. A total of 269 households were surveyed in both 2011 and 2015 (approximately 10% of the total 2,717 households; see Table 14.2).

The primary-data collection comprised three parts: (1) semi-structured interviews of sampled households from all sites; (2) semi-structured interviews of RWA members, community leaders and community members; and (3) semi-structured interviews of government officials, academics and NGO staff working on the seven BSUP sites (see Table 14.3 for a detailed description of the fieldwork). During the fieldwork, we realised that tracing the households from the previous research (2011) would be very difficult, despite having a detailed list of the residential units allotted to resettled households in 2011. Although locating households from the list was indeed difficult, 82.5% of households could be traced with the help of local community leaders and NGO staff.

Table 14.4 gives an overview of the 2015 survey of households. Of the original 269 households interviewed in 2011, some 222 could be located, and these participated in our 2015 survey. The remaining 47 households either could not be traced or declined participating. Among these, 18 had rented out their residential unit to other people, 14 had returned to their previous (slum) locations or another part of the city, nine households were not willing to participate in our survey and five households were not available during the survey period. Most households (167 of 222) permitted the recording of their interview, thereby facilitating the transcription and validation of the data.

14.4 ANALYSIS OF CHANGES IN RISKS OF IMPOVERISHMENT

This section provides a comparison* of the changes between 2011 and 2015. The analysis focuses on identifying and understanding factors, according to eight impoverishment indicators, that are affecting resettled households' lives in significant ways (see Table 14.1).

---

* The statistical analysis was done per settlement and in total. For the total analysis, as settlement sizes vary, the socio-economic data of all resettled households were analysed as a single group.

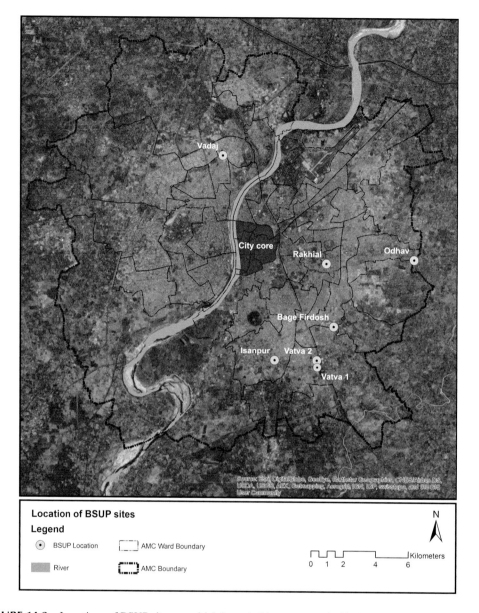

FIGURE 14.2 Locations of BSUP sites on which households were resettled in 2011.

14.4.1 LANDLESSNESS

Land and its location are important factors in understanding the risk of impoverishment after resettlement. The restructuring of households' livelihoods depends upon contextual opportunities for education, livelihood and social amenities. It is also important to understand the influence of distance from socio-economic amenities. Increased distance from socio-economic amenities such as health facilities, schools and work locations can have a negative impact on a household's monthly expenditure and reduce contextual opportunities for improving a household's life after resettlement.

Table 14.5 shows that the average distance travelled to schools increased for all but one BSUP site (Odhav). Vadaj was the only site with a decline, albeit relatively small (<10%). Thus, students from the surveyed households had to travel further for their education in 2015, and the increased travel costs added to the burden of tight household budgets. Many students who were studying in

TABLE 14.2
Residence Status on BSUP Sites and Sampling Size for Household Survey, October 2015

	Residence Status			
BSUP Sites	**Residences Constructed**	**Residences Allotted**	**Residences Occupied**	**No. of Sampled Households (10%)**
Isanpur	384	197	172	17
Vadaj	576	465	395	39
Rakhial	704	479	422	42
Bag-e-Firdosh	672	472	459	46
Odhav	320	164	121	12
Vatva	2768	674	533	51
Vatva 2	2224	824	615	61
Total	**7648**	**3275**	**2717**	**269**

Source: Patel et al. (2015).

primary schools are now in secondary school and have to travel much longer distances due to the lack of secondary and higher secondary schools in the surroundings of BSUP areas. This increases travel costs and monthly expenditures.

Proximity to a market for daily food supplies has a considerable impact on travelling time and costs. Overall, 75% of the households had access to markets within 2 km of their location. However, as Table 14.6 shows, for three BSUP sites (Isanpur, Vadaj and Odhav), many households have to travel far greater distances, for some even more than 10 km.

TABLE 14.3
Detailed Description of Data Collection, October 2015

Primary Data Collection	No. of Interviews Done	Description of Interviews
Household interviews	222	222 households were traced out of original 269. Detailed break-down of households' interviews are shown in Table 14.4
Settlement level checklists	7	To understand the condition of social amenities (e.g. maintenance)
Discussions with RWA members and community leaders	7	Three interviews and discussions took place with RWA members on the BSUP sites of Vadaj, Rakhial and Bag-e-Firdosh. Four interviews and discussions took place with community leaders and community members
Discussions with NGOs	4	Two discussions took place with the head of Mahila Housing Sewa Trust (MHT) and NGO staff on BSUP sites at Rakhial, Bag-e-Firdosh and Vatva 1. Another two discussions took place with the head of SAATH and other NGO staff
Discussions with government officials	6	Discussions took place with officials of the Housing and Slum networking project department, a city engineer, an assistant city engineer (BSUP projects), a technical supervisor, an assistant city engineer (BSUP finance department), an AMC senior town planner and an engineer from the Affordable Housing Mission, Government of Gujarat
Discussions with academics	2	Associate professor, Faculty of Planning, CEPT University, Ahmedabad Anthropologist, PhD candidate, University of Jyvaskla, Finland

TABLE 14.4

Number of Households Surveyed at Each BSUP Site, October 2015

BSUP Site	Households Surveyed, 2011	Households Surveyed, 2015	Reason for Missing Households, 2015				Total Missing Households
			Rented Out to Others[a]	Migrated to Other Locations[a]	Not Willing to Participate in Interview[a]	Not Available for Interview[a]	
Isanpur	17	17	0	0	0	0	0
Vadaj	39	28	3	3	2	2	10
Rakhial	42	36	2	1	1	2	6
Bag-e-Firdosh	46	42	3	1	0	0	4
Odhav	13	9	2	2	0	0	4
Vatva 1	51	42	5	2	2	0	9
Vatva 2	61	48	3	5	4	1	13
Total	**269**	**222**	**18**	**14**	**9**	**5**	**47**

[a] Missing households.*

* Indicates 38 households missing out of 269 households which were not traced back during the 2015 survey and nine households that were not willing to participate in the fieldwork survey.

TABLE 14.5

Proportion of Households on Each BSUP Site Travelling Various Distances to School, 2011 and 2015

BSUP Site	<2 km (%)		2 to 4 km (%)		4 to 10 km (%)		>10 km (%)	
	2011	2015	2011	2015	2011	2015	2011	2015
Isanpur	92	50	–	20	8	30	–	–
Vadaj	90	81	7	14	–	5	3	–
Rakhial	83	59	9	24	8	14	–	3
Bag-e-Firdosh	51	44	30	21	17	33	2	2
Odhav	20	50	20	25	–	–	60	25
Vatva 1	69	40	–	3	–	28	31	29
Vatva 2	80	58	15	24	5	18	–	–
Total	**75**	**56**	**12**	**19**	**6**	**19**	**7**	**6**

14.4.2 HOUSING AND BASIC SERVICES (WATER AND SANITATION)

Under the BSUP programme, households can obtain ownership after paying USD 1,044 (67,860 INR) to the AMC, with USD 120 (7,860 INR)* as a first instalment and the remaining USD 923 (60,000 INR) through loans from a bank (according to AMC). However, we found that four years after resettlement, none of the surveyed households formally owned the house they occupy. Many had paid only USD 50 (3,260 INR) for the first instalment, with only a few from Vatva and Vadaj paying the first instalment in full.

At the time of the 2015 fieldwork, it was not clear how and when the resettled households should pay the remaining amount to AMC, nor were the consequences of not doing so known. We found

* In 2015, currency value was calculated based on an exchange rate of 1USD = INR 65 in October 2015, when the fieldwork was conducted (http://www.xe.com/currencytables/?from=INR&date=2015–10-08).

TABLE 14.6

Proportion of Households on Each BSUP Site and Distances to Markets, 2015

BSUP site	<2 km (%)	2–4 km (%)	4–10 km (%)	>10 km (%)
Isanpur	12	44	44	–
Vadaj	51	3	41	5
Rakhial	92	3	5	–
Bag-e-Firdosh	100	–	–	–
Odhav	–	100	–	–
Vatva 1	92	3	5	–
Vatva 2	98	–	2	–
Overall	**75**	**9**	**15**	**1**

that none of the households had started paying the outstanding debt to AMC. Furthermore, AMC had not identified any financial institutions where households could get a loan to pay the remaining amount. Most commercial banks will not issue loans without adequate documentation of the house, ownership documents and evidence of creditworthiness. As a result, four years after resettlement, many households expressed their fear of yet another displacement if they are not able to pay the required amount. Such feelings of insecurity and uncertainty have been previously found to be important risks associated with DIDR processes (Patel et al., 2015).

Table 14.7 shows that resettlement has significantly improved the quality of housing and provision of water and sanitation for most households: in 2015, 83% had a permanent abode with an individual water connection and toilet. These conditions were significantly better when compared to slum areas. Surprisingly, the level of access to water and sanitation has dropped slightly since 2011 from 88% to 83% in 2015, suggesting that some households were unable to maintain such basic infrastructure.

14.4.3 JOBLESSNESS

Table 14.8 shows that the percentage of workers experiencing unemployment increased from 12% in 2011 to 14% in 2015. Households in BSUP sites at Odhav, Vadaj and Vatva 1 and 2 were most badly affected by unemployment: it increased from 18% to 40%, with households whose livelihood depended on day-labour for their income suffering most. A lack of skills development programmes in resettlement sites and the inability to find suitable work were the main reasons given for the increase in unemployment since 2011.

During the household interviews, we found that households that had changed their occupation after resettlement were more vulnerable to employment loss. In Vatva 1, one interviewee told us

TABLE 14.7

Proportion of Households with a Permanent Abode and Basic Services, Before Resettlement in 2011 and in 2015

Households	Permanent Abode	Water Connection	Toilet
Households in slum (< 2011)	8	32	17
Households in BSUP 2011	76	88	88
Households in BSUP 2015	83	83	83

TABLE 14.8

Proportion of Workers Experiencing Unemployment by BSUP Site, 2011–2015

BSUP Site	Proportion Unemployment 2011 (%)	Proportion Unemployment 2015 (%)
Isanpur	10	7
Vadaj	8	22
Rakhial	3	–
Bag-e-Firdosh	14	4
Odhav	18	40
Vatva 1	11	14
Vatva 2	20	25
Overall	**12**	**14**

that '*before I had a fast food stall at the law garden and was earning enough for my family. After resettlement, due to travel cost, I cannot travel more than 20 km to work every day, so now I am working in the surrounding area as a construction labourer and the work is not stable, especially during the monsoon.*'

The root cause of unemployment lies in the relocation of slum dwellers to resettlement sites without properly understanding their existing livelihood conditions (Patel et al., 2015). Even after resettlement (2011–2015), unemployment increased as many workers had difficulties in finding work in surrounding areas because of their lack of skills. Consequently, they often had to change their occupation.

To sustain their livelihoods, households were travelling more in 2015 than in 2011. Overall, distance travelled to work increased by 2%, from an average of 7.2 km in 2011 to 7.5 km in 2015 (see Table 14.9). The average distance to work decreased slightly for four BSUP sites (Isanpur, Rakhial, Odhav and Vatva 2) but increased in three (Vatva 1, Bag-e-Firdosh and Vadaj).

Table 14.9 also shows that many workers switched from public transport to private motorised transport. All but one BSUP site, Odhav, shows a decrease in the use of public transport for work-related trips, owing to insufficient and irregular provision of service. For two sites, Rakhial and Bag-e-Firdosh, the use of private motorised transport had increased by 2015 by 129% and 138%, respectively.

TABLE 14.9

Change in Distance to Work and Mode of Transportation by BSUP Site, 2011–2015

BSUP sites	Average Distance to Work, 2015 (km)	Change 2011–2015 (%)	Proportion of Work Travel Switch to Public Transport, 2015 (%)	Proportion of Work Travel Switch to Private Motorised Transport, 2015 (%)
Isanpur	7	–4	–20	0
Vadaj	6	4	0	–56
Rakhial	5	–3	–58	129
Bag-e-Firdosh	7	7	–47	138
Odhav	9	–3	9	–67
Vatva 1	10	8	–38	15
Vatva 2	8	–6	–27	0
Overall	**7.5**	**2**	**–32**	**13**

14.4.4   Access to Education and Health Facilities

New primary health centres and primary education centres had been constructed in all BSUP sites since 2011. However, not all were fully functioning in 2015 (see Table 14.10). In Isanpur, Odhav and Vatva 2, the primary health centres were found abandoned, and in Isanpur and Odhav the primary education centres were also not functioning. AMC claimed that they were unable to start these services in Isanpurn, Odhav and Vatva 2 due to a lack of medical and teaching staff. Some households in Vatva 2 and Isanpur use the abandoned buildings for illegal activities, such as gambling, drinking alcohol and storing their personal belongings.

Table 14.11 shows that the number of student dropouts decreased overall by 45% from 2011 to 2015; only in Vatva 1 did student dropouts increase – by 27%. These dropouts are mainly girls, and the reasons for dropping out given by households were concerns about safety and a lack of funds to send children to school. One interviewee mentioned that 'father doesn't want to send our daughter to school. The school is not safe for girls due to criminal activity happening at the school. We don't have the money to get admission to a private school'. Another interviewee remarked: 'After resettlement I lost my job. I would like to send my son to a school. To earn money for the family, he has to work so at the end of a day we can fulfil our basic needs'.

The average distance travelled to school increased from 2.3 km in 2011 to 3.5 km in 2015 due to a lack of higher education institutions in the surroundings of the BSUP sites. Table 14.12 shows that the cost of school travel increased by 192%, up from an average monthly travel expenditure of USD 3 in 2011 to USD 12 in 2015. Not only do students generally travel longer distances than in 2011, they also increasingly use more expensive motorised transport. The shift to private transport is particularly strong in Isanpur, Rakhial and Odhav, where access to public transport is poor and frequency of service is low.

Four years after resettlement, access to health facilities had not improved. Although several primary health centres had originally been constructed, only in two BSUP sites were these fully functioning (Vadaj and Vatva). In the other sites, the health centres were either only partially operational or a doctor visits once a week (Bag-e-Firdosh).

Figure 14.3 shows the accessibility of AMC multi-specialist hospitals from BSUP sites. The average distance is 5.7 km, but only three BSUP sites are well connected to these hospitals. These hospitals are run by AMC and provide free medical treatment to the urban poor, making the demand for services high.

14.4.5   Health Risks (Morbidity, Mortality and Food Security)

Table 14.13 shows that the proportion of relocated households with a Below Poverty Line (BPL) card increased from 14% in 2011 to 36% in 2015. Many households had a BPL card based on their previous residential address, which meant that they could not access fair price shops near their current place of residence. In 2015, we found that 56% of households had a social security card (Adhar card), providing access to subsidised cooking gas and other benefits. Households having neither a BPL nor an Adhar card have to buy cooking fuel and gas at market prices, which increases their monthly expenditures. For poor households, such situations may lead to reduced nutrition.

Table 14.14 shows that the condition of sewerage infrastructure had not improved much since 2011; although the overflowing problem reduced, the proportion of households reporting problems with blocked sewers had increased by 31%. This is an indicator of the quality of maintenance of this important infrastructure, which is vital for good hygiene and public health and is in need of improvement.

Table 14.15 shows that the frequency of solid waste collection varied for each BSUP site, both in 2011 and 2015. We observed on all sites that many households disposed of their solid waste in open spaces due to a lack of dustbins. In total, about half the households reported daily collection, while most others reported irregular collection. The sanitary conditions on many BSUP sites was

TABLE 14.10

Status of Community Facilities at BSUP Sites in 2015 in Comparison with Status* in 2011

BSUP Sites	Residential Welfare Association Formed	Primary Health Centre Constructed	Primary Health Centre Functioning	Primary Education Centre Constructed	Primary Education Centre Functioning	Primary School Centre Constructed	Primary School Centre Functioning
Isanpur	In progress	Yes	No/Mobile health van	Yes	No	No	No
Vadaj	Yes	Yes	Yes	Yes	Yes	No	No
Rakhial	Yes	Yes	Yes (doctor visits 2 h/day)	Yes	Yes	No	No
Bag-e-Firdosh	Yes	Yes	Yes (doctor visits once a week)	Yes	Yes	No	No
Odhav	In progress	Yes	No	Yes	No	No	No
Vatva 1	In progress	Yes	Yes	Yes	Yes	Yes/Common for all Vatva BSUP sites	Yes/Common for all Vatva BSUP sites
Vatva 2	No	Yes	No/Mobile health van	Yes	Yes	Yes/Common for all Vatva BSUP sites	Yes/Common for all Vatva BSUP sites

* Shaded boxes show a change in the status of community facilities from 2011 to 2015.

TABLE 14.11

School Dropouts in 2011 and 2015

BSUP Site	No. of Student Dropouts in 2011	No. of Student Dropouts in 2015	Change in Dropouts (%)
Isanpur	5	0	−100
Vadaj	6	6	0
Rakhial	11	4	−64
Bag-e-Firdosh	12	7	−42
Odhav	20	6	−70
Vatva 1	11	14	27
Vatva 2	2	0	−100
Overall	**67**	**37**	**−45**

therefore problematic, with the accumulation of solid waste increasing the risk of disease and undermining the visual quality of the neighbourhood.

For their portable water, all households have an underground bore well. In 2011, the quality of water was unfit for drinking and daily use (Patel et al., 2015). In 2015, only Vatva had a functional water distribution plant. The Odhav and Isanpur sites, located near industrial areas, had the worst water quality, with many households mentioning that typhoid, gastroenteritis and hepatitis were common diseases on these sites.

14.4.6 MARGINALIZATION

In 2011, displaced households experienced marginalization through the loss of economic and social capital during the displacement and resettlement process (Patel et al., 2015). By 2015, little improvement could be observed in their situation. Households resettled near relatively good neighbourhoods reported negative social interactions from the surrounding community. Many neighbouring communities refer to the BSUP housing as 'Ganda Vasvato' (slum area) or 'Maftiya para' (an area where people do not want to pay for housing and municipal services). During our fieldwork, an interviewee said that 'Whenever AMC workers come to distribute books or medicines for children, they treat us like beggars and AMC has just left us here to die'. Many households reported a lot of frustration about the loss of employment and lack of basic services (e.g., water and sanitation) on BSUP sites.

TABLE 14.12

Distance to School and School Travel Expenditures by BSUP Site, 2011 and 2015

BSUP sites	Average Distance to School in 2015 (km)	Change in Distance to School (%)	Proportion of Switch to Private Motorised Transport for School Travel (%)	Change in Expenditure for School Travel (%)
Isanpur	3	84	500	−29
Vadaj	2	79	0	10
Rakhial	3	75	700	61
Bag-e-Firdosh	4	48	29	88
Odhav	6	92	100	138
Vatva 1	7	51	−100	17
Vatva 2	3	102	−40	124
Overall	**3.5**	**57**	**38**	**192**

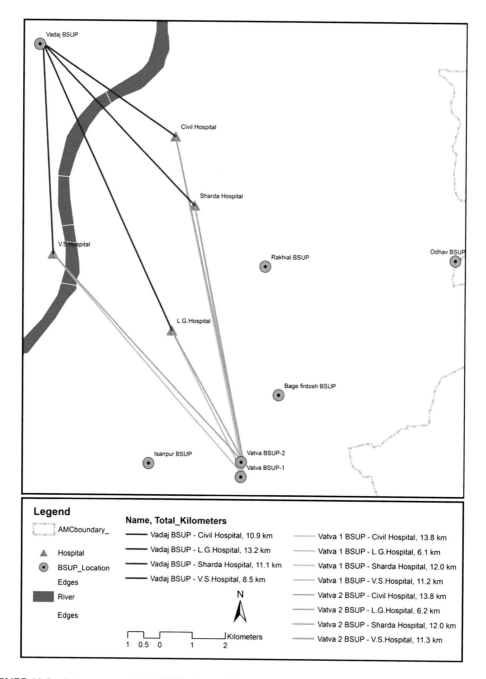

FIGURE 14.3   Distance matrix for BSUP sites and hospitals.

Table 14.16 shows that, overall, the proportion of households having access to formal financial services increased; only the proportion of households participating in community savings[*] schemes decreased slightly, from 6% to 4% in 2011 and 2015, respectively.

---

[*] Community savings is a small umbrella set up by NGOs in poor urban areas to strengthen their financial condition by organizing the saving of a small amount of money every month. The urban poor can get a loan at very low interest rates. Most community savings schemes are run by a women's group at each site. In Ahmedabad, Mahila Housing Trust is working to improve the urban poor's financial situation through community saving (d'Cruz & Mudimu, 2013; Obino, 2013).

TABLE 14.13

Changes in Access to BPL and Social Security Cards

BSUP Sites	Proportion with BPL Card in 2011 (%)	Proportion with BPL Card in 2015 (%)	Proportion with Adhar Card (Social Security card) (%)
Isanpur	0	52	94
Vadaj	3	31	82
Rakhial	24	36	89
Bag-e-Firdosh	7	26	88
Odhav	8	44	78
Vatva 1	18	45	81
Vatva 2	21	33	88
Overall	**14**	**36**	**56**

TABLE 14.14

Proportion of Households with Sanitation Problems by BSUP Site, 2011 and 2015

BSUP Site	Proportion with Sewerage Infrastructure Problems, 2011(%)			Proportion with Sewerage Infrastructure Problems, 2015 (%)		
	Over-flowing	Blocked	No issues	Over-flowing	Blocked	No issues
Isanpur	–	–	100	–	65	35
Vadaj	–		100		17	83
Rakhial	100	–	–	2	6	92
Bag-e-Firdosh	4	2	94	–	19	81
Odhav	92	8	–	–	67	33
Vatva 1	77	2	21	–	64	36
Vatva 2	11	2	87	–	19	81
Overall	**36**	**3**	**61**	**0.5**	**30.5**	**69**

TABLE 14.15

Solid Waste Management by BSUP Site, 2011 and 2015

BSUP Site	Solid Waste Management, Households in 2011 (%)			Solid Waste Management, Households in 2015 (%)		
	Every day	Irregular	No collection	Every day	Irregular	No collection
Isanpur	100	–	–	35	65	–
Vadaj	–	100	–	28	59	13
Rakhial	–	100	–	86	14	–
Bag-e-Firdosh	96	4	–	76	24	–
Odhav	No data	No data	No data	89	11	–
Vatva 1	–	100	–	40	60	–
Vatva 2	89	3	8	29	71	–
Overall	**46**	**52**	**2**	**52**	**46**	**2**

TABLE 14.16

Proportion of Households (%) with Access to Banking, Community Saving and Insurance Services, 2011 and 2015

BSUP site	Bank Account 2011	Bank Account 2015	Community Saving 2011	Community Saving 2015	Insurance 2011	Insurance 2015
Isanpur	41	47	0	0	6	2
Vadaj	41	66	3	17	14	28
Rakhial	17	42	2	3	3	17
Bag-e-Firdosh	49	81	7	2	20	34
Odhav	56	78	11	11	11	22
Vatva 1	43	50	5	2	21	17
Vatva 2	25	83	13	0	13	27
Overall	36	64	6	4	14	23

14.4.7 SOCIAL DISARTICULATION

Table 14.17 shows that 65% of households had a positive perception about their community composition in 2015. Despite this quite positive outcome, however, applications to transfer to other BSUP sites increased from 11% to 23% between 2011 and 2015, with as many as 89% of the surveyed households in Odhav indicating a desire to relocate. We observed that many community leaders try to resolve various community issues and were promoting the foundation of resident welfare associations (RWAs) within their communities. On some BSUP sites, households were able to form an RWA with help from NGOs and AMC. There were clear signs that where RWAs had not been formed, households had problems maintaining common property and community harmony.

In addition, households with apartments on the ground floor had started to encroach on common property for their personal use. We even found that some elderly households, with apartments on the uppermost floors, had started to encroach on common ground and build temporary shelters there to live in during periods of illness: their physical condition made them unable to access their allocated housing units. Frequent conflicts were reported between neighbours over encroachment of common

TABLE 14.17

Proportion of Households with a Positive Perception of their Community's Composition and Proportion Applying for Relocation, 2011 and 2015

BSUP sites	Positive Perception About Community Composition (%)		Applied for Transfer to Other BSUP Site (%)	
	2011	2015	2011	2015
Isanpur	18	59	12	12
Vadaj	8	48	–	7
Rakhial	5	62	7	3
Bag-e-Firdosh	2	85	15	24
Odhav	39	100	100	89
Vatva 1	31	86	10	33
Vatva 2	8	35	1.6	33
Overall	13	65	11	23

properties and disputes about maintenance funds for common property. Before relocation in 2011, slum dwellers had a very strong sense of community despite their belonging to different castes and living in slums. Between 2011 and 2015, households had been unable to create a similar sense of community at various BSUP sites owing to differences in beliefs, food habits and cultural practices among them. This situation was found to be worse at BSUP sites with mixed religious communities.

14.4.8   THE DYNAMICS OF IMPOVERISHMENT

Between 2011 and 2015, improvements in the lives of households were limited to certain aspects, such as provision of better infrastructure, decreasing student dropout rates, increased accessibility to a BPL card, access to bank accounts and insurances facilities and residence in an authorised housing settlement. Most households (60%) stated that after resettlement they were provided with good quality housing. However, increased risk of impoverishment was also observed, the result of a lack of social amenities and income opportunities at BSUP locations. For example, between 2011 and 2015, distances travelled for work and schooling had increased, in turn increasing monthly expenditures. Some workers also reported having to change their occupations due to inability to find employment in the surrounding area.

The absence or poor quality of basic institutions/facilities, such as RWAs, primary health centres and primary schools, contribute significantly to increased risk of impoverishment. In addition, inadequate access to water, sewerage and solid waste management have led to further deprivation. Our analysis did, however, also identify some improvements in Vadaj, Rakhial and Bag-e-Firdosh. Ultimately, changes in households' lives also depend upon the location of their BSUP site. Those households that were resettled near an industrial location have faced more challenges than others that were not. In addition, households resettled near industrial areas are more exposed to further impoverishment, such as joblessness, lack of access to community facilities, health risks and social disarticulation due to the relative isolation of the BSUP sites.

By 2015, four years after being resettled, 23% of households stated that their lives had become worse, and they voiced their frustration towards AMC for not helping them to reach a basic standard of living. Despite their problems, most households were, however, doing their best to cope with the location and available opportunities.

14.5   DISCUSSION

The basic aim of the BSUP programme is to improve the socio-economic conditions of slum dwellers' lives by improving the infrastructure in the settlements in which the urban poor live. After resettlement, households' lives should show signs of improvement and the ability to change their socio-economic situation. However, despite being provided with basic infrastructural facilities, many inhabitants of BSUP settlements still face many sorts of deprivation in their lives. Our study shows that structural improvements in the lives of the urban poor do not just depend on having access to basic infrastructure; it is also necessary to strengthen the socio-economic situation by creating conditions that provide a sustainable livelihood (Baud et al., 2008, 2009; Paulo et al., 2007; Richards and Thomson, 1984; Sufaira, 2013; World Bank, 2007). Some resettled households experienced a deterioration of conditions due to the lack of a supporting social and economic environment at their relocation site (Patel and Mandhyan, 2014; Perlman, 2010). Resettled households often face increased impoverishment. It is very difficult for them to cope with their new location and its surroundings and adjust to the many challenges, which affect all household members in different ways.

Even after four years, many households experienced high levels of uncertainty about their future, which contributed to their impoverishment. Many resettled households still face difficulties in paying the remaining beneficiary share for their BSUP unit. Due to a lack of access to financial institutions and an unclear policy on monthly instalments, many resettled households have not started to pay off their remaining debt. This is significant, as many resettled households are unwilling to contribute to the maintenance of common infrastructural measures because they fear being displaced in the future, which would mean a loss of scarce personal capital.

Our study shows that 12% of the surveyed households had migrated to other parts of the city due to their inability to cope on their designated BSUP site. In total, 23% of surveyed households want to shift to other BSUP sites in the belief that livelihood opportunities would be better there. For all BSUP sites, many households were forced to change their occupation because of increased travel expenditures. Though we found that households resettled near industrial areas had more income opportunities, increased travelling distances to schools, health services and markets led to an increase in their monthly expenditure. In the case of Odhav and Vatva 2, we found that unemployment had increased despite being located near heavy industrial areas, and in Vadaj, which is located near a residential area, unemployment rose to 40% in 2015, up from 8% in 2011. Many resettled households in Vadaj have had to change their occupations due to their new surroundings. Such issues contribute to feelings of uncertainty and stress.

BSUP housing policy has a clear remit to provide basic social amenities such as common ground, primary health centres, primary education centres and pump station rooms for the water supply system. Although AMC constructed the infrastructure for social amenities on all BSUP sites, only in Vadaj and Rakhial are they fully functioning. Their lack at other BSUP sites means long travelling distances and higher expenditure to access distant social amenities.

Another critical issue for successful relocation is a strong RWA. However, by 2015 only Vadaj, Rakhial and Bag-e-Firdosh had RWAs. The absence of an RWA reduces a community's ability to build social capital and to maintain common infrastructure. The BSUP sites at Vadaj and Rakhial had very active and well-developed communities due to their RWAs. At other BSUP sites, resettled households depend, instead, on community leaders to resolve issue. On most sites, we observed a lack of awareness about the roles and responsibilities of RWAs.

Resettled households were positive about their dwellings and the presence of basic infrastructure, but they were not satisfied with the quality of infrastructure provided and its maintenance. In Vadaj, which AMC presents as a model BSUP site, one community leader mentioned that '*We know we have a good house and access to basic infrastructure. After resettlement, AMC are not willing to pay for our deteriorating houses and common infrastructures. We do not have any idea who is going to pay for repair and maintenance. We do not want to invest our money because we do not have legal documents for the houses.*'

After more than four years, households only have a very fuzzy, uncertain perception about their future, owing to AMC's unclear policies and deficient implementation. Residents have a very negative perception of AMC for not helping to improve their livelihoods after resettlement. These households are also unhappy with the defunct RWAs, the poor condition of their settlements' infrastructure and safety concerns at sites located near industrial areas. According to a newspaper report, 'EWS (economic weaker section) households mention that slum rehab needs jobs, not a new place' (Yagnik, 2016).

Table 14.18 summarises the impoverishment risks we observed. In general, the study indicates that resettled households continue to experience a high risk of impoverishment. Mostly, resettled households are affected by landlessness, joblessness, restricted access to infrastructural facilities and social disarticulation. The increased travelling distances for many aspects of life is a major burden for low-income households.

14.6   CONCLUSIONS

For our study, we revisited resettled households in Ahmedabad to analyse whether impoverishment had increased or improved between 2011 and 2015. In general, BSUP policy has focused on improving slum dwellers' lives. However, four years after resettlement, no BSUP site shows a fully positive

TABLE 14.18

Summary of Changes in Parameters of Impoverishment Risk, 2011–2015

Impoverishment Form	Factors Affecting Resettled Households in 2015	Effect of Factor in 2015 (-ve/+ve)
Landlessness	Distance to school	-ve
	Distance to market	-ve
	Access to public transport	-ve
Joblessness	Unemployment	-ve
	Average distance to work	-ve
	Travel to work on public transport	-ve
	Travel to work on private transport	-ve
	Monthly income	+ve
	Monthly expenditures	-ve
Homelessness	Uncertainty about eviction	-ve
	Migration to another location	-ve
	Perception related to quality of housing	+ve
Marginalization	Access to banking facilities	+ve
	Community savings scheme	-ve
	Access to insurance	+ve
Health risks (food security, morbidity, and mortality = health risk interpreted as combination of two categories)	Access to BPL card	+ve
	Access to fair price shop	-ve
	Access to social security card	+ve
	Sewerage network	+ve
	Solid waste management	+ve
	Quality of water	-ve
	Access to primary and higher-order health facilities	-ve
Loss of access to common facilities	Access to social amenities, RWAs	-ve
	Access to social amenities, health	-ve
	Access to social amenities, primary education	-ve
	School dropout rates	+ve
	Average distance to school	-ve
	School travel shifts to private or motorised mode of transport	-ve
	travel expenditure	-ve
	Access to hospital in emergency	-ve
Social disarticulation	Perception of community composition	+ve
	Applications for transfer to other BSUP sites	-ve
	Households' perception of resettlement-related improvement	0

+ve sign shows positive effect on resettled households' lives in 2015; -ve sign shows negative effect on resettled households' lives in 2015; and 0 sign shows no effect on resettled households' lives in 2015.

transformation into a well-serviced, functional and authorised housing area. Because of the poor socio-economic environment at BSUP sites, many aspects of the lives of resettled households had worsened by 2015. This is a worrying observation in view of the large scale and high ambitions of BSUP policy.

We noticed that RWAs and NGOs could have an important role to improve resettled households' lives. The BSUP sites where RWAs had been formed were able to provide basic social amenities such as health services and community spaces at their sites with the help of NGOs. We observed that these BSUP sites had relatively less social disarticulation issues than other BSUP sites. However, at the level of individual households, RWAs were not improving access to economic opportunities. Even for NGOs, it was difficult to implement social development programmes and build a sense of community without the presence of RWAs. The social fabric became more vulnerable at BSUP sites between 2011 and 2015. Households were positive about having a permanent dwelling and access to basic infrastructural facilities such as water supply, sewerage and solid waste management. However, they were negative about the quality of public services and the deteriorating infrastructure on all BSUP sites. In many respects, lives of relocated households deteriorated due to a lack of a stable livelihood.

Many of the risks of impoverishment are spatial in nature. For example, BSUP sites located near residential areas are less impoverished than other BSUP sites. The lack of local job opportunities, insecure livelihoods and poor access to public transportation are major factors contributing to further impoverishment. Increased distances to all public services make monthly expenditures higher. Households are also affected by the nature of surrounding neighbourhoods. Those resettled near residential areas face fewer impoverishment risks than those near industrial areas. Location also affects a household's ability to maintain employment and generate income over time. Furthermore, we found that elderly people, children, women and day-labourers are most affected by the lack of hospitals, primary school and public transportation at BSUP sites and their surroundings. To ensure the success of BSUP resettlements, it is crucial to pay specific attention to the most vulnerable members of households.

REFERENCES

Alexandrescu, F. (2013). Mediated risks: The Roşia Montană displacement and a new perspective on the IRR model. *Canadian Journal of Development Studies/Revue canadienne d'études du développement, 34*(4), 498–517. doi:10.1080/02255189.2013.845549

Baud, I. S. A., Sridharan, N., & Pfeffer, K. (2008). Mapping urban poverty for local governance in an Indian mega-city: The case of Delhi. *Urban Studies, 45*(7), 1385–1412. doi:10.1177/0042098008090679

Baud, I. S. A., Pfeffer, K., Sridharan, N., & Nainan, N. (2009). Matching deprivation mapping to urban governance in three Indian mega-cities. *Habitat International, 33*(4), 365–377. doi:10.1016/j.habitatint.2008.10.024

Cernea, M. (1997a). *African involuntary population resettlement in a global context* (Vol. 45). Environment Department working papers; no. 45. Social assessment series*Social Development papers; no. SDP 18. Washington, DC: World Bank. http://documents.worldbank.org/curated/en/495981468767432101/pdf/multi-page.pdf

Cernea, M. (1997b). The risks and reconstruction model for resettling displaced populations. *World Development, 25*(10), 1569–1587. doi:10.1016/S0305-750X(97)00054-5

Cernea, M. (2008). Compensation and benefit sharing: Why resettlement policies and practices must be reformed. *Water Science and Engineering, 1*(1), 89–120. doi:10.1016/S1674-2370(15)30021-1

d'Cruz, C., & Mudimu, P. (2013). Community savings that mobilize federations, build women's leadership and support slum upgrading. *Environment and Urbanization, 25*(1), 31–45. doi:10.1177/0956247812471616

Desai, R. (2014). *Municipal politics, court sympathy and housing rights: A post-mortem of displacement and resettlement under the Sabarmati Riverfront Project, Ahmedabad.* CUE Working Paper 23, CEPT, Ahmedabad, India.

Heggelund, G. (2006). Resettlement programmes and environmental capacity in the three Gorges Dam Project. *Development and Change, 37*(1), 179–199. doi:10.1111/j.0012-155X.2006.00474.x

Hooper, M., & Ortolano, L. (2012). Confronting urban displacement: Social movement participation and post-eviction resettlement success in Dar es Salaam, Tanzania. *Journal of Planning Education and Research, 32*(3), 278–288. doi:10.1177/0739456X12439066

Jackson, S., & Sleigh, A. (2000). Resettlement for China's Three Gorges Dam: Socio-economic impact and institutional tensions. *Communist and Post-Communist Studies, 33*(2), 223–241. doi:10.1016/S0967-067X(00)00005-2

Kaida, N., & Miah, T. M. (2015). Rural–urban perspectives on impoverishment risks in development-induced involuntary resettlement in Bangladesh. *Habitat International, 50*, 73–79. doi:10.1016/j.habitatint.2015.08.008

Muggah, H. C. R. (2000). Conflict-induced displacement and involuntary resettlement in Colombia: Putting Cernea's IRLR model to the test. *Disasters, 24*(3), 198–216. doi:10.1111/1467-7717.00142

Obino, F. (2013). *Housing finance for poor working women: Innovations of the self-employed women's association in India.* Retrieved from Manchester, UK: http://www.wiego.org/sites/default/files/publications/files/Obino-Home-Based-Workers-India-SEWA-Housing-WIEGO-PB14.pdf

Patel, S., & Mandhyan, R. (2014). Impoverishment assessment of slum dwellers after off-site and on-site resettlement: A case of Indore. *Commonwealth Journal of Local Governance, 15*(July), 104–127. doi:10.5130/cjlg.v0i0.4065

Patel, S., Sliuzas, R., & Mathur, N. (2015). The risk of impoverishment in urban development-induced displacement and resettlement in Ahmedabad. *Environment and Urbanization, 27*(1), 231–256. doi:10.1177/0956247815569128

Paulo, M., Rosário, C., & Tvedten, I. (2007). *'Xiculungo' – social relations of urban poverty in Maputo, Mozambique.* Retrieved from Bergen, Norway: https://www.cmi.no/publications/2930-xiculungo-social-relations-of-urban-poverty-maputo

Perlman, J. (2010). *Favela: Four decades of living on the edge in Rio de Janeiro.* New York: Oxford University Press.

Quetulio-Navarra, M., Niehof, A., Van der Horst, H., & van der Vaart, W. (2014). Short-term risk experience of involuntary resettled households in the Philippines and Indonesia. *Habitat International, 41*, 165–175. doi:10.1016/j.habitatint.2013.07.013

Richards, P. J., & Thomson, A. M. (Eds.). (1984). *Basic needs and the urban poor: The provision of communal services.* Sydney, Australia: Routledge.

Robinson, W. C. (2003). *Risks and rights: The causes, consequences, and challenges of development-induced displacement.* Retrieved from Washington, DC, USA:

Sufaira, C. (2013). Socio economic conditions of urban slum dwellers in Kannur municipality. *Journal of Humanities and Social Science, 10*(5), 12–24.

UN-Habitat. (2003). *The challenge of slums – global report on human settlements* (E. P. Ltd Ed.). London, UK: Earthscan.

UN-Habitat. (2015). *Habitat III Issue Papers: Informal settlements* (Vol. 22). New York, USA.

World Bank. (2007). *Bangladesh Dhaka: Improving living conditions for the urban poor.* Retrieved from Washington, DC, USA: http://documents.worldbank.org/curated/en/587231468007834055/Bangladesh-Dhaka-Improving-living-conditions-for-the-urban-poor

Yagnik, B. (2016). Ahmedabad let's get smart: Slum rehab needs jobs, not new place. *Times of India.* Retrieved from https://timesofindia.indiatimes.com/city/ahmedabad/Ahmedabad-lets-get-smart-Slum-rehab-needs-jobs-not-new-place/articleshow/50861298.cms

15 Planning for Transit Oriented Development (TOD) Using a TOD Index

*Yamini Jain Singh, Johannes Flacke, Mark Zuidgeest,
and Martin van Maarseveen*

CONTENTS

15.1 Introduction 267
15.2 Methodology 269
    15.2.1 Identification of Indicators...271
    15.2.2 Computing Indicators and TOD Indices..272
        15.2.2.1 Actual TOD Index 272
        15.2.2.2 Potential TOD Index 274
15.3 Results..277
15.4 Inferences and Recommendations 278
    15.4.1 Actual TOD Index 278
    15.4.2 Potential TOD Index...279
15.5 Conclusions...280
References..280

15.1 INTRODUCTION

Transit-oriented development (TOD) has various definitions, including those given by Calthorpe (1993), Dittmar and Poticha (2004), Boarnet and Crane (1997), Schlossberg and Brown (2004), Centre for Transit-Oriented Development (2009) and many more. Most commonly, TOD is defined as a relatively high-density form of urban development that features a balanced mix of land uses and a walkable or cyclable urban environment that encourages people to walk, cycle or use various public forms of transit instead of their cars. As such, the TOD concept is an approach to sustainable development that hinges on the integration of land use and transport.

TOD plans, which primarily aim at bringing about a modal shift from transit by car to forms of transit with a lesser carbon footprint, have been made on a wide range of scales, varying from train and bus stations, and urban areas through to entire regions; see, for example, case studies found in Schlossberg and Brown (2004), Balz and Schrijnen (2009), Cascetta and Pagliara (2009), Yang and Lew (2009), Hoffman (2006), McKone (2010) and Cervero and Murakami (2009). Typically, these case studies also discuss and evaluate TOD plans, including their success or failure. The discussions in these studies tend to be very qualitative in nature, with little quantitative analysis involved. We argue in this chapter that quantitative analyses not only offer the means with which to compare different approaches and findings but also help in establishing measurable performance thresholds that can be set as goals for better performance. To aid this quantitative evaluation, a number of indicators for evaluating TOD plans have been proposed by Renne (2009), Evans and Pratt (2007) and Belzer and Autler (2002). In addition to evaluating TOD projects, Evans and Pratt (2007) emphasised the need to be able to assess and measure the degree to which a development fulfils TOD criteria. To accomplish this objective, we propose to develop a 'TOD index' that quantifies key TOD characteristics and combines them into a score of a TOD index. An index can help in assessing existing

TOD characteristics and in identifying reasons for poor TOD performance, so that specific policies, programmes and fiscal interventions can be made to improve TOD performance. Such a TOD index, when calculated over different points in time, can also help to gauge whether an area is moving towards meeting its aspirational (TOD) threshold values or slipping back. To plan more effectively for TOD or to evaluate TOD projects, a TOD index could, therefore, be a powerful tool.

In this chapter we develop a TOD index to facilitate planning for transit-oriented development, elaborate our methodology and discuss its application in a case study. We argue that planning for TOD must not be restricted to just making any development more transit-oriented: it should also include providing transit in places where development has already been highly transit-oriented, yet access to transit is still lacking. Hence, our approach towards planning for TOD has the following two objectives to fulfil (Singh et al., 2014):

1. To improve TOD conditions at locations where transit connectivity is available but where conditions for TOD are poor.
2. To improve transit connectivity at locations or in areas where conditions for TOD exist but access to transit connectivity is either absent or poor.

The TOD index elaborated in this chapter provides solid ground for achieving the above planning objectives. For the first objective, TOD needs to be measured around existing transit nodes and recommendations made at that scale to improve TOD conditions. For the second objective, TOD needs to be measured over an entire region such that areas in the region can be identified where TOD conditions are good but provision of transit connectivity is poor (Singh et al., 2014). When measuring TOD at a regional scale, TOD is characterised by typical urban development characteristics, such as urban densities, land-use diversity and others. However, when measuring for TOD at a local scale around an existing transit node, TOD is also affected by transit-related conditions such as frequency of service, travel comfort and node access, as these aspects influence people's decisions to use transit. Thus, for our first objective, development and transit-related characteristics need to be measured, while for the second objective, only development-related TOD characteristics need to be measured. Since we measure different aspects of TOD for each of these objectives, one index is not enough. We therefore developed two TOD indices, one for each objective: an 'Actual TOD index' and a 'Potential TOD index'. The Actual TOD index measures TOD around existing transit nodes, and the Potential TOD index measures TOD over an entire region, to identify areas that are potential locations for transit connectivity (Singh et al., 2014). These indices differ in terms of where they are measured and what they measure. Since TOD is an inherently spatial concept, its planning and evaluation needs to be done using spatial analyses, so both indices are measured using Multiple Criteria Analysis (MCA) or a Spatial MCA (SMCA) tool, as applicable.

BOX 15.1   Case Study Area

Arnhem-Nijmegen

At the time of the study (2012–2014), the City Region **Arnhem–Nijmegen**, located in the Province of Gelderland, was the third largest of the eight city regions in the Netherlands. The City Region covered more than 1,000 km² and was home to the two large cities – Arnhem–Nijmegen. Total population of the City Region was about 735,000 in 2012, of which about 40% was residing in its two main cities, Arnhem and Nijmegen. The region had 20 municipalities and was served by a rail-based national and regional transit system. Regional and urban trains and a neighbourhood bus system also operated within the Region. There were 21 train stations in the

City Region at the time of our research; a couple more having been added since, but they are not included in this study.

To stimulate a modal shift from cars to transit, the City Region wanted to make public transport as competitive and attractive as private modes of transport and encourage its use for regional travel. It saw TOD as a powerful tool that could help in achieving this. The City Region Arnhem–Nijmegen was therefore a natural choice as a case study for investigating the use of TOD indices for planning for TOD.

We chose the City Region Arnhem–Nijmegen (the Netherlands) as a case study for this work because its vision for sustainable growth involved a modal shift from cars to transit. The City Region was spread over 1,000 km², includes 20 municipalities and had a population of around 735,000, 40% of which resides in its main cities of Arnhem and Nijmegen. The Region is striving to become the second-most important region in the Netherlands in economic terms, after the Randstad region. Figure 15.1 shows the City Region and its location within the Netherlands.

15.2 METHODOLOGY

BOX 15.2 Methods Applied in the Chapter

This chapter presents a methodology for calculating TOD indices using a combination of quantitative GIS methods and statistical analyses. Specifically, the GIS tools used for this study included ArcGIS™ 10.1 and the (Spatial) Multiple Criteria Assessment (MCA) platform in ILWIS™ (ITC, 2007). While ArcGIS was used to analyse individual indicators that contribute to TOD indices, ILWIS was used to calculate TOD index values. For the study, a variety of spatial and non-spatial data was required, which was collected from the City Region, Statistics Netherlands (CBS), the website of transit service, ESRI Top 10NL™ and Open Street Map (OSM)™. Administrative boundaries, the rail network, rail/bus station locations, road networks and land-use data (including building footprints) were all available as spatial data layers in vector format. No primary data was collected for this study, but site visits were made to, for example, train stations, to check some data and information collected through secondary sources. The methodology adopted to produce TOD indices is logical, straightforward, easy to implement and highly transferable to different geographies. If used before and after implementation of TOD-related plans, the indices can be used to decide whether an area is moving towards meeting its aspirational (TOD) threshold values or slipping back.

The methodologies adopted for the calculation of Actual and Potential TOD indices are similar; it is in their interpretation that they differ. For the calculation of both indices, first an area of analysis needs to be defined for which each index will be calculated. Each index is measured at a different scale. The Actual TOD index must be measured around an existing transit node (in accordance with various definitions described in the literature), extending outwards to what is considered the walkable limit from the node. According to various references (Calthorpe, 1993; Schlossberg and Brown, 2004; CTOD, 2009; Cervero and Murakami, 2009; City of Calgary, 2004), this distance can range from 400 m to 800 m. For the particular case of the Netherlands, Molster and Schuit (2013) have defined a comfortable ten-minute walking distance to be 800 m. Accordingly, we defined the area

FIGURE 15.1   The City Region Arnhem–Nijmegen (the Netherlands).

to be considered for transit-oriented development as that within a radius of 800 m of a transit node, and all indicators for the Actual TOD index were measured in this area.

The Potential TOD index needs to be measured over an entire region to see how index values vary from one location to another and whether there are some areas where levels of transit-oriented development are already high. Thus, for this index we divided the City Region into a number of grid cells such that a (potential) TOD index could be calculated for each grid cell. By applying tessellation of space, different grids of 100 m × 100 m, 200 m × 200 m, 300 m × 300 m and 500 m × 500 m were created. Ultimately, the 300 m × 300 m grid was chosen because a smaller cell size had a considerably negative influence on computational performance, while a larger size could lead to a loss of accuracy. Laying this grid over the City Region's 1,000 km² resulted in more than 12,000 grid cells for which a Potential TOD index was calculated.

It is also important to identify the transit system and nodes that we considered in this case study. The City Region is served by a rail-based national and a regional transit service, which also includes inter-urban and intra-urban bus services. There is discussion in the literature (Newman and Kenworthy, 2007; Newman, 2009) that for TOD to be successful, a high-quality transit service is required that has a high-traffic volume, higher speeds than cars and is not affected by traffic conditions on the roads. Only rail-based transit services and Bus Rapid Transit (BRT) offer such advantages (Hale and Charles, 2006; Hoffman, 2006; McKone, 2010; Newman and Kenworthy, 2007), as regular buses can be slower than cars: they share the same road space as cars and are thus vulnerable to adverse traffic conditions.

Transit in the City Region Arnhem–Nijmegen is rail-based with 21 train stations providing connections. A BRT system was in its early planning stages and thus could not be studied, so indicators for the Actual TOD index calculations were measured in the area around each of the 21 train stations. The Potential TOD index is independent of the location of transit nodes, so indicators had to be measured for all grid cells. However, once our calculations were complete, we were able to identify cells with high Potential TOD index values for which access to rail transit was at that time nevertheless low.

15.2.1 Identification of Indicators

For both indices, relevant indicators needed to be identified for measurement. There are some typical physical characteristics relating to urban development that define transit-oriented development: high urban densities, high levels of land-use diversity, a walkable environment and attractive urban design, to name a few. Thus, to measure TOD one must measure development-related indicators based on the 3Ds-concept – Density, Diversity and (urban) Design – as discussed by Cervero and Murakami (2009), since these factors are expected to affect travel demand and mode choice (Cervero and Kockelman, 1997), as well as reducing peak-hour congestion levels (Zhang, 2010). Since transit is core to TOD, it is also important to measure those characteristics of transit that affect travel behaviour. These could include the frequency of transit services, access to transit services and to opportunities provided by them, user-friendliness of the system, travel comfort and so on. Transit characteristics can, however, only be measured around existing transit nodes, so they were measured along with the urban development characteristics being measured for the Actual TOD index. By contrast, the Potential TOD index is measured on the basis of grid cells, independently of the location of transit nodes, which is why only development characteristics need to be measured for this index. This helps to identify areas that, with development, would be highly suitable for transit. This will help in identifying those areas where the development has better TOD characteristics. Later, the level of access of these areas to high quality transit is assessed.

Various indicators proposed by Schlossberg and Brown (2004), Renne (2009), Evans and Pratt (2007), Belzer and Autler (2002) and others were also studied in detail to finalise a list of quantifiable TOD indicators for the measurement of the urban development and transit characteristics. These indicators can be grouped according to broad criteria, as shown in Table 15.1. The indicators are not specific to our case study and can be used to measure TOD around any transit node in any city or country.

For our list of indicators for both indices, we needed a variety of spatial and non-spatial data, which we acquired from the City Region's authorities, Statistics Netherlands (CBS), ESRI TOP 10 NL, Open Street Map (OSM) and the website of transit services. The spatial data collected were all in vector format and minor issues relating to incomplete map coverage, inconsistent administrative boundaries, conflicting land-use classifications and different map projections were addressed by comparison of the data amongst each other and with the help of a third-party reference such as Google Maps. Although we were able to collect data for most of the indicators, there were some that were deleted from both lists for want of data: specifically, indicators for tax earnings of municipalities and employment levels. The indicator 'basic amenities at station' considered the presence of waiting rooms, benches to sit on, elevators and/or access for the disabled and ticket machines at the stations. Since all these amenities were found to be present at all stations, this indicator was dropped as it did not contribute to TOD comparisons for any stations. 'Safety' at a station can be measured in a number of ways. We based our assessment on the presence of lighting and the presence of other people at a station. Since all stations were found to be properly lit day and night, we focused on the number of people at stations, since the presence of 'many eyes' in an area adds to the feeling of being safe (Jacobs, 1961). However, we had neither the data for this indicator nor the resources to collect the data ourselves. We therefore measured safety using 'number of shops/

eateries/takeaways' as a proxy for presence of other people. In the absence of these shops, fellow passengers can be expected to arrive only shortly before departure and leave immediately after arrival, reducing numbers of people around a station.

15.2.2 Computing Indicators and TOD Indices

Broadly speaking, the calculation of our two indices involves four steps: calculation of individual indicators; their standardization to account for differing units of measurement; weighting; and, finally, calculation of index values in MCA/SMCA (as applicable).

15.2.2.1 Actual TOD Index

For this index, we measured a total of 18 indicators related to urban development and transit (see Table 15.1), some of which are spatial in nature and others non-spatial. The development-related indicators were spatial and were measured in ESRI ArcGIS software using the statistical data on various urban densities, the City Region's administrative boundaries, land-use data and road network data. While the computation of the indicators is quite straightforward, we show below the equations used to calculate land-use diversity and mix of land use. These two indicators sound very similar but are quite different. Diversity of land use is a measure of how diverse the mix of all land uses is in an area and was calculated using the entropy index (Cheng et al., 2013):

$$LU_d(i) = \frac{-\sum_i Q_{lui} \times \ln(Q_{lui})}{\ln(n)} \tag{15.1}$$

$$\sum_i Q_{lui} \times \ln(Q_{lui}) = Q_{ai} \times \ln(Q_{ai}) + Q_{bi} \times \ln(Q_{bi}) + \ldots + Q_{ni} \times \ln(Q_{ni}) \tag{15.2}$$

$$Q_{lui} = \frac{S_{lui}}{S_i} \tag{15.3}$$

where:
lui = land-use class within the analysis window i
S_{lui} = total area of the specific land use within the analysis window i
S_i = total area of the analysis window i
n = total number of land use classes within the analysis window i

A high degree of land use diversity is desirable for TOD as this helps in utilizing off-peak directions/hours of transit capacity (City of Calgary, 2004), as well as adding to the liveliness of a place. The mix of land use, on the other hand, measures how well residential land use is mixed with other land uses in an area and is important because a greater mix means that residents can walk or cycle when undertaking non–work-related trips (Cervero and Kockelman, 1997; Thorne-Lyman et al., 2011; Zhang and Guindon, 2006). Thus, this indicator is used to measure the walkability or cyclability of an area and is calculated using the formula (Zhang and Guindon, 2006):

$$MI(i) = \frac{\sum_{\cap i} S_c}{\sum_{\cap i} (S_c + S_r)} \forall i \tag{15.4}$$

where:
i = area of analysis
Sc = sum of the total area under non-residential urban land uses within i
Sr = sum of the total area under residential land use within i.

TABLE 15.1
Criteria and Indicators for Actual and Potential TOD Indices

No.	Criteria	Indicators for Actual TOD Index	Indicators for Potential TOD Index
1.	Area around transit node should have transit-supportive densities.	Population density (No. persons/km²)	Population density (No. persons/km²)
		Commercial density [a] (No. commercial enterprises/km²)	Commercial density (No. commercial enterprises/km²)
			Employment density [c] (No. employees/km²)
2.	Land-use diversity is essential for effective utilization of transit in off-peak hours.	Land-use diversity	Land-use diversity
3.	Area around transit node should be walkable and cyclable.	Mix of residential land use with other land uses	Mix of residential land use with other land uses
		Total length of walkable/cyclable paths (km)	Total length of walkable/cyclable paths (km)
		Intersection density (No. of intersections/km²)	Intersection density (No. intersections/km²)
		Impeded pedestrian catchment area (IPCA)	
4.	Greater economic development in area around transit leads to more transit-oriented development.	Density of business establishments [b] (number of business establishments/km²)	Density of business establishments (number of business establishments/km²)
		Tax earnings of municipalities in previous year	Tax earnings of municipalities in previous year
		Employment levels	Employment levels
5.	Transit system's capacity should be utilised to optimal levels.	Passenger load at peak hours	n.a.
		Passenger load at off-peak hours	
6.	Transit node should be user-friendly and visually attractive.	Safety of commuters at transit stop	n.a.
		Basic amenities	
		Information display systems	
7.	Transit nodes should be accessible for passengers and provide good accessibility to a range of destinations.	Frequency of transit service (No. trains operating/hr)	n.a.
		Interchange to different routes of same transit mode (No. of routes)	
		Interchange to other transit modes	
		Access to job opportunities within walkable distance of transit node (No. of jobs) ('cumulative opportunities' or 'contour measure' (Geurs and Van Wee, 2004))	
8.	Transit nodes should provide optimum parking supply for different modes.	Parking supply–demand for cars/four-wheeled vehicles	n.a.

[a, b] To avoid double counting of same data in 'commercial density' and 'density of business establishments', 'commercial establishments' were counted for 'commercial density' and 'non-commercial establishments' were counted as business establishments.

[c] Employment density was not measured for Actual TOD index because the same data were used to measure 'access to job opportunities around the station' and we did not want to double-count the number of jobs.

n.a. = not applicable.

The transit-related indicators are non-spatial and were calculated accordingly. A combined measurement of these indicators for each train station had to be done to arrive at the value of the Actual TOD index; this was done using MCA in ILWIS software (ITC, 2007). MCA allows computation of the index using indicators that have different units of measurement by standardizing them. In the standardization process, the highest value of an indicator among all stations was fixed as 1 and, proportionately, all other indicator values were standardised. After the standardization process, the effect of each indicator's value on the final value of the TOD index was also specified. For example, an increase in densities listed in Table 15.1 should lead to an increase in the Actual TOD index value: this relationship between the indicator and the TOD index was specified for each indicator. Most of the indicators have a similar directly proportional relationship with the TOD index value, except for mix of land use. For this indicator, the directly proportional relationship remains until a mid-value of 0.5 is reached, which implies a balanced mix of residential use with other land uses. Beyond this value (0.5), any increase in an indicator's value decreases the Actual TOD index value because it implies a less-balanced mix of uses. Four more indicators – i.e. passenger load during peak hours and off-peak hours, and parking utilization by four-wheeled vehicles and cycles – also exhibit a similar relationship with TOD index values. For these indicators, the relationship remains directly proportional up to a value of 0.9, i.e. 90% utilization of transit capacity or parking supply, after which the relationship becomes inversely proportional. That is because a utilization level of 90% or more can be quite uncomfortable for users and signals the need for additional capacity.

The next stage in MCA requires weighting of indicators, and for this step we held a workshop in the City Region's offices, with the heads of all 20 municipalities invited to attend. They were chosen to participate because they are the decision-makers for a variety of planning policies within their municipalities, as well as advising officials of the City Region on regional planning policies. For other case studies, however, the group(s) of stakeholders chosen could vary considerably. Before the workshop, all participants were sent information about the project and the purpose of the workshop. During the workshop, the interim results of our study were presented and the focus of the workshop was discussed and explained. A questionnaire (also available online; Singh, 2013) was then physically distributed so that the participants could rank criteria and indicators based on perceptions of their importance for realization of transit-oriented development. After the workshop, the ranks of criteria and indicators were compiled using the Borda Count method (Reilly, 2002) and input into ILWIS, where ranks were converted into weights using a rank–sum method (ITC, 2007). These weights (see Table 15.2) were used in the final calculations of the Actual TOD index.

15.2.2.2 Potential TOD Index

For the Potential TOD index, we measured eight indicators for the more than 12,000 grid cells that cover the City Region. All indicators for this index are spatial in nature and were, therefore, calculated with ArcGIS software using the statistical data on various urban densities, administrative boundaries and land-use and road network data. The formulae used for computing land-use diversity and mix of land use were the same as for Equations 15.1 and 15.3, although there was a difference in the way they were calculated. As mentioned before, the size of a grid cell is 300 m × 300 m, so this size covers about 9 ha of area, which is not big enough to assess land-use diversity nor mix of land use. For this reason, we used a 'window of analysis' around each grid cell that covered an area bound by the eight cells surrounding the 'central' cell. The indicators for Potential TOD were calculated over this area, although the result was assigned to the central grid cell only. This ensures that these indicators also consider land use surrounding the grid cell, which is a better measure.

As a result of measuring the indicators, a map was produced for each indicator and combined into a TOD index. The maps were input into ILWIS for SMCA. For this, too, all indicators were standardised and the relationship of each indicator with the Potential TOD index was specified in the same manner as for the Actual TOD index. During the workshop held to obtain the weights of criteria and indicators, the municipal heads were also requested to rank criteria and indicators for the Potential TOD index. The ranks were compiled and input into ILWIS, where they were

TABLE 15.2

Weights of Criteria and Indicators for Actual and Potential TOD Indices

No.	Weights for Actual TOD Index				Weights for Potential TOD Index			
	Criteria	Weights	Indicators	Weights	Criteria	Weights	Indicators	Weights
1.	Area around transit node should have transit-supportive urban densities	0.15	Population density Commercial density	0.67 0.33	Area around transit node should have transit-supportive urban densities	0.3	Population density Commercial density Employment density	0.5 0.33 0.17
2.	Land-use diversity is essential for effective utilization of transit	0.03	Land-use diversity	1	Land-use diversity is essential for effective utilization of transit	0.2	Land-use diversity	1.0
3.	Area around transit node should be walkable and cyclable	0.06	Mix of residential land use with other land uses Total length of walkable/cyclable paths Intersection density Impeded Pedestrian catchment area (IPCA)	0.1 0.4 0.2 0.3	Area around transit node should be walkable and cyclable	0.1	Mix of residential land use with other land uses Total length of walkable/cyclable paths Intersection density	0.17 0.50 0.33
4.	Higher economic development around transit area leads to more transit-oriented development	0.22	Density of business establishments	1	Higher economic development around transit area leads to more transit-oriented development	0.4	Density of business establishments	1.0
5.	Transit system's capacity should be utilised at optimal levels	0.19	Passenger load at peak hours Passenger load at off-peak hours	0.67 0.33	n.a.	n.a.	n.a.	n.a.
6.	Transit node should be user-friendly and visually attractive	0.11	Safety of commuters at transit stop Information display systems	0.5 0.5				

(Continued)

TABLE 15.2 (CONTINUED)
Weights of Criteria and Indicators for Actual and Potential TOD Indices

| | | Weights for Actual TOD Index | | | Weights for Potential TOD Index | | |
No.	Criteria	Weights	Indicators	Weights	Criteria	Weights	Indicators	Weights
7.	Transit nodes should be accessible to passengers and provide good accessibility to a range of destinations	0.15	Frequency of transit service	0.4				
			Interchange to different routes of same transit mode	0.3				
			Interchange to other transit modes	0.2				
			Access to job opportunities within walkable distance from transit station	0.1				
8.	Transit nodes should provide optimum parking supply for different modes	0.08	Parking supply–demand for cars/four-wheeled vehicles	0.67				
			Parking supply–demand for cycles	0.33				

converted to weights that are tabulated in Table 15.2. The indicators were then weighted to arrive at a combined Potential TOD index value for each grid cell in the City Region.

15.3 RESULTS

The Actual TOD index values for all 21 railway stations in the City Region Arnhem–Nijmegen are shown in Figure 15.2, for which, on a scale to 1, the maximum index value is 0.76 for Arnhem railway station and minimum index value is 0.16 for Wolfheze railway station. Since Arnhem and Nijmegen are the biggest urban centres in the region, it was to be expected that they would have high TOD scores. Similarly, the quiet villages of Zetten-Andelst, Molenhoek-Mook and others like them were expected to score very low. Thus, the predictive performance of the Actual TOD index is well demonstrated.

We subjected our results to a sensitivity analysis to make sure they were robust enough. For sensitivity analysis, we varied the weights of each criterion one at a time by $\pm 10\%$ (Malczewski, 1999) and assessed its effect on the final TOD index values.

For the Actual TOD index there were eight criteria, so 16 scenarios of weights were created by changing each criterion by $\pm 10\%$. For the sensitivity analysis, the Actual TOD index was recalculated for all 16 scenarios: the top five ranking stations remained the same and the lowest and highest Actual TOD index values in the region ranged from 0.152 to 0.162 and from 0.76 to 0.77, respectively, which makes them close to the original values. Thus, the Actual TOD index was confirmed to be robust and reliable. The inferences that can be drawn from the results of the Actual TOD index are discussed in next section.

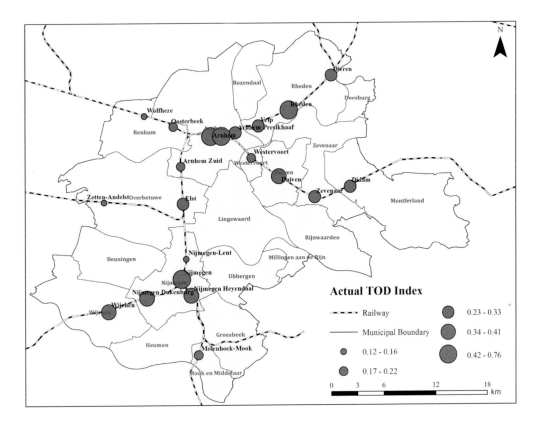

FIGURE 15.2 Actual TOD index values for 21 train stations in the City Region Arnhem–Nijmegen.

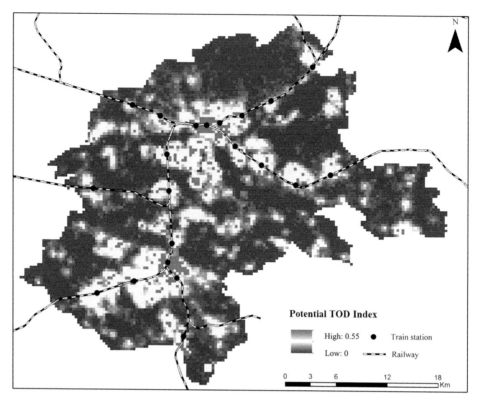

FIGURE 15.3   Potential TOD index values throughout the City Region Arnhem–Nijmegen.

The Potential TOD index map for the City Region Arnhem–Nijmegen is shown in Figure 15.3; the maximum value in the region was 0.55 on a scale of 0 to 1. Potential TOD index values are generally low throughout the region: the mean index value for the region is only 0.044. Similar to Actual TOD index results, the Potential TOD index shows the highest TOD scores around the biggest stations of the Arnhem–Nijmegen region, which matches the pattern expected for this index.

A sensitivity analysis was also carried out for the Potential TOD index. We measured four criteria for this index, so eight scenarios of weights were created by changing weights of each criterion by ± 10%. Potential TOD index values were recalculated for all eight scenarios for the sensitivity analysis. The results showed a marginal change in the spread of Potential TOD index values: the maximum index value only ranged from 0.55 to 0.56, with the mean index value also ranging from 0.0432 to just 0.0461. Thus, the results of Potential TOD index were confirmed to be robust.

15.4   INFERENCES AND RECOMMENDATIONS

High TOD index values in an area imply that transit is a core concern that needs to be taken into account when planning for development. Proper inferences are required to be able to make recommendations for effective TOD planning.

15.4.1   ACTUAL TOD INDEX

On a scale of 1, the highest value of the Actual TOD index was 0.76, which seems like a good score. However, as this study is the first of its kind, we do not have any benchmarks or reference values to infer in any other way. Even if there were other, similar case-studies, their results might not be comparable to ours since each case has different characteristics: their weights would differ, making a whole lot of difference in the results. It was expected that at the regional scale Actual TOD scores would be highest for Arnhem and Nijmegen's railway stations and that scores would become gradually lower as distance from these cities increased, just as the degree of urbanization declines with distance. However, no such pattern is visible in Figure 15.3, except for areas to the south of Nijmegen station. At the regional level, planners may like to first work towards improving the TOD levels at neighbouring stations to Arnhem and Nijmegen, where the scores drop abruptly. Yet even these top-ranking stations have some scope for improvement of TOD levels, and this may be investigated further.

To understand the reasons behind a good or a bad score in terms of index values, we created web charts for all stations. The web charts of the three best scoring stations – Arnhem, Arnhem Velperpoort, Nijmegen – are shown in Figure 15.4. For Arnhem Velperpoort station, for example, even though the Actual TOD index value is one of the highest in the region, it clearly scores very low on user-friendliness, the urban densities access to and from the station and parking utilization. Low parking utilization indicates the need to redirect investment in any extra supply of parking to other uses that support transit while still keeping enough supply to accommodate existing and forecast parking demand. At Arnhem station, user-friendliness is highest of all stations, but it can still be improved. The web charts make clear which criteria need to be improved upon to enhance conditions for TOD around each station. Planners and decision-makers could take this information into account before making specific planning proposals on, for example, how to improve urban densities or land-use diversity around stations.

15.4.2 POTENTIAL TOD INDEX

Like the Actual TOD index, there are no benchmarks or reference values to guide us on whether a Potential TOD index value of 0.55 is high or not. It can, however, be safely inferred that overall the region's scores are low since its mean index value is 0.0447, with nearly 86% of grid cells scoring below 0.30 and 40% grid cells scoring below 0.15. Figure 15.3 shows the spread of Potential TOD index values over the City Region, but this does not help in identifying TOD hot-spots. Thus, spatial statistical analyses were needed to carve out clusters or hot-spots of high TOD index values. Global Moran's *I* statistic confirmed that statistically significant clustering of similar values is possible within the region. Subsequently, statistical analysis using Anselin Local Moran's *I* was carried out to identify clusters or hot-spots of similarly high TOD index values. Of these hotspots, those within 800 m of existing train stations were omitted since these could be expected to have access to transit and have also been studied when Actual TOD index values were being calculated. The remaining hot spots were those areas that had high Potential TOD index values but at that time poor access to high-quality transit modes; see Figure 15.5. These areas can be considered by the planners for better transit connectivity. We would not recommend connecting these areas by rail since it is a highly cost-intensive mode. Rather, the more cost-effective option of BRT could be implemented for these areas. Conversion of existing regular bus routes to BRT routes would also be an option worth considering.

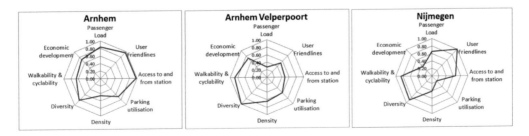

FIGURE 15.4   Web charts of indicators of the Actual TOD index for the three highest-scoring stations in the City Region Arnhem–Nijmegen.

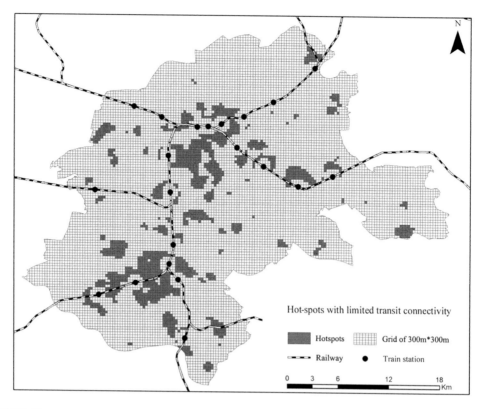

FIGURE 15.5   Hot-spots that have high Potential TOD index scores and poor access to transit.

15.5   CONCLUSIONS

We demonstrate in this chapter how a TOD index can be a useful tool in pursuing the TOD planning objectives we set out early in this chapter. The methodology adopted to calculate TOD indices is logical, straightforward and easy to implement, which, together with its high degree of suitability across different geographies, are the strengths of this approach. With the sort of results presented here, urban planners are able to make more accurate planning proposals, knowing what exactly is missing around each transit node and what can be changed to improve conditions for TOD. The ambiguity of qualitative assessments can thus be avoided by replacing them with quantitative assessments.

Once TOD plans have been drawn up and implemented, the same index can be used to measure the conditions for TOD later and so gauge the success of TOD initiatives. Identification of hot-spots in a region also helps to highlight areas that would benefit from better access to high quality transit. For the first time in a case study, the views of decision-makers have also been incorporated in defining and measuring TOD. TOD index results from other cases would also be extremely helpful to urban planners when making specific TOD planning proposals at regional and local scales. Application of a TOD index to different contexts would further improve this tool, perhaps leading to ready-to-implement assessment and evaluation guidelines that would help in optimizing planning of TOD.

REFERENCES

Balz, V., & Schrijnen, J. (2009). From Concept to Projects: Stedenbaan, the Netherlands. In C. Curtis, J. L. Renne & L. Bertolini (Eds.), *Transit Oriented Development: Making It Happen* (pp. 75–90). Ashgate e-Book.

Belzer, D., & Autler, G. (2002). *Transit Oriented Development: Moving from Rhetoric to Reality*. The Brookings Institution Center on Urban and Metropolitan Policy and The Great American Station Foundation.

Boarnet, M., & Crane, R. (1997). L.A. Story – A Reality Check for Transit-Based Housing. *Journal of the American Planning Association, 63*(2), 189–204.

Calthorpe, P. (1993). *The Next American Metropolis – Ecology, Community and the American Dream*. Canada: Princeton Architectural Press.

Cascetta, E., & Pagliara, F. (2009). Rail Friendly Transport and Land-Use Policies: The Case of the Regional Metro System of Naples and Campania. In C. Curtis, J. L. Renne & L. Bertolini (Eds.), *Transit Oriented Development: Making It Happen* (pp. 49–64). Ashgate e-Book.

Cervero, R., & Kockelman, K. (1997). Travel Demand and the 3Ds: Density, Diversity, and Design. *Transportation Research Part D: Transport and Environment, 2*(3), 199–219. doi:http://dx.doi.org/10.1016/S1361–9209(97)00009-6

Cervero, R., & Murakami, J. (2009). Rail and Property Development in Hong Kong: Experiences and Extensions. *Urban Studies, 46*(10), 2019–2043. doi:10.1177/0042098009339431

Cheng, J., Bertolini, L., Clercq, F. l., & Kapoen, L. (2013). Understanding Urban Networks: Comparing a Node, a Density and an Accessibility-Based View. *CITIES – The International Journal of Urban Policy and Planning, 31*, 165–176.

CTOD. (2009). Why Transit-Oriented Development and Why Now? Reconnecting America – Centre for Transit Oriented Development (CTOD).

Dittmar, H., & Poticha, S. (2004). Defining Transit-Oriented Development: The New Reginal Building Block. In H. Dittmar & G. Ohland (Eds.), *The New Transit Town – Best Practices in Transit Oriented Development* (pp. 19–39). Island Press.

Evans, J. E., & Pratt, R. H. (2007). Transit Oriented Development. *Transit Cooperative Research Program (TCRP) Report 95: Traveler Response to Transportation System Changes Handbook* (3 ed., pp. 17(11)–17(138)). U.S.A.: Transport Research Board of the National Academies.

Geurs, K. T., & Van Wee, B. (2004). Accessibility Evaluation of Land-Use and Transport Strategies: Review and Research Directions. *Journal of Transport Geography, 12*, 127–140.

Hale, C., & Charles, P. (2006). *Making the Most of Transit Oriented Development Opportunities*. Paper presented at the 29th Australasian Transport Research Forum Gold Coast, Queensland.

Hoffman, A. (2006). *Transit: Can it Sustain TOD?* The Forum at Redwood City. Retrieved from http://www.missiongrouponline.com/sitebuildercontent/sitebuilderfiles/HoffmanTODpt2a.pdf www.missiongrouponline.com

ITC. (2007). ILWIS 3.3 User Manual: International Institute for Geo-Information Science and Earth Observation, Enschede, the Netherlands. Retrieved from http://www.itc.nl/ilwis

Jacobs, J. (1961). *The Death and Life of Great American Cities*. New York: Random House.

Malczewski, J. (1999). *GIS and Multicriteria Decision Analysis*. United States of America: John Wiley & Sons, Inc.

McKone, J. (2010). Cities in Focus: Curitiba. Retrieved Nov. 15, 2011, from http://thecityfix.com/blog/cities-in-focus-curitiba/?utm_source=feedburner&utm_medium=feed&utm_campaign=Feed%3A+thecityfix%2Fposts+%28TheCityFix%29

Molster, A., & Schuit, S. (2013). *Voetsporen Rond het Station*. Paper presented at the Het Nationaal Verkeerskunde Congres, Netherlands.

Newman, P. (2009). Planning for Transit Oriented Development: Strategic Principles. In C. Curtis, J. L. Renne & L. Bertolini (Eds.), *Transit Oriented Development: Making it Happen* (pp. 13–22). Ashgate e-Book.

Newman, P., & Kenworthy, J. (2007). Greening Urban Transportation. In L. Starke (Ed.), *State of the World 2007: Our Urban Future (Digital Edition)* (pp. 66–89). New York etc.: W.W. Norton & Company.

Reilly, B. (2002). Social Choice in the South Seas: Electoral Innovation and the Borda Count in the Pacific Island Countries. *International Political Science Review, 23*(4), 355–372. doi:10.1177/0192512102023004002

Renne, J. L. (2009). Measuring the Success of Transit Oriented Development. In C. Curtis, J. L. Renne & L. Bertolini (Eds.), *Transit Oriented Development: Making it Happen* (pp. 241–257). Ashgate e-Book.

Schlossberg, M., & Brown, N. (2004). Comparing Transit-Oriented Development Sites by Walkability Indicators. *Transportation Research Record: Journal of the Transportation Research Board* (1887), 34–42.

Singh, Y. J. (2013). Ranking the Stars! Prioriteren van criteria en indicatoren voor twee TOD indices (gebiedsbreed en OV knooppunt specifiek). Retrieved 2013, 2013 from https://www.surveymonkey.com/s/TODonderzoek

Singh, Y. J., Fard, P., Zuidgeest, M. H. P., Brussel, M., & van Maarseveen, M. F. A. M. (2014). Measuring Transit Oriented Development: A Spatial Multi-Criteria Assessment Approach for the City Region Arnhem and Nijmegen. *Journal of Transport Geography, 35*(0), 130–143. doi:http://dx.doi.org/10.1016/j.jtrangeo.2014.01.014

Singh, Y. J., He, P., Flacke, J., & van Maarseveen, M. F. A. M. (2014). *Measuring Transit Oriented Development over a Region Using an Index.* Paper presented at the 12th Annual Transport Practitioners' Meeting London, U.K.

The City of Calgary. (2004). Transit-Oriented Development: Best Practices Handbook. The City of Calgary.

Thorne-Lyman, A., Wood, J., Zimbabwe, S., Belzer, D., Breznau, S., Fogarty, N., Brennan, T., Tumlin, J., & Yake, C. (2011). Transit-Oriented Development Strategic Plan/Metro TOD Program. Metro TOD Program of Portland, Oregon.

Yang, P. P., & Lew, S. H. (2009). An Asian Model of TOD: The Planning Integration in Singapore. In C. Curtis, J. L. Renne & L. Bertolini (Eds.), *Transit Oriented Development: Making it Happen* (pp. 91–108). Ashgate e-Book.

Zhang, M. (2010). Can Transit-Oriented Development Reduce Peak-Hour Congestion? *Transportation Research Record: Journal of the Transport Research Board* (2174), 148–155. doi:10.3141/2174-19

Zhang, Y., & Guindon, B. (2006). Using Satellite Remote Sensing to Survey Transport-related Urban Sustainability. Part 1: Methodologies for Indicator Quantification. *International Journal of Applied Earth Observation and Geoinformation, 8*, 149–164.

16 Performance, Preferences and Policies in Urban Water Supply in Yogyakarta, Indonesia

Yohannes Kinskij Boedihardja, Mark Brussel,
Frans van den Bosch, and Anna Grigolon

CONTENTS

16.1 Introduction 283
16.2 Literature Review 284
    16.2.1 Perception and Behaviour of Water Users...............................284
    16.2.2 Stated Choice Experiments..285
16.3 Water Supply in Yogyakarta...285
16.4 Methodology...286
16.5 Stated Choice Experiment on Water Provision Choice 289
16.6 Results...290
16.7 Implications of Findings and Recommendations 292
References..294

16.1 INTRODUCTION

Many cities in the Global South struggle to provide a continuous and reliable supply of good quality, potable water to all inhabitants. In most cases, a variety of infrastructural systems co-exist, based on different sources of water supply and with different characteristics and levels of service. Common sources of water are piped supply systems, local (private or communal) wells, local streams and water vendors.

In urban water supply, piped water systems are often the most preferred type of delivery as they are usually associated with higher levels of safety and reliability. Nevertheless, piped water-supply systems face a number of challenges, such as provision to the whole population within a city, minimizing the gap between coverage and service, enhancing water quality and preserving water supplies for current and future requirements (Alegre et al., 2006). If these challenges are not met, problems such as an ineffective network (leakage), failing continuity of supply and poor quality of water may arise (Whittington et al., 2002).

Other systems commonly found also have their disadvantages. Groundwater from wells, from which the water is extracted using pumps, often has problems with pollution from human faecal material (Pang et al., 2004), especially in highly urbanised areas. Besides, at some point ground-water levels and the quantity that can be extracted will gradually fall below renewable levels. The option of supply through private water vendors, i.e. distribution of water by trucks to neighbour-hoods, is unattractive because of very high costs and a lack of regulation of water quality.

Indonesia, as a developing country, faces all these problems. In most Indonesian cities, the local PDAM (*Perusahaan Daerah Air Minum*, or Piped Water Supply Company) provides drinking water. Large portions of the population, however, do not use these services, and for a variety of reasons depend on groundwater sources instead.

BOX 16.1   Case Study Area

Yogyakarta is located in central Java (Indonesia), at the heart of D.I. Yogyakarta Province, and is also referred to as the Special Region of Yogyakarta. Yogyakarta city has 14 districts and 45 urban villages, together covering an area of 32.5 km². In 2008, it had a population of around 412,000 inhabitants (BPS Indonesia, 2015).

Based on discussions with experts from the University of Gadjah Mada, we decided to study three urban villages in the district Tegalrejo, selected for their variation in socio-demographic characteristics and types of water supply, thus allowing us to capture variation in preferences. For our study, we selected the urban villages of Bener, Kricak and Karangwaru (Figure 16.1).

Competition between the use of groundwater and piped water for water supply is complex. Both are physically and economically communicating vessels: reducing use of one source leads to an increased use of the other and an increase in spending on that source. From a public goods perspective, it is important that the number of users of the local PDAM increases, to increase the chances of survival of the piped service and to create a more sustainable water supply system. Competition between water sources forms the context of this chapter.

Our objective was to evaluate the choice behaviour of users of different types of water supply in Yogyakarta (Indonesia) and so gain understanding of links between supply characteristics, consumer choice behaviour and supply performance of different sources, as well as to estimate the potential influence of policy measures on water-supply choice. To this end, a stated choice experiment was conducted to understand water-supply choice in relation to characteristics and performance attributes of piped water and groundwater.

16.2   LITERATURE REVIEW

16.2.1   PERCEPTION AND BEHAVIOUR OF WATER USERS

The perception of water users is important for determining the behaviour of people in their choice of water supply. This is affected by several factors, including risk perception, attitudes towards chemical content, familiarity with specific water properties, trust in suppliers, past problems attributed to water quality and information provided by the mass media and interpersonal sources (Doria, 2010). In addition, Rahut et al. (2015) mentioned that affordability affects people's choice and behaviour, and that wealthier households have better access to and can afford the installation and monthly rates of piped water, as opposed to poorer households. In addition, affordability, quantity and continuity are also determinants when choosing a mode of water supply.

Water quality is also a crucial factor in choice of water supply. Doria et al. (2009) state that perception of water quality, especially for drinking water, is influenced by the water's colour, odour and flavour. Furthermore, perception of chemical characteristics (lead, chlorine and hardness) is a major variable in determining water safety. By taking perceptions of chemical characteristics into account, people decide if water is safe to be consumed or not (Doria et al., 2009).

16.2.2 STATED CHOICE EXPERIMENTS

To analyse the connection between supply-system performance and the behaviour of water users when choosing a mode of water supply, we developed a stated choice experiment. Aimed at understanding and predicting an individual's preferences and choices, stated choice methods in the form of discrete choice experiments (DCEs) are being increasingly used to identify and evaluate the relative importance of aspects of decision-making. According to Hensher (1994), a stated choice experiment is generally classed as behavioural research aiming to identify respondents' behaviour towards their choice for a given combination of attributes. Stated choice data have several advantages: for instance, they provide reliable estimates of the relative importance of each attribute; they enable the testing of new products or attribute levels that do not currently exist; and they enrich the choice model by easily matching choice behaviour with socio-demographics. The main drawback with this type of experiment is that it is notorious for its complexity, and results are sometimes not realistic if choice sets are not comprehensive and exhaustive enough (Hensher et al., 2005).

Stated choice experiments have been used in various disciplines, such as transport (e.g. Louviere and Hensher, 1982) and marketing (e.g. Louviere and Woodworth, 1983), due to the need to predict demand for new products that were potentially expensive to produce. To the best of our knowledge, applications in water research are more recent and scarcer.

MacDonald et al. (2005) measured customer service standards for urban water in Adelaide, Australia. Attributes used in their experiment were continuity, connection interruptions per month, information about connection interruptions, alternative water supply and annual water bill. It was known that price sensitivity was an important component affecting user perceptions. In this research, users tended to choose a combination that benefited them based on their annual water bill.

In Kenya, Brouwer et al. (2015) compared rural and urban respondents' willingness to pay for improved drinking-water quality. Four attributes were used: flow rate, storage capacity, level of diarrhoea prevalence and price. Comparison of urban and rural responses indicated that respondents from both urban and rural areas were mostly concerned with the water quality in relation to prevalence of diarrhoea. Willingness to pay was also an important driver in their choices. Respondents living in urban areas were found to have a higher willingness to pay for improved water quality than respondents living in rural areas, who were more price-sensitive, mainly due to lower household incomes. This may also be due to the variety of sources of water available. In their study, Brouwer et al. (2015) found that most respondents living in urban areas had tap water only, whereas most of the respondents living in rural areas collected water both from streams and protected or unprotected springs and wells.

16.3 WATER SUPPLY IN YOGYAKARTA

Yogyakarta city is situated on the island of Java (Indonesia) and is located at the heart of D.I. Yogyakarta Province; it is also referred to as the Special Region of Yogyakarta. Yogyakarta city consists of 14 districts and 45 urban villages, together covering an area of 32.5 km². In 2008, it had a population of around 412,000 inhabitants (BPS Indonesia, 2015).

In Yogyakarta city, piped water is managed by PDAM (Regional Water Company Tirtamarta). Data from BPS Indonesia (2015) shows that PDAM Tirtamarta had 33,871 households as consumers out of a total of 144,137 households in the city; the rest (thus, 77%) was served by wells/springs. Of the total 15.6 million m³ of water distributed each year, 70% was sold to consumers, while the remaining 30% was lost due to leakage. As these figures make clear, in this situation it is very difficult to operate a feasible piped-water supply system, as all parts of the city are connected to the piped network but consumer use is limited and leakage is very high.

To fulfil its annual water production targets, PDAM Tirtamarta draws water from four sources: springs, shallow wells, deep wells and surface water (Pemkot Yogyakarta, 2001). Before being distributed, all water from wells and rivers undergoes treatment for aeration, coagulation, flocculation, filtration and chlorination to ensure the quality of supply.

Different rates apply according to type of building use: social, non-commercial, commercial, industrial and special. Rates also differ according to levels of water consumption, which is checked monthly by PDAM Tirtamarta (2014).

The process of water provision from groundwater wells in Yogyakarta follows a different procedure. First, well users need to acquire a well permit, the application procedures for which are laid down in Regulation No. 28, Year 2013. According to Pemkot Yogyakarta (2013), this regulation not only gives information on how to apply for a well permit, but also describes the drilling and monitoring procedures. When the permit is issued by the licensing bureau, a field survey needs to be conducted as part of the drilling procedure. If the location is found to be suitable, a licensed driller will construct the well according to the instructions provided by the licensing bureau.

Large-scale drilling of wells has had a significant impact on groundwater levels: the water table falls by 20–30 cm annually (Jakarta Post, 2014). If this situation continues, the groundwater aquifers beneath Yogyakarta will soon be depleted, with possible land subsidence following. Recently, the Indonesian government has tried to remedy this situation by suspending the '2004 Law on Water Resource' that allowed participation of private equity in water provision activities. The 2004 law was not compliant with the National Statutes of 1945, which stipulate that all national resources should be managed only by the state. As a consequence, water distribution and revenue collection should only be conducted by PDAM (Jakarta Post, 2015). Under these new conditions, the first step to be taken in applying the policy is that PDAM will cooperate with ESDM (Ministry of Energy and Minerals) at the provincial level to regulate the construction and use of groundwater wells and to enhance the current supply of piped water.

In addition to the issues related to water quantity discussed above, water quality is also an important factor in determining water supply choice. Private wells in Yogyakarta are usually sunk to different depths, depending on individual wishes, the individual's financial resources and local conditions of the soil and aquifer. Although most of these private wells are technically sound, the quality of groundwater obtained can be substandard because of aquifer pollution, particularly in areas with poor sanitation practices. Salendu (2010) reported that 93 groundwater samples collected during a study in the Yogyakarta area at well depths varying from 5 m to 30 m showed very high levels of *E. coli*, pointing to faecal contamination and indicating that water taken from these wells was not fit for direct consumption. In addition, for those water users who have the possibility to use either piped water or groundwater, the majority preferred groundwater, mainly because they preferred its smell and taste over the chlorinated piped water.

16.4 METHODOLOGY

BOX 16.2   Methods Applied in the Chapter

This chapter describes a choice model for water supply with regards to the two main types of water supply provision found in Yogyakarta: piped water and well water. The spatial distribution of water users was visualised using ArcGIS™, and cluster sampling was applied to select households for taking part in the experiment. A discrete choice model was used to gain insight into which attributes respondents consider important when choosing a mode of water provision; Biogeme™ software (Bierlaire, 2003) was used for model estimation. The primary data set used for the analysis came from a stated choice experiment conducted among 500 households during the fieldwork. The experiment required respondents to choose between piped or well water, the characteristics of which varied in terms of price, quality and continuity of supply.

The experiment incorporated nine choice profiles, each with three choice options. Secondary data collected consisted of maps of the area and reports on drinking water supply in the city.

Choice modelling is a method that has been applied successfully in areas such as market analysis, health services and transportation. The method does have several limitations, however: it is a complex method, and its results are sometimes unrealistic if the choice sets provided are not sufficiently well understood by respondents, or if they lack the knowledge/experience to be able to appreciate the choice options provided, or if the choice options provided are not exhaustive enough.

Three urban villages were chosen for the study, based on discussions with experts from Gadjah Mada University in Yogyakarta (see Figure 16.1). Our intention was to select areas with enough variation in socio-demographic characteristics and types of water provision for us to capture variation in preferences. At first, the experts suggested urban villages from Umbulharjo and Tegalrejo Districts as study areas. Both districts have variation in socio-demographic characteristics and in types of water provision. However, since Umbulharjo District is bigger than and farther away than

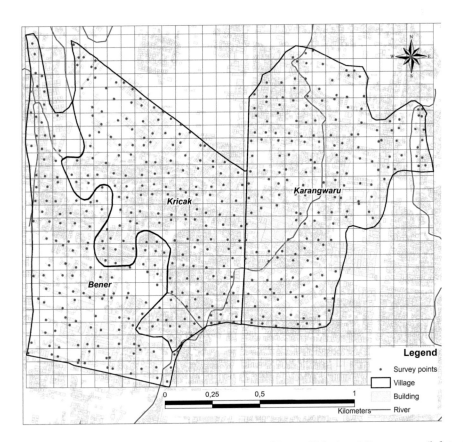

FIGURE 16.1 Cluster sampling in the study area , villages of Bener, Kricak and Karangwaru (left to right).

Tegalrejo District, which would lead to inefficiencies in time and cost, experts finally suggested that we select urban villages within Tegalrejo District for our study. As indicated in BPS Indonesia (2015), Tegalrejo is a district comprising several urban villages with differing socio-economic characteristics. We selected the villages of Bener, Kricak and Karangwaru for the study. Households were selected using a cluster sampling method (Kumar, 2005; Martinez, 2015). Cluster sampling is a random sampling method, meaning that the choice of any one respondent is not influenced by the choice of other respondents, since they are considered to be independent from one another. Moreover, Kumar (2005) mentions that random/probability sampling has advantages compared to other methods, such as: samples can represent the total sampling population and the collected random sample can be subjected to statistical tests.

According to Martinez (2015), the first step in cluster sampling that uses satellite images is to divide the satellite images into grid cells. The grid cells can be selected using simple random or systematic sampling. Then, all dwellings within those grids are numbered and selected at random. In the last step, individual households are selected from each dwelling unit. In our study, satellite images derived from Martinez (2015) were used. All buildings within the three selected urban villages were included and 500 households were selected at random (Figure 16.1).

Table 16.1 describes the sample of households. It shows that most households consisted of four or more persons, at least one of whom had an occupation. Almost half of those sampled reported

TABLE 16.1
Sample Profile

Sampling Variables	Classes	Proportion (%)
Number of persons in household	1	1.2
	2	7.6
	3	19.8
	4	39
	More than 4	32.4
Number of persons with an occupation	1	46.28
	2	38.43
	More than 2	15.29
Head of household's occupation	No occupation	3.4
	Self employed	46.6
	Military/police	3.6
	Government officer	20.2
	Other	26.2
Head of household's education level	No formal education	0.8
	Primary school	3
	Junior high school	10.2
	Senior high school	39
	University	47
Own/rent house	Own	91.6
	Rent	6.4
	Other	2
Own motorcycle	Yes	98
	No	2
Number of motorcycles	1	22.04
	2	47.35
	3	24.08
	> 3	6.53

FIGURE 16.2   Distribution of water users in the study.

being self-employed. In terms of education, almost half the respondents had completed a university degree, which is a very high proportion. Almost all households owned the house they were living in and all reported having at least one motorcycle.

Figure 16.2 shows the location of water users in the study. Of the 500 households, 163 (33%) reported using mainly PDAM (tap) water, while 260 (52%) mainly extract water from the ground. Of the 500 households, 77 (15%) of them use tap water and groundwater in equal shares.

16.5   STATED CHOICE EXPERIMENT ON WATER PROVISION CHOICE

In this section, the details of the design of the stated choice experiment and the estimated coefficients of the choice model of water supply will be discussed. The variables, attributes and attribute performance levels of the experiment were determined after a review of the literature. Choice of water supply was modelled for two variables: piped water and well water (groundwater). The attributes chosen for these variables were price, water quality and continuity of supply. For each of these attributes, three performance levels (high, medium, low) were established to allow for sufficient variation and to make the experiment realistic. We chose to use relative attribute performance levels for the variables price and quality – these levels refer to the current situation, with which respondents are familiar – and absolute levels for continuity, which was considered more appropriate because of the variation in continuity of supply in the study area. Table 16.2 summarises the experimental set-up.

For the two variables (piped and well), and three attributes (price, quality, continuity) varying among three attribute performance levels, the full factorial design comprises 729 combinations (6 variables with 3 levels each, thus 3^6). An orthogonal fraction of this full factorial design, consisting of 18 choice sets, was constructed. In addition, the 18 choice sets were randomised by the addition of one extra variable to block the design (Hensher et al., 2005) into two, meaning that instead of answering the full choice set of 18 questions, each respondent would only answer nine choice sets.

TABLE 16.2
Variables, Attributes and Performance Levels of the Choice Model for Water Supply

	Piped Water	Well Water (groundwater)
Price	(1) More than you pay now for piped water	(1) More than you pay now for groundwater
	(2) Same as you pay now for piped water	(2) Same as you pay now for groundwater
	(3) Less than you pay now for piped water	(3) Less than you pay now for groundwater
Quality	(1) Drinkable	(1) Better than now, but not drinkable
	(2) Same as now	(2) Same as now
	(3) Worse than now	(3) Worse than now
Continuity	(1) Available the whole day	(1) Available the whole day
	(2) Difficulties at peak hours (06.00–09.00 and 17.00–21.00)	(2) Difficulties at peak hours (06.00–09.00 and 17.00–21.00)
	(3) Only available 2 hours in the day	(3) Only available 2 hours in the day

This approach was chosen to avoid placing too high a burden on respondents while still providing enough variation for responses by the respondent.

The final step was to construct the stated choice questionnaire. Clear instructions were given to interviewers about how to explain the overall purpose of the research and, if necessary, how to instruct respondents on filling in the questionnaire. Respondents were required to choose between piped or well water, the characteristics of which were varied in terms of price, quality and continuity of supply. Respondents also had the option of choosing neither piped nor well water if the attributes were not attractive enough for them. Although previous studies have tested the effect of including an opt-out option in discrete choice experiments and found small differences between the forced and unforced choice models (e.g. Veldwijk et al., 2014), our decision to include the 'none of these' option was intended to make the task more realistic, thus prompting respondents to perform the experiment more thoughtfully. Figure 16.3 shows an example of a choice set in the experimental task.

16.6 RESULTS

A discrete choice model was used to analyse the responses from the stated choice experiment that we developed for this study. It models the choices made by people among a finite set of alternatives and postulates that 'the probability of individuals choosing a given option is a function of their socioeconomic characteristics and the relative attractiveness of the option' (Ortúzar and Willumsen, 2011, p. 227). The relative attractiveness can be represented using the concept of utility maximization, which assumes that individuals choose the alternative that maximises their utility or benefit.

Quality	Drinkable	Better than now, but not drinkable	
Continuity	Only available for 2 hours	Difficulties on peak hours (06.00 -09.00 and 17.00-21.00	
Price	More than you pay now	More than you pay now	
Choice	○	○	○ None of these

FIGURE 16.3 Example of a choice set.

The estimated coefficients can then be used to predict probability of an alternative being chosen among a set of discrete alternatives.

According to Hensher et al. (2005), there are three main formulations for estimating discrete choice models: multinomial logit (MNL), nested logit and mixed logit. MNL models are the most popular due to the simplicity of their formulation and because they treat responses from the same respondents as independent observations in the estimation.

The formulation of our model follows the traditional multinomial logit model, in which the utility of an individual i associated with alternative j may be described as in Equation 16.1:

$$V_{ij} = \beta_{0i} + \beta_{1i} f\left(X_{1i}\right) + \beta_{2i} f\left(X_{2i}\right) + \beta_{3i} f\left(X_{3i}\right) + \ldots + \beta_{ki} f\left(X_{ki}\right) \tag{16.1}$$

where

V_i	is the utility of an individual choosing an alternative i (pipe, well, none)
β_{ki}	is the parameter associated with attribute X_k and alternative i (price, quality, continuity)
β_{0i}	is an alternative-specific constant not associated with any of the observed attributes
k	is the number of attributes from 1 until k (in our study, $k = 3$).

The β coefficients estimated by the MNL model can be used to calculate probabilities of an alternative being chosen (Equation 16.2) and express the

probability of an individual choosing alternative i out of the set of j alternatives is equal to the ratio of the (exponential of the) observed utility index for alternative i to the sum of the exponentials of the observed utility indices for all J alternatives, including the ith alternative.

(Hensher et al., 2005)

$$\text{Prob}_i = \frac{\exp V_i}{\sum_{j=1}^{J} \exp V_j}; \ j = 1, \ldots, i, \ldots, J \ i \neq j \tag{16.2}$$

With Biogeme software (Bierlaire, 2003), maximum likelihood was used to estimate the parameters of the MNL model, which are displayed in Table 16.3. To interpret the results, it is important to realize that the attribute levels were effect-coded; consequently, the estimated utilities for each attribute add up to zero across the levels of that attribute. The t-statistics of each part-worth utility indicate any significant differences against the mean utility of that attribute. To measure the goodness-of-fit of the MNL model, McFadden's ρ^2 (Rho-squared) index is usually used. For this model, the value was 0.389, which according to Louviere et al. (2000) is indicative of good model fit. Most of the estimated coefficients are statistically significant, as indicated by the p-values.

Table 16.3 shows the estimated coefficients for the variables (PDAM (piped water), well water (groundwater)), attributes (price, quality, continuity) and their performance levels (for ease of interpretation, high, medium, low); a detailed description of the performance levels can be found in Table 16.2. Because the choice 'none of these' was used as a reference, coefficients estimated for the constants 'PDAM' and 'well' indicate a deviation from this reference. This means that 'none' was the most chosen option by respondents, followed by 'pipe' and then 'well'. This was also confirmed by the descriptive statistics of the data, which show that 67% of the responses chose 'none', with 17% choosing piped water and 15% well water. This may indicate that respondents did not answer questions in a thoughtful manner or that it was indeed difficult to trade off between attributes. Future research should debrief respondents to discover the reasons for choosing the opt-out option.

Table 16.3 also indicates that coefficients are positive for PDAM, reflecting that respondents prefer the ideal condition of a piped water supply (low price, high quality and high continuity), which is a realistic finding. Results for the supply of well water indicate a higher preference for low cost

TABLE 16.3

Parameters of the Choice Model for Water Provision

Attribute and Level	Coefficient	p-value
Constant PDAM (α)	−1.768	0
Constant Well (α)	−2.122	0
Constant None*	0	–
PDAM price High	−0.310	0
PDAM price Medium	0.026	0.68
PDAM price Low*	0.283	–
PDAM quality High	0.797	0
PDAM quality Medium	0.222	0
PDAM quality Low*	−1.019	–
PDAM continuity High	1.100	0
PDAM continuity Medium	−0.438	0
PDAM continuity Low*	−0.662	–
Well price High	−0.535	0
Well price Medium	0.046	0.54
Well price Low*	0.488	–
Well quality High	−0.649	0
Well quality Medium	1.360	0
Well quality Low*	−0.711	–
Well continuity High	1.320	0
Well continuity Medium	−0.745	0
Well continuity Low*	−0.575	–

Note: * = reference level.

and high continuity, but – unexpectedly – only a medium level of water quality. One reason for this could be that the respondents were unable to imagine a better quality for groundwater, which in turn would affect how realistic the choices were for respondents.

The estimated coefficients can also be interpreted in terms of the importance of each attribute for overall utility. The importance of each attribute can be indicated by the percentage from relative ranges (range between the highest and the lowest utility of the attribute). Figure 16.4 shows that quality and continuity have a higher degree of importance for the total utility of piped (43% and 42%, respectively) and well water (around 40% for both attributes), whereas price only represents 15% of the total contribution to the utility of piped water and 20% to well water. This is an important finding as it indicates that users of both piped and well water attach higher importance to the quality and continuity of the water provided and are less sensitive to the costs involved. It is worth exploring in future studies the issue of willingness to pay more for a service if, in turn, users are provided with better water in terms of quality and continuity.

16.7   IMPLICATIONS OF FINDINGS AND RECOMMENDATIONS

People's preferences with regard to the two main types of water supply – piped water and well water – were investigated to enable formulation of a more focused policy for water supply improvements. The method of discrete choice modelling that we applied has been useful for gaining insight into which attributes respondents consider to be important when choosing their mode of water supply. Notwithstanding this, the results also indicate several issues that will be addressed in the remainder of this section.

Perhaps surprisingly, the attributes of quality and continuity were considered more important by respondents than the attribute price. The model shows that the likelihood of people shifting to PDAM is high, provided water quality and continuity are brought to a higher level.

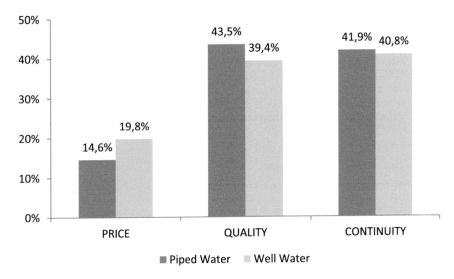

FIGURE 16.4 Importance of attributes in choosing mode of water provision.

The current PDAM municipal water supply is operating at a loss and has problems in generating sufficient revenue. The results indicate that the vicious circle of reducing levels of service and declining income can be broken if PDAM is able to improve water quality and continuity of supply.

To achieve this, investments in physical improvements to the piped water supply system are probably necessary, leading to improved water quality and a reduction of losses due to leakage. If possible, this should be done in combination with a reduction of chlorination levels, thus making the water more drinkable.

In addition, water supply from wells could also be made more unattractive by incorporating the environmental cost and damage associated with falling water tables in the pricing of well permits. An annual levy could be considered in which these costs are taken into account. Even though price is considered the least important attribute by respondents, it is the only one that policy-makers can directly influence. The other attributes – quality and continuity – do not lend themselves well to intervention. To make well water unattractive in terms of quality would imply that the government would take steps to reduce the quality levels of groundwater, while making wells unattractive in terms of continuity would imply reducing the availability of electricity for driving pumps. A better option for reducing groundwater use altogether probably lies in stricter enforcement of existing rules and regulations and a fairer pricing system for permits.

In the implementation of the stated choice experiment, a few difficulties appeared. The high level of respondents who chose the opt-out option was unexpected and undesirable. This may have been caused by the fact that people found it difficult to trade off between the two choices offered. To be able to analyse this situation, we recommend splitting the experiment into two parts: one half of the sample would be given the opt-out option and the other half would not, so that the influence of this option can be tested. In the case that most of the respondents favour the opt-out option, they would need to be approached again to discover their reasons for doing so.

Research in the future may consider a larger and more segmented body of respondents, which would allow choice behaviour to be evaluated across different socio-economic groups and possibly geographic areas, thus providing more spatially targeted improvements of the water supply system as a result.

REFERENCES

Alegre, H., Hirner, W., Baptista, J. M., Parena, R., Cubillo, F., Cabrera Jr., & E. Duarte. (2006). *Performance indicators for water supply services* (2nd ed.). London: IWA Publishing.

Bierlaire, M. (2003). *BIOGEME: A free package for the estimation of discrete choice models.* Proceedings of the 3rd Swiss Transportation Research Conference, 1–27. Ascona, Switzerland. Retrieved from http://infoscience.epfl.ch/record/117133

BPS Indonesia. (2015). *Statistics Bureau of Indonesia.* Retrieved from https://www.bps.go.id/

Brouwer, R., Job, F. C., Kroon, B. van der, & Johnston, R. (2015). Comparing willingness to pay for improved drinking-water quality using stated preference methods in rural and urban Kenya. *Applied Health Economics and Health Policy*, 13(1), 81–94. http://doi.org/10.1007/s40258-014-0137-2

Doria, M. F. (2010). Factors influencing public perception of drinking water quality. *Water Policy*, 12(1), 1–19. http://doi.org/10.2166/wp.2009.051

Doria, M. F., Pidgeon, N., & Hunter, P. R. (2009). Perceptions of drinking water quality and risk and its effect on behaviour: A cross-national study. *Science of the Total Environment*, 407(21), 5455–5464. http://doi.org/10.1016/j.scitotenv.2009.06.031

Hensher, D. A. (1994). Stated preference analysis of travel choices: The state of practice. *Transportation*, 21(2), 107–133. http://doi.org/10.1007/BF01098788

Hensher, D. A., Rose, J. M., & Greene, W. H. (2005). *Applied choice analysis: A primer* (1st ed.). New York, USA: Cambridge University Press.

Jakarta Post. (2014). Over-exploitation of Water a Threat to Yogyakarta. Retrieved June 1, 2015, from http://www.thejakartapost.com/news/2014/10/18/over-exploitation-water-a-threat-yogyakarta.html

Jakarta Post. (2015). Court Bans Monopoly on Water Resources. Retrieved June 2, 2015, from http://www.thejakartapost.com/news/2015/02/20/court-bans-monopoly-on-water-resources.html

Muryanto, B.. (2014). Over-exploitation of water a threat to Yogyakarta. *Jakarta Post.* Retrieved from http://www.thejakartapost.com/news/2014/10/18/over-exploitation-water-a-threat-yogyakarta.html

Sundaryani, F. S. (2015). Court bans monopoly on water resources. *Jakarta Post.* Retrieved from http://www.thejakartapost.com/news/2015/02/20/court-bans-monopoly-on-water-resources.html

Kumar, R. (2005). *Research methodology: A step-by-step guide for beginners* (3rd ed.). London, England: Sage Publications.

Louviere, J. J., & Hensher, D. A. (1982). Design and analysis of simulated choice or allocation experiments in travel choice modeling. *Transportation Research Record*, (890), 11–17. Available online at https://trid.trb.org/view.aspx?id=189334.

Louviere, J. J., & Woodworth, G. (1983). Design and analysis of simulated consumer choice or allocation experiments: an approach based on aggregate data. *Journal of Marketing Research* 20(4), 350–367. http://doi:10.2307/3151440.

Louviere, J. J., Hensher, D. A., & Swait, J. D. (2000). *Stated choice methods: analysis and application* (1st ed.). Cambridge, England: Cambridge University Press.

MacDonald, D. H., Barnes, M., Bennett, J., Morrison, M., & Young, M. D. (2005). Using a choice modelling approach for customer service standards in urban water. *Journal of the American Water Resources Association*, 41(3), 719–728. http://doi.org/10.1111/j.1752-1688.2005.tb03766.x

Martinez, J. (2015). Lecture: Sampling [PowerPoint Slides]. Enschede, the Netherlands: ITC – University of Twente.

Ortúzar, J. D., & Willumsen, L. G. (2011). *Modelling Transport* (4th ed.). Chichester, UK: John Wiley & Sons, Ltd.

Pang, L., Close, M., Goltz, M., Sinton, L., Davies, H., Hall, C., & Stanton, G. (2004). Estimation of septic tank setback distances based on transport of E. coli and F-RNA Phages. *Environment International*, 29(7), 907–921. http://doi.org/10.1016/S0160-4120(03)00054-0

PDAM Tirtamarta. (2014). *Price for water in PDAM* [In Bahasa]. Retrieved from http://pdamkotajogja.co.id/berita-tarif-belangganan.html

Pemkot Yogyakarta. (2001). *Profile of Yogyakarta city* [In Bahasa]. Retrieved from http://upik.jogjakota.go.id/

Pemkot Yogyakarta. (2013). *Public law about licensing to extract groundwater* [In Bahasa]. Retrieved from http://upik.jogjakota.go.id/

Rahut, D. B., Behera, B., & Ali, A. (2015). Household access to water and choice of treatment methods: empirical evidence from Bhutan. *Water Resources and Rural Development*, 5, 1–16. http://doi.org/10.1016/j.wrr.2014.09.003

Salendu, B. (2010), Quality assessment and interrelations of water supply and sanitation: a case study of Yogyakarta City, Indonesia. Unpublished MSc thesis, University of Twente. Retrieved from: https://webapps.itc.utwente.nl/librarywww/papers_2010/msc/upm/salendu.pdf

Veldwijk, J., Lambooij, M. S., de Bekker-Grob, E. W., Smit, H. A., & de Wit, G. A. (2014). The effect of including an opt-out option in discrete choice experiments. *PLoS One*, 9(11), e111805. http://doi.org/10.1371/journal.pone.0111805

Whittington, D., Pattanayak, S. K., Yang, J. C., & Kumar, K. C. B. (2002). Household demand for improved piped water services: evidence from Kathmandu, Nepal. *Water Policy*, 4(6), 531–556. http://doi.org/10.1016/S1366-7017(02)00040-5

17 Simulating Spatial Patterns of Urban Growth in Kampala, Uganda, and Its Impact on Flood Risk

Eduardo Pérez-Molina, Richard Sliuzas, and Johannes Flacke

CONTENTS

17.1 Introduction 295
17.2 Background..296
17.3 Problem Conceptualization 299
17.4 Projecting Drivers of Urban Growth 300
17.5 Urban Growth Scenarios: Development and Comparison..302
    17.5.1 Patterns of Future Urban Development 302
        17.5.1.1 Evaluation of Metropolitan Urban Growth..303
    17.5.2 Flood Impacts of Metropolitan Dynamics in Upper Lubigi....................................304
17.6 Conclusions...305
References..305

17.1 INTRODUCTION

Local flooding is a problem often aggravated by urban growth. Urban growth can be guided by land-use planning systems, which then become tools for reducing flood risk. The most general question suggested by framing the problem in this way is how should land-use planning systems be designed and deployed so that they have a real impact on reducing flood risks – and, more generally, natural hazards?

The case study reported in this chapter approaches this question by adopting a geo-computational approach: it details the use of existing urban growth (Pérez-Molina et al., 2017) and flood models (Sliuzas et al., 2013a) of Kampala (Uganda) for designing future urban growth scenarios and testing their flooding impact. Specifically, the scenarios seek to reproduce and operationalise building densification options available for inclusion in the Kampala Physical Development Plan (KPDP), commissioned by the Kampala Capital City Authority in 2012 (ROM Transport et al., 2012). With this in mind, our research aims were to: (1) identify building densification options proposed in the KPDP; (2) formalise these options within the existing urban growth model (of Pérez-Molina et al., 2017); and (3) apply the existing urban growth/flood models (Pérez-Molina et al., 2017; Sliuzas et al., 2013a) to predict future built-up patterns and their impact on flooding (Figure 17.1).

BOX 17.1   Case Study Area

Kampala is Uganda's largest city and its major centre of commerce, as well as being the largest city in East Sub-Saharan Africa (UN-Habitat, 2014). The Kampala metropolitan area covers an area of approximately 325 km². The city proper is divided into five districts: Kampala Central Division, Kawempe, Makindye, Nakawa and Rubaga, although a significant amount of new development occurs beyond the city boundaries.

The case study area develops an urban growth model for the Kampala metropolitan region – the city proper and its surrounding area – and evaluates the impact of flooding of the Upper Lubigi sub-catchment, an area that roughly coincides with the Kawempe division. The Upper Lubigi has been the focus of several studies on flooding and urban growth because of the recurring flooding that the large informal settlement of Bwaise suffers; Kampala is itself prototypical of a group of large cities in Sub-Saharan Africa that exhibit rapid growth and physical expansion in a complicated context and a weak institutional setting (see African Planning Association, 2014). Kampala is located near the equator, north of Lake Victoria, in a hilly terrain that also contains large areas of wetlands. Its tropical weather and soil infiltration properties already lead to large runoff volumes, a main cause of recurrent flooding, which has been exacerbated by urban growth.

The use of scenarios was chosen as an approach for exploring the potential consequences of land-use plans because, as argued by Steinitz et al. (2005), they enable the assessment and comparison of different options (irrespective of their feasibility) faced by decision-makers, allowing local actors to make informed choices.

Scenarios are also an efficient way of dealing with uncertainty. Sliuzas, Flacke and Jetten (2013b) had already noted a number of uncertainties present in Kampala that preclude the possibility of using typical design processes for reaching the goal of designing an ideal city for the future. Kampala is undergoing rapid population growth and urban development, much of which is informal and unplanned. To this can be added that its planning institutions are weak – which creates uncertainty about their ability to control development – and the available budget for public investment is highly dependent on funds from the World Bank. Moreover, and most importantly, key data for flood modelling (e.g. data on rainfall events and soils) are often unavailable.

17.2   BACKGROUND

Land systems are complex phenomena that involve multiple interactions between physical and human elements, with dynamics that cannot be easily analysed. This inherent complexity commonly leads to counter-intuitive responses in what would otherwise seem to be straightforward situations (White and Engelen, 1993). Yet the management of human settlements on the landscape requires, as a basic democratic principle, input from citizens, including many people who are not trained in dealing with such complexities (Pearce, 2003). The tension this introduces has long been a methodological problem, usually tackled through participation in the definition

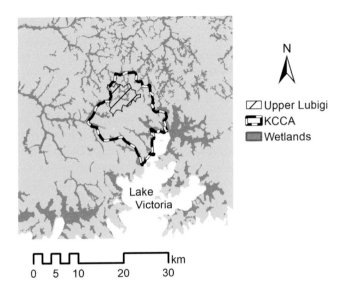

FIGURE 17.1   Case study: Kampala metropolitan area, Kampala City Council Authority and Upper Lubigi sub-catchment.

of planning instruments (Arias et al., 1999). The problem has been compounded as scientific understanding of planning practice and its consequences have driven planners towards future-oriented approaches such as scenario planning (Hulse et al., 2004; Vargas-Moreno and Flaxman, 2012).

Scenarios have been proposed as a way to create workable methods that take into account both science research and the values of citizens in land-use planning. As argued by Xiang and Clarke (2003), scenarios can play a simultaneous, double role by acting both as a catalyst for broadening thinking on concrete planning problems and as a bridge between modelling (science) and planning (the practice of policy). This duality plays out if scenario designs are deliberately organised to fulfil these roles, which has implications in terms of what is to be simulated. Xiang and Clarke (2003) call for unexpected yet plausible scenarios to challenge stakeholders' unexamined beliefs and, ideally, to trigger discussion on new strategies: 'out of the box thinking', so to speak. The consequences of these strategies can be systematically examined through the modelling in order to assess their impact on planning options.

Couclelis (2005), in the context of a futures-oriented planning practice, viewed scenario development more comprehensively. For her, there are three possibilities for scenarios of an urban system: what *may* be – 'the system's internal contingent futures, conditional upon actions and events that the planners generally have some control over' (p. 1362); what *should* be, a reflection of the normative aspect of planning, i.e. the desirable future states of the system and the paths leading to them; and what *could* be – questions that 'transcend the planned system itself and interrogate the broader environment within which that system is embedded' (p. 1362).

A key feature of land system scenarios is that they must be structured around specific questions about the landscape. These questions can be general features, common to the management of land-use systems (see Steinitz, 2012) or specific to a case study. Scenario development reveals consequences, but only if the process (which starts by setting up models of phenomena that are the object of interest) has incorporated the elements that are to be questioned. Thus, scenario development begins by a conceptualization of the problem and, from that, creates process models and evaluation models to assess each scenario.

Interestingly, the practice of modelling the flooding impacts of land-use change remains limited in exploring the full range of possibilities. Prototypical examples (Lin et al., 2007) make use of ready-made land models, rather than customizing case-specific urban growth models for examining the flooding process in greater detail. As a consequence, the range of scenarios that can be modelled is constrained by the original design of the land model.

Land use scenarios for the city of Kampala (Uganda) have been recently developed to address at least two pressing urban problems: Vermeiren et al. (2012) analysed urban growth of the Kampala metropolitan region with the ultimate goal of studying social inequality (Vermeiren et al., 2016); and Sliuzas et al. (2013a) created scenarios for exploring the phenomenon of rapid flooding in response to urban development trends in the Upper Lubigi sub-catchment.

Vermeiren et al. (2012) based their scenarios on population growth and population density, design constraints and new infrastructure (new major roads). They view population growth as a major driver, since it determines how much new urban space will be required. Further, they consider scenarios of increased population density in central areas and contrast them with sparse development of the urban fringe. The latter is affected by new infrastructure, which modifies the suitability of the fringe for urban development. Design constraints – such as zoning laws that prohibit new built-up land in wetlands and existing open space – also modify suitability for development.

The land scenarios of Sliuzas et al. (2013a), which were restricted to Kampala's Upper Lubigi sub-catchment area, were based on a static spatial logistic regression analysis (with development in the neighbourhood of each location, distance to urban centralities and slope as the determinants); Vermeiren et al. (2012) use a similar statistical approach. However, unlike Vermeiren et al. (2012) – for which such considerations were not relevant – Sliuzas et al. (2013a) also considered specific measures that would affect the impact of flooding: grass paths along secondary drains, constraints on flooding zones (and the evacuation of buildings therein) and the greening of random locations. As these examples illustrate, the fundamental hydraulic problem of the Upper Lubigi is to slow down runoff and allow infiltration of rainfall upstream, in order to reduce water volumes reaching the main drainage channel, because, as was proven by analysing trend conditions, the expansion of this main drainage channel will not be able to mitigate flooding under current conditions, let alone in the future, when climate change could lead to higher rainfall and heavier, more frequent storms.

This chapter builds on the work of Sliuzas et al. (2013a) and on the development of new methods that have further expanded this framework (Pérez-Molina et al., 2017); these methods were specifically designed to better incorporate the uncertainty of spatially random human behaviour into the urban growth model, resulting in a better reflection of past trend conditions of human and physical dynamics. The new methods also consider land dynamics, rather than static suitability for urban development, and are more flexible. Specifically, they allow us to simulate a spatially explicit potential supply of urban development (representing the actions of developers and planners), as well as a spatially explicit suitability of land for development (which reflects the human locational preferences for urban space of residents) – the simulation then proceeds by assuming the supply can match the suitability, so residents choose from the available supply, taking first locations with higher suitability scores.

In the following sections, we present a synthesis of the research problem (interaction between urban growth and flooding) as well as a brief description of the modelling approach, including a discussion on key drivers of urban growth (and flooding) and the methods used to project them. Several scenarios of urban development for the Kampala metropolitan area are generated; they are then assessed in terms of their urban growth patterns and of the flooding patterns they cause in the Upper Lubigi sub-catchment.

17.3 PROBLEM CONCEPTUALIZATION

The models and simulations we present in this chapter view Kampala from the perspective of a coupled human and natural system (Alberti et al., 2011), specifically a rainfall–runoff approach in which runoff increases due to human causes (Sene, 2010). The landscape is thought of as an area over which rain falls, and part of this rain infiltrates the soil (the soil's rate of infiltration determines the volume). The fraction of rainfall that infiltrates is larger for vegetated areas and zero for impervious areas (built-up areas and paved roads). Further, another fraction of the rainfall is intercepted by vegetation. The fraction of rainfall that remains – the runoff – moves downslope, eventually reaching drainage channels. The capacity of these drainage channels may, in turn, be exceeded, which causes overflow into adjacent areas – in other words, flooding. The human impact on this natural system comes through urban construction, the creation of impervious areas. Floods, in turn, could influence locational preferences for urban land in a number of ways. For example, they could drive most urban agents away because of the risk of flood losses, but they could also attract the poorest urban agents, since the risk of flooding makes these otherwise desirable locations unattractive for others; see Bathrellos et al. (2017).

BOX 17.2 Methods Applied in the Chapter

In this chapter, we describe a cellular automata model for generating prospective urban growth scenarios that reflect possible policy-driven building densification options. The results were then masked and used as inputs for a flood model of the Upper Lubigi sub-catchment to calculate the hydrological impacts associated with each scenario. The cellular automata model was implemented in ArcGIS™ 10.5 and the results were used as inputs for the openLISEM rainfall–runoff/flood model of the Upper Lubigi (Sliuzas et al., 2013a). Spatial data of the Kampala metropolitan region included: land-cover fractions of built-up surface, vegetation, bare soil and water for 2015 (derived from Landsat imagery by linear spectral unmixing); spatial factors that determine urban growth (distance from main roads, travel time to city centre, wetlands location); and formalizations of possible urban growth envisaged by the KPDP. The database for openLISEM is described in Sliuzas et al. (2013a) and includes data on soil infiltration, rainfall and drainage channels. The general framework integrates spatially explicit land and hazard models, so it can be used to evaluate other hazards that interact with settlement patterns (e.g. sedimentation, landslides) and other landscapes exposed to human-driven change by incorporating other land models (statistical, rule-based; agricultural, deforestation, etc.). The methodology assumes physical settings that have been/are being modified by human intervention. It reflects aggregate human behaviour – and, therefore, cannot deal with micro-level effects.

The modelling suite we used is fully described in Pérez-Molina et al. (2017). In a nutshell, we coupled a cellular automata–based model of urban growth with an openLISEM flood model of the Upper Lubigi sub-catchment area. Cellular automata were chosen to model the urban system because of their capability to replicate complex urban patterns. As noted by White and Engelen (1993), a cellular automata model can be thought of as an *array of cells* (in this case study, a uniform array of square cells – the pixels of a raster). Each cell has a *state* that characterises it at a moment in time (in this case study, the built-up fraction is the state for a given year). The state of a cell changes from one period to the next depending on the states of its surrounding cells (in this case study, the higher the average built-up fraction of the adjacent cells, the higher the probability that the cell in

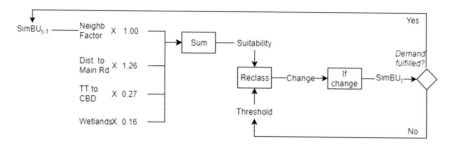

FIGURE 17.2   Cellular automata model of urban growth.

question experiences an increase of its built-up fraction). In addition to the state of adjacent cells, other factors can be incorporated to define the probability of change – the combination of these with the state of neighbouring cells defines the *transition rules* that govern whether or not a cell undergoes change from one period to the next (White, 1998).

The cellular automata model (see Figure 17.2) makes use of a map of an index representing the suitability of land for urban development, a weighted summation of accessibility and physical and neighbourhood factors. The neighbourhood factor is dynamic (it is updated at each time step, i.e. one year), whereas all other factors are static. The scenarios developed in the following sections (17.4 and 17.5) use the version of this urban growth model that was calibrated for Kampala's metropolitan area (Pérez-Molina et al., 2018), since this scale of the phenomenon better represents the human drivers of urban growth (travel time to the city centre, distance to nearest main road, average built-up area in the immediate vicinity of the location in question) and their relation to changes in the built-up fraction.

Based on the suitability map referred to above, for any given time step, the best locations for urban development are successively allocated as developed land until an exogenous demand for that period is satisfied. The process then repeats itself for the next time step, using as input the simulated map. This assumes, as noted in section 17.2, that urban agents seeking land rationally choose the best locations provided to them by developers. How much land demand is required by all urban agents in a given period is a function of the expected demographic change (how many people will require new housing, which also assumes new business areas are proportional to this growth) and the expected gross population density, hence its exogeneity. The amount of development per cell is a spatially explicit model that can reflect the spatially random character of observed data – as in the Kampala case (see Pérez-Molina et al., 2017) – or it may also simulate the guidance of land-use policy (as for the case study of Rwampara in Kigali (Rwanda), reported in Pérez-Molina et al. 2016). It captures the aggregate behaviour of suppliers of buildings and urban land, possibly including the extent to which regulators can influence their activities. Once the land-cover scenarios were generated, these results were then clipped to the area of the Upper Lubigi sub-catchment, and the flood model generated by Sliuzas et al. (2013a) was applied to assess the impacts on flooding.

17.4   PROJECTING DRIVERS OF URBAN GROWTH

The scenarios in our study build on three sets of main drivers. As stated in Vermeiren et al. (2012), population growth is a very important driver because it determines the total demand for land for new development. Previous scenarios have proven to be very sensitive to this figure (Pérez-Molina et al., 2017). Spatial factors that determine land use, as a reflection of human behaviour in the choice of urban locations, are also key drivers. The relative importance of each factor is pre-determined by calibrating the cellular automata model (Pérez-Molina et al., 2018). In the absence of deeper behavioural models (such as agent-based models), it is not possible to experiment with the effect of policy at this micro level (i.e. the cumulative effect of decisions of many individuals). However, changes in the physical elements that determine these (individual) preferences may have potential consequences (e.g. the introduction of new roads, as Vermeiren et al. 2012 simulated). The third element is land-use planning and the constraints and opportunities such policies represent (see Pérez-Molina et al. 2016 for a case study of Kigali (Rwanda) that incorporates this element).

Scenarios were generated for the target year 2025 from the baseline year 2015. Data on population numbers for Kampala are limited and of questionable reliability. While population distributions per district are available for census years, information on the evolution of this population and on its vital statistics are unavailable. To generate future scenarios, we adopted the population projections of the United Nations (United Nations, 2014). Demand for land was estimated as population growth (the expected number of new residents for Kampala, 2015–2025, which is the sum of births and immigrants minus deaths) divided by gross population density (total estimated population of Kampala into the total built-up area for that year, calculated with data of 2015).

The core of the results we report in this chapter refer to the implementation of land-use planning and its potential to mitigate flooding problems through increasing population and urban activity densities of central locations. Development scenarios are operationalised through models of potential supply, as follows:

- The baseline scenario (S0) makes use of the results of Pérez-Molina et al. (2017): potential supply follows a spatially random pattern. The S0 map generated is the result of the multiplication of values taken from a random-values map (with values between 0 and 1), the average per cell growth of 2010–2015, the time elapsed since the baseline (number of time periods since $t=0$) and an expansion factor that controls the level of aggregation. The S0 map is adjusted so that for all cells the resulting built-up fraction (after urban growth has occurred) cannot exceed 0.85 (Figure 17.3).
- The densification scenario (S1) allows for greater population densities in the urban core. Specifically, based on the KPDP proposal, two densification zones were proposed: for the city core (Zone 1), a greater gross population density was applied when calculating the potential housing supply; for infill development by building densification of areas within the city of Kampala (Zone 2), half that density was assumed (these two zones are shown in Figure 17.3). To operationalise this principle, the baseline gross population density (calculated, as noted, with data from 2015) was multiplied by a factor greater than 1.0 (which changes depending on the scenario but which is also, by definition, larger for Zone 1 than for Zone 2).

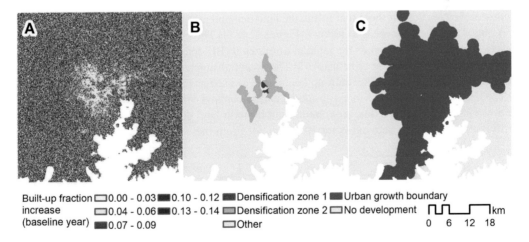

Built-up fraction ☐0.00 - 0.03 ■0.10 - 0.12 ■Densification zone 1 ■Urban growth boundary
increase ☐0.04 - 0.06 ■0.13 - 0.14 ☐Densification zone 2 ☐No development
(baseline year) ■0.07 - 0.09 ☐Other

0   6   12   18 km

FIGURE 17.3 Potential supply model for 2025. A: Potential supply of urban development. B: Densification zones. C: Urban growth boundary.

- The growth control scenario (S2) incorporates the same basic pattern as that of S1, to which a hypothetical urban growth boundary is added. Beyond this boundary, no further increase of the built-up fraction is allowed (Figure 17.3).
- For all scenarios, the non-built fraction is split up into vegetation and bare soil, based on the proportions of these types of land cover in the base year 2015.

17.5 URBAN GROWTH SCENARIOS: DEVELOPMENT AND COMPARISON

17.5.1 Patterns of Future Urban Development

The land demand required for 2025 was calculated to be 156.7 km². The United Nations (2014) reported a population of 1,935,654 for 2015 and 3,054,592 for 2025. From available geo-data, the former population required a total built-up area of 271.1 km² (note that this estimate may be smaller than other data reported in the literature since it was calculated based on built-up fractions per cell rather than the usual classification by categories, such as those used by Vermeiren et al. 2012). The total projected land demand for 2015–2025 can be expressed as:

$$\text{Land Demand}_{2015-2025} = \frac{\text{Population Growth}_{2015-2025}}{\text{Gross Density}_{2015}} = \frac{3054592 - 1935654}{1936564 / 27112.4\,\text{ha}} = 15{,}672.8\,\text{ha}$$

The cell size of all model inputs was 90 m. Therefore, land demand is equivalent to 19,349 cells of 90 m × 90 m for the ten-year period that is to be simulated.

Computationally, the cellular automata model consists of a model calling two sub-models. The first sub-model creates the potential supply map, based on the map of built-up land cover for t and on the built-up land cover of 2015; this model is recalculated at each time step to enforce the limit of 0.85 as the maximum possible built-up fraction. The selected expansion factor for the model is one-fifth of the calibrated factor reported in Pérez-Molina et al. (2018), to account for the fact that the calibration in their study was performed for a five-year period, whereas each iteration performed for the simulations in this chapter corresponds to a one-year period.

The second sub-model is an allocation algorithm: it computes the neighbourhood factor, defined as the average built-up fraction within a moving window of 3 × 3 cells; then the sub-model combines it with other factors (accounting for the calibrated weights of each one) to generate a suitability map; and it then reclassifies the suitability map into growth (higher suitability values) or no change (lower suitability values). This reclassification is based on a threshold parameter value; the algorithm searches for the threshold that fulfils the land demand requirement for any given time step.

A model run, then, proceeds as follows (Figure 17.4): (1) The built-up land-cover fraction map of the baseline year (2015) is set as the input at time step 0 (BU denotes this built-up land-cover fraction map; BU at $t=0$ is the baseline map). (2) The potential supply map is generated for the baseline year and, together with the 2015 built-up land cover, is used as input for the allocation sub-model. (3) The third input of this allocation sub-model is land demand, which is set to the time step 1 (since we are calculating the first time step) multiplied by total land demand (19,349 cells) and divided by 10 (because the land demand is for 10 years). (4) From these inputs, the allocation sub-model

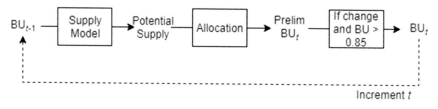

FIGURE 17.4 Computational flow chart of the urban growth model.

produces the simulation for the first time step, i.e. for 2016. For the second period, this simulated land cover is used instead of that of the baseline year, and the cellular automata model starts from (1) again, using built-up land-cover fractions from this map as input – albeit with the added difference that, in (3), land demand would be $2 \times 19{,}349/10$, because the second time step is being calculated. The model runs for ten time steps, ultimately generating a simulation for 2025.

The increase in building density, reflected by the possibility of accommodating greater population (which is to say the *computational* assumption of a greater population density), postulated by Scenarios 1 and 2 is obtained by multiplying all cells in the potential supply map of densification Zone 1 by 3 (for S1) or 7 (for S2), equivalent to increases in the potential of these areas to accommodate gross population densities, relative to S0, by a factor of 3 and 7, respectively. Similarly, all cells in the potential supply map of Zone 2 were multiplied by 1.5 or 3.5 (for S1 and S2, respectively), again increasing their potential to accommodate higher gross population densities (the conceptual assumption here is that the most desirable area for greater building densities, from a policy perspective, is the city centre – Zone 1 – followed by the relatively central neighbourhoods of Zone 2). These modified potential supply maps are used to allocate demand (note, though, that at this point in the process the fraction of built-up land cover can be higher than 1.00 in some cases, to allow for the possibility of multi-storeyed buildings; in practice, this means the potential supply is expressed as floor area rather than building footprint, although it would be incorrect to equate this figure to housing demand, since it also includes business developments in proportion to land use of the 2015 baseline); once the final simulation has been completed, a further assumption – that all predicted built-up fractions (which in the case of Zones 1 and 2 are actually floor space estimates) produce a building footprint of, at most, $0.85 \times$ cell area – is enforced by reclassifying the value of any cell undergoing growth that results in a built-up fraction > 0.85 back to the value 0.85.

17.5.1.1 Evaluation of Metropolitan Urban Growth

The simulations generated from the models are shown in Figure 17.5. Clearly, future demand for urban activities is very high in Kampala. Furthermore, the simulated scenarios show a trend that suggests the city is highly susceptible to sprawling development, especially associated with the proximity of main roads that branch out from Kampala into the rest of the country and the neighbouring Wakiso district.

An important finding from the comparison of these scenarios is the role of building densification in the implementation of smart growth policies. As can be clearly seen in Figure 17.5, the differences introduced by only building densification (Scenario 1) relative to the trend (Scenario 0) are not too significant. Both scenarios lead to substantial peri-urban development. By contrast, when coupled with a regulatory constraint (the urban growth boundary, Scenario 2), sprawling development is noticeably smaller. However, to obtain such a pattern, the population densification levels assumed to meet the needs for building densification were implausibly high (a factor of 7

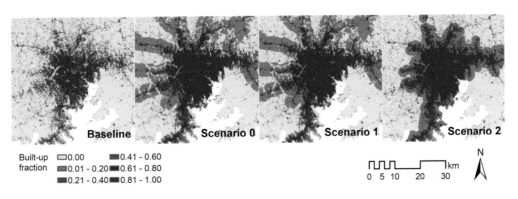

FIGURE 17.5   Land cover projections for 2025.

for central locations and 3.5 for inner neighbourhoods). The use of lower values, however, causes convergence problems in the model because there would not be enough land supply to fulfil land demand. The ultimate conclusion is that to control sprawl, targeted increases of building density are not sufficient. Infill development would have to be systematically promoted at much greater building density levels than is currently common practice in Kampala – in addition to effective land use controls.

While the challenge thus revealed is very large, it must also be noted that opportunities for systematic building densification of large areas exist. In particular, Kampala lacks a mass transportation system (a Bus Rapid Transit system is now being developed for Kampala). The implementation of such a system may guide future development processes towards greater building densities along main public transport corridors, and this new pattern could prove useful in controlling scattered development at the peri-urban interface of the metropolitan region.

The second noticeable issue arising from a comparison of scenarios concerns open space, specifically the wetlands. Despite wetlands being a negative factor in determining suitability for urban development, all scenarios predict wetland areas in Kampala to be fully developed by 2025. This is most noticeable for the Lubigi and Nalukolongo wetlands, located in western Kampala, which are clearly undeveloped in the baseline map of 2015 but are full by 2025 for all scenarios. On the one hand, this speaks for the efficiency of existing land-use controls, since in fact most wetlands have remained undeveloped despite the difficult institutional context of weak enforcement capabilities by the local government. Yet it must also serve as a warning, in the sense that wetland locations are highly desirable for urban development because of their accessibility. Any breakdown in existing land-use controls, even temporarily, may have serious consequences for flood risk and other natural hazards in the city.

17.5.2 Flood Impacts of Metropolitan Dynamics in Upper Lubigi

To assess the scenarios chosen, the resulting built-up land cover projections for 2025 were used as inputs in the flooding model developed in Sliuzas et al. (2013a). The hydrological results produced by the model are summarised in Table 17.1.

Table 17.1 shows very similar outcomes for all three scenarios, with slightly less flooding under Scenario 0 (the most diffuse development pattern). This outcome follows on from the assumptions made about development: that a denser urban fabric (in Scenarios 1 and 2) accommodates more people within the city, but with more impervious surfaces in these same relatively central locations, of which Upper Lubigi is one. The lesson, therefore, is that building densification by itself will not reduce flood risk. Redevelopment of the city must be accompanied by substantial greening of the urban landscape, in order to increase infiltration by both less impervious areas (with no infiltration) and slower runoff, thus decreasing floods by reducing the volume of runoff.

TABLE 17.1

Hydrological Outcomes in the Upper Lubigi Sub-Catchment Under the Scenarios Developed for 2025

Hydrological Outcomes	Scenario 0	Scenario 1	Scenario 2
Total area flooded to depth > 25 cm (ha)	146.0	155.9	155.4
Built-up area flooded to depth > 25 cm (ha)	119.9	129.7	132.5
Peak discharge (m³/s)	41.7	41.7	41.3
Total discharge (10³ m³)	408.9	416.3	435.5
Discharge/Rainfall (%)	39.4	40.7	42.9

17.6 CONCLUSIONS

Looking back at the framework on scenarios for land-use planning, the scenarios we present in this chapter do not exhaust the possibilities. Regarding Couclelis (2005), our scenarios concentrate on exploring what *(perhaps) should* be: assessing the possibilities of land use planning to change urban growth patterns (and developing also, counterfactually, a scenario of what *may be* – Scenario 0, or the current trend). They are not meant to be broadening scenarios, so the outlook produced is only partial. Specifically, long-run hypotheses involving climate change or massive population displacements are not explored. However, the immediate challenges are already so large that any work done towards increasing the system's resilience against short-run trends would yield enormous benefits, even in the face of a potentially worsening overall context.

What is investigated with these scenarios is the previously unexamined assumption that building densification in Kampala is an appropriate goal for smart growth, in particular with regard to the goal of reducing flood risk. Our results suggest the need for an effective and efficient system that promotes infill densification of the urban fabric (i.e. more and higher buildings) and, as well, controls – in other words, a system that restricts peri-urban expansion. Further, building densification not only requires higher buildings but also introduces the need for greening (as has been noted, for example, by Nadraiqere 2014). Even with such a system in place, the reality of Kampala is that much of its future expansion will be greenfield development on its periphery, resulting from a complex interaction between land availability and accessibility constraints. There is, therefore, also a need to expand and improve the road network of the metropolitan region.

Finally, and consistent with the previous considerations, flooding in the Upper Lubigi sub-catchment is not very sensitive to regional urban-growth policies. Indeed, it could be argued that by directing growth away from central locations the trend of peri-urban expansion causes less flood risk than efforts to increase the occupation of central locations through infill redevelopment. One should be cautious, though, not to act aggressively based solely on simulations of the Upper Lubigi sub-catchment. In this sense, the need to expand flood risk assessment to a regional level is very clearly justified, since what the analysis suggests is that, even with well-designed policies and a system for enforcing them, there is likely to be a trade-off between infill densification of the urban fabric and peri-urban expansion at low population and building densities in terms of local impacts of flooding.

REFERENCES

African Planning Association. (2014). *The state of planning in Africa. An overview.* Nairobi, Kenya: APA and UN-Habitat.

Alberti, M., Asbjornsen, H., Baker, L. A., Brozovic, N., Drinkwater, L. E., Drzyzga, S. A., Jantz, C. A., Fragoso, J., Holland, D. S., Kohler, T. A., Liu, J., McConnell, W. J., Maschner, H. D. G, Millington, J. D. A., Monticino, M., Podestá, G., Pontius Jr., R. G., Redman, C. L., Reo, N. J., Sailor, D., & Urquhart, G. (2011). Research on coupled human and natural systems (CHANS): Approach, challenges, and strategies. *The Bulletin of the Ecological Society of America, 92*(2), 218–228. doi:10.1890/0012-9623-92.2.218

Arias, E. G., Eden, H., Fischer, G., Gorman, A., & Scharff, E. (1999, December). *Beyond access: Informed participation and empowerment.* Paper presented at the 1999 Conference on Computer Support for Collaborative Learning, Palo Alto, CA. Retrieved from https://dl.acm.org/citation.cfm?id=1150242

Bathrellos, G. D., Skilodimou, H. D., Chousianitis, K., Youssef, A. M., & Pradhan, B. (2017). Suitability estimation for urban development using multi-hazard assessment map. *Science of the Total Environment, 575,* 119–134. doi:10.1016/j.scitotenv.2016.10.025

Couclelis, H. (2005). 'Where has the future gone?' Rethinking the role of integrated land-use models in spatial planning. *Environment and Planning A, 37*(8), 1353–1371. doi:10.1068/a3785

Hulse, D. W., Branscomb, A., & Payne, S. G. (2004). Envisioning alternatives: Using citizen guidance to map future land and water use. *Ecological Applications, 14*(2), 325–341. doi:10.1890/02-5260

Lin, Y. P., Hong, N. M., Wu, P. J., & Lin, C. J. (2007). Modeling and assessing land-use and hydrological processes to future land-use and climate change scenarios in watershed land-use planning. *Environmental Geology, 53*(3), 623–634. doi:10.1007/s00254-007-0677-y

Nadraiqere, E. (2014). *Urban Design With Sustainable Drainage Systems. 'A Case Study for Kampala's Expansion'*. (M.Sc. Thesis, University of Twente, Enschede, the Netherlands).

Pearce, L. (2003). Disaster management and community planning, and public participation: How to achieve sustainable hazard mitigation. *Natural Hazards, 28*(2–3), 211–228. doi:10.1023/A:1022917721797

Pérez-Molina, E., Sliuzas, R., van Maarseveen, M. F. A. M., & Jetten, V. (2016). Urban flood management under rapid growth in Kigali, Rwanda: Developing perspectives on land use planning and the future by exploring the case of the Rwampara wetland. In Schmitt, H. C., Danielzyk, R., Greiving, S., Gruehn, D., Thinh, N. X., & Warner, B. (Eds.), *Proceedings of Dortmund conference 2016: Spatial patterns-structure, dynamic, planning* (pp. 351–365). Dortmund, Germany: Klartext Verlag.

Pérez-Molina, E., Sliuzas, R., Flacke, J., & Jetten, V. (2017). Developing a cellular automata model of urban growth to inform spatial policy for flood mitigation: A case study in Kampala, Uganda. *Computers, Environment and Urban Systems, 65*, 53–65. doi:10.1016/j.compenvurbsys.2017.04.013

Pérez-Molina, E., Sliuzas, R., van Maarseveen, M. F. A. M., & Jetten, V. (2018). *The use of Bayesian statistics to calibrate cellular automata models of urban growth: Deriving and validating transition rules for Kampala, Uganda, 2001–2016*. Unpublished manuscript.

ROM Transport, Shapira Hellerman Planners, Aberman Associates, & Tzamir Architects and Planners. (2012). *Kampala Physical Development Plan. Updating Kampala Structure Plan and Upgrading the Kampala GIS Unit*. Kampala, Uganda: Kampala City Council Authority.

Sene, K. (2010). *Hydrometeorology. Forecasting and applications*. Heidelberg: Springer.

Sliuzas, R., Jetten, V., Flacke, J., Lwasa, S., Wasige, J., and Pettersen, G. (2013a). *Flood Risk Assessment, Strategies and Actions for Improving Flood Risk Management in Kampala*. Enschede, the Netherlands: ITC and UN-HABITAT.

Sliuzas, R., Flacke, J., & Jetten, V. (2013b). Modelling Urbanization and Flooding *in Kampala, Uganda*. Paper presented at N-AERUS XIV/GISDECO: Urban Futures. Multiple Visions, Paths and Constructions?, Enschede, the Netherlands. Retrieved from http://n-aerus.net/web/sat/workshops/2013/PDF/N-AERUS14_sliuzas%20et%20al%20Final_FINAL.pdf

Steinitz, C., Faris, R., Flaxman, M., Vargas-Moreno, J. C., Canfield, T., Arizpe, O., & Lambert, C. D. (2005). A sustainable path? Deciding the future of La Paz. *Environment: Science and Policy for Sustainable Development, 47*(6), 24–38. doi:10.3200/ENVT.47.6.24-38

Steinitz, C. (2012). *A framework for geo design: Changing geography by design*. Redlands, CA: ESRI.

United Nations. (2014). *World urbanization prospects, the 2014 revision*. New York: United Nations, Department of Economic and Social Affairs.

Vargas-Moreno, J. C., & Flaxman, M. (2012). Participatory climate change scenario planning and simulation modeling: Exploring future conservation challenges in the Greater Everglades Landscape. In Karl, H., Scarlett, L., Vargas-Moreno, J. C. & Flaxman, M. (Eds.), *Restoring lands – coordinating science, politics and action* (pp. 27–56). Dordrecht, the Netherlands: Springer.

Vermeiren, K., Van Rompaey, A., Loopmans, M., Serwajja, E., & Mukwaya, P. (2012). Urban growth of Kampala, Uganda: Pattern analysis and scenario development. *Landscape and Urban Planning, 106*(2), 199–206. doi:10.1016/j.landurbplan.2012.03.006

Vermeiren, K., Vanmaercke, M., Beckers, J., & Van Rompaey, A. (2016). ASSURE: A model for the simulation of urban expansion and intra-urban social segregation. *International Journal of Geographical Information Science, 30*(12), 2377–2400. doi:10.1080/13658816.2016.1177641

White, R., & Engelen, G., (1993). Cellular automata and fractal urban form: A cellular modelling approach to the evolution of urban land use patterns. *Environment and Planning A, 25*(8), 1175–1199. doi:10.1068/a251175

White, R. (1998). Cities and cellular automata. *Discrete Dynamics in Nature and Society, 2*(2), 111–125. doi:10.1155/S1026022698000090

Xiang, W. N., & Clarke, K. C. (2003). The use of scenarios in land-use planning. *Environment and Planning B, 30*(6), 885–909. doi:10.1068/b2945

18 Volunteered Geographic Information (VGI) for the Spatial Planning of Flood Evacuation Shelters in Jakarta, Indonesia

Adya Ninggar Laras Kusumo, Diana Reckien, and Jeroen Verplanke

CONTENTS

18.1 Introduction 307
    18.1.1 Background..307
    18.1.2 Case Study Area, Flooding History and Response 309
18.2 Data and Methods..311
    18.2.1 Data Retrieval and Processing..311
        18.2.1.1 Twitter Data 311
        18.2.1.2 Primary Data on Residents' Preferences on Evacuation Shelters..............312
        18.2.1.3 Secondary Data on Evacuation Shelters 312
    18.2.2 Data Analysis and Methods...312
        18.2.2.1 Location of Twitter Users in or Near Evacuation Shelters.........312
        18.2.2.2 Spatial Pattern of Twitter Users in or Near Evacuation Shelters 313
        18.2.2.3 Shelter Preferences Among Twitter Users 314
18.3 Results..315
    18.3.1 Location of Twitter Users in or Near Evacuation Shelters 315
    18.3.2 Spatial Pattern of Twitter Users in or Near Evacuation Shelters..........316
    18.3.3 Preferences for Evacuation Shelters Among Twitter Users...................316
    18.3.4 Comparison of Sites of Official and Unofficial Evacuation Shelters........318
18.4 Discussion..319
18.5 Conclusion 321
References...321

18.1 INTRODUCTION

18.1.1 BACKGROUND

During disasters and crises, social media have been acknowledged as key communication channels (Wendling et al., 2013). Authorities and emergency response agencies use social media as a valuable source of information, as well as a useful platform for the rapid delivery of it (Kreiner and Neubauer, 2012). Social media assist the response to and management of disasters by, for example, providing a platform for sending alerts, identifying critical needs and focusing responses (Carley et al., 2015). Inhabitants use social media to request help during crises, share views and experiences on the events and criticise responses of government agencies and other organizations (Takahashi et al., 2015). Social media thereby allow people to participate in disaster response and management (Goodchild, 2007).

Using social media, people often voluntarily provide data about their own locations – known as volunteered geographic information (VGI). VGI can be made accessible by harnessing tools for assembling and disseminating these geographic data (Goodchild, 2007). In this form, VGI can aid disaster response and management (Takahashi, Tandoc and Carmichael, 2015) by increasing the speed of interaction between victims and relief organizations. Some social media applications, including Twitter™, Flickr™ and Open Street Map™, incorporate VGI by offering a geo-location feature (Schade et al., 2011). All these can be used in situations requiring near real-time disaster response and management (Carley et al., 2015; Kreiner and Neubauer, 2012; Takahashi et al., 2015), both before and during a disaster. VGI is, however, rarely used to address spatial planning problems arising after an extreme weather event or disaster that relate to mitigation of future disasters and/ or climate change adaptation. VGI has considerable potential for such applications and appears to offer advantages over traditional methods. For example, VGI could be used for the planning of sites for evacuation shelters, enabling residents' knowledge and preferences for shelters to be captured in a much faster, more timely and more comprehensive fashion than is possible with, for example, questionnaires and other types of surveys.

The research discussed in this chapter investigates the advantages and disadvantages of using VGI for capturing residents' preferences for (official and unofficial) flood evacuation shelters and explores the usefulness of this information for urban planning. We were motivated to carry out this study by the observation that official evacuation shelters provided by the authorities are frequently not used by residents – a fact known, for example, within Jakarta's city planning department for a long time (one of us works there) and established by research at least 30 years ago (Perry, 1979). A variety of factors determine people's use of official and non-official evacuation shelters (Stein et al., 2010). According to Rahman et al. (2014), these include aspects such as drainage capacity, soils, technical feasibility, shelter capacities, basic facilities present, environmental impact, accessibility, land availability and levels of maintenance. Additional reasons are to be uncovered by our research. We used Twitter data related to flooding in Jakarta, Indonesia to address this objective and the role of VGI.

BOX 18.1   Case Study Area

The province of **Jakarta** or 'Special Capital Region of Jakarta' (DKI Jakarta) is the capital and largest city in Indonesia. Jakarta has an estimated population of over 10 million people in 2016, up from 9,607,787 recorded during the 2010 Census. It is, furthermore, the largest city in Southeast Asia and one of the most populous urban agglomerations on Earth. Jakarta has been selected as case study because of 1) its recurrent problems with flooding and, consequently, its need for a functioning system of flood evacuation shelters; and 2) the popularity of using Twitter among the residents of Jakarta. Flooding affects Jakarta on an almost annual basis and has been an issue since the colonial era. As a response to flood emergencies, the

Jakarta government's Jakarta Disaster Management Agency (BPBD Jakarta), in collaboration with the SMART Infrastructure Facilities and Twitter, recently provided an online resource known as 'Peta Jakarta'. Peta Jakarta (@petajkt) is a system that utilises social media to gather, sort and display information about flood events in Jakarta in real time (BPBD Jakarta, 2015). In this respect, authorities in Jakarta are among the forerunners in the use of social media for coordinating emergency flooding responses.

18.1.2 CASE STUDY AREA, FLOODING HISTORY AND RESPONSE

The province of Jakarta or 'Special Capital Region of Jakarta' (DKI Jakarta) is the capital of Indonesia (Figure 18.1). Jakarta Province has a total area of 662 km² and comprises five administrative cities on the mainland and one administrative coastal region that covers the marine area and islands to the north of the mainland. Only the five administrative cities, with their 267 sub-districts, have been considered in our research (see panel on right hand side of Figure 18.1). Jakarta not only is the largest city in Indonesia and the largest city in Southeast Asia but also is one of the most populous urban agglomerations on Earth. Its estimated population was in 2016 greater than 10 million, up from 9,607,787 recorded during the 2010 Census. Jakarta is now considered a global city and home to one of the fastest growing economies in the world (World Population Review, 2018).

Jakarta was selected as a case study because of 1) its recurrent problems with flooding and the need for a functioning system of flood evacuation shelters; and 2) the popularity of using Twitter among Jakarta's residents.

Flooding affects Jakarta almost annually and has been an issue since the colonial era. Historical records show major floods occurred in 1654, 1872, 1909 and 1918 (Team Mirah

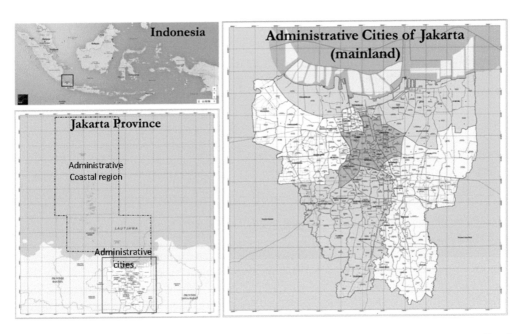

FIGURE 18.1   Map of Jakarta Province (5°19'12"–6°23'54" S, 106°22'42"–106°58'18" E). (Sources: Google (2015), Jakarta Capital City Government (2014)).

Sakethi, 2010). Currently, flooding occurs nearly every year. In 2002 and 2007, Jakarta was severely affected by two '50-year' floods' (i.e. floods with a statistical probability of occurring once every 50 years). According to Firman et al. (2011), the 2002 flood covered about one-fifth of Jakarta's total area. Hundreds of thousands of people were made homeless, 68 persons were killed, 190,000 people suffered from flood-related illnesses and about 422,300 had to be evacuated. Flood losses were estimated at 9 trillion Indonesian Rupiahs (USD 998 million) (Akmalah and Grigg, 2011).

As a flood emergency response, the Jakarta Disaster Management Agency (BPBD Jakarta), in collaboration with the SMART Infrastructure Facilities and Twitter, recently launched an online resource known as 'Peta Jakarta'. Peta Jakarta (@petajkt) utilises social media to gather, sort and display information about flood events in Jakarta in real time (BPBD Jakarta, 2015). Jakarta's residents can also use Peta Jakarta to report on situations and conditions in their neighbourhood: information on flooding, evacuation processes, traffic jams and other flood-related problems. One of the reasons behind the development of the Peta Jakarta system was the enormous volume of volunteered geographic information (VGI) being generated by residents of Jakarta through their use of social media, particularly tweets. A tweet is any message posted through Twitter and may contain photos, videos, links and up to 140 characters of text (see http://www.twitter.com). After analysing a sample of 10.6 billion public tweets posted by 517 million Twitter users, Semiocast (2012), social media intelligence consultants, attributed more than 2% of those tweets to Twitter accounts in Jakarta. This means that Jakarta holds first place as the city most actively using Twitter among all cities worldwide. And, judged by the number of people visiting the Twitter site per month on a country basis (Smart Insights, 2015), Indonesia is recognised as the country with the most Twitter users. In this respect, the authorities in Jakarta are forerunners in the use of social media by using Twitter to coordinate flood emergency response. Given that it can provide a robust sample of Twitter users and its regular exposure to flood emergencies, Jakarta is an appropriate and interesting case for studying the potential of VGI.

BOX 18.2   Methods Applied in the Chapter

The analysis in this chapter investigates the advantages of using volunteered geographic information (VGI) for capturing residents' preferences for (official and unofficial) flood evacuation shelters and explores the usefulness of VGI for related urban planning tasks. A geographic information system (GIS) was used to buffer the actually recorded location of tweets – to offset inaccuracies during the transmission process and to spatially join the sites of residents' unofficial shelters with the locations of official shelters. These geo-processing tasks were conducted with ESRI ArcGIS™ (Version 10.2.2.). Data include secondary data in the form of (1) geo-located Twitter data retrieved from the 'Digital On-line Life and You' (DOLLY) archive using the Twitter Application Program Interface (API); (2) locations of official evacuation shelters and land use categories supplied by the Jakarta City Planning Department (DPK); and (3) the distribution of flood areas supplied by the Jakarta Disaster Management Agency (BPBD Jakarta).

The method could also be applied in cases of other extreme events for which there is a need for an emergency response, for example heat waves and earthquakes. The method has the advantage of relatively easily processing large amounts of VGI and residents' preferences; one disadvantage is the limited insight to be gained from short tweets. It makes sense, therefore, to use VGI in combination with supplementary methods that can provide more insight into residents' location preferences.

18.2 DATA AND METHODS

To establish the usefulness of VGI data for planning the location of evacuation shelters, first we determined the location of Twitter users in or near evacuation shelters. The second step was to determine the spatial pattern of these users in or near specific types of evacuation shelters, followed by the third and last step, which was to determine their preferences regarding the use of these particular types of shelters. Our study employs both secondary and primary data, which was analysed using several methods (Figure 18.2). A description of the retrieval, processing and analysis of the data to carry out the three steps mentioned above follows hereafter.

18.2.1 Data Retrieval and Processing

Our study makes use of both primary and secondary data. Primary data was collected to determine the preferences of individuals – Twitter users – regarding shelters. Initially, the preferences were thought to be elicited by way of a questionnaire sent to people that had previously tweeted about their flood experience. However, due to a low response rate for the questionnaire, secondary data had to be used. Secondary data comprised information on Twitter use as well as land-use categories and other spatially explicit GIS data from the statistical offices, i.e. the Jakarta City Planning Department (DPK) and the Jakarta Disaster Management Agency (BPBD Jakarta).

18.2.1.1 Twitter Data

Twitter data was retrieved from the 'Digital On-line Life and You' (DOLLY) archive – the massive database of geo-located Twitter data. Developed by the Floating Sheep collective, the DOLLY Project is a repository of billions of geo-located tweets that allows real-time research and analysis. Building on top of existing open source technology, the Floating Sheep collective has created a back-end that ingests all geo-tagged tweets (~8 million a day) and does basic analysis, indexing and geocoding to allow real-time search of the entire database (3 billion tweets since December 2011 (Zook et al., 2016)) – using the Twitter Application Program Interface (API). An API is a set of routines, protocols and tools for building software applications. According to Durahim and Coşkun

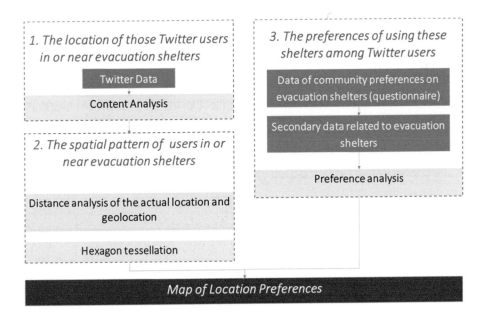

FIGURE 18.2 Analytical framework.

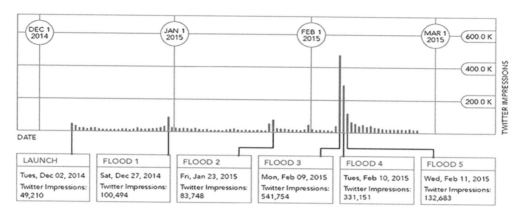

FIGURE 18.3   Twitter impressions during the 2014/15 flooding period (Holderness and Turpin, 2015).

(2015), Twitter API is the means most commonly used to gather data from Twitter. We requested geo-tagged information and received a random sample of 1% of all geotagged tweets with and without keywords sent in Jakarta. Spatially, the study includes data within the Jakarta bounding box, i.e. between 5.20166° and 6.37248°S and 106.390266° and 106.974274°E. As this bounding box does not exactly represent the administrative boundary of Jakarta Province, the data frame was clipped.

Temporally, the study focuses on tweets that were sent during two periods of severe flooding that were categorised by BPBD Jakarta (2015) as events demanding an emergency response. The flood periods analysed were from December 2013 to March 2014 and from December 2014 to March 2015 (Figure 18.3). The retrieved data contain one or more relevant hashtags and keywords, for example #banjir, #banjirjkt, #evakuasi, #logistik, #relawan, pengungsi, korban, and @petajkt. These hashtags and keywords were chosen by the authors (the first author is a native Indonesian living in Jakarta) in cooperation with Mrs Fitria Sudirman from Peta Jakarta (http://www.petajakarta.org). Peta Jakarta was the official operational consultant of DKI Jakarta responsible for managing Twitter reports from residents. The retrieved data therefore represents the locations of Twitter users in Jakarta Province commenting about the floods at the time the flooding was taking place, i.e. during the two periods of major flooding in 2013/14 and 2014/15.

18.2.1.2   Primary Data on Residents' Preferences on Evacuation Shelters

Primary data were initially aimed to capture residents' preferences regarding shelter locations by way of a questionnaire. Residents were selected based on their tweeting behaviour regarding evacuation-shelter locations and their apparent role in the flooding disaster, i.e. residents were identified as evacuees (see Figure 18.4). The questionnaire was designed in Survey Monkey. It contained a mix of open and closed questions about respondents' use of evacuation shelters during previous flood events. The link to the questionnaire was sent to residents through their Twitter accounts.

18.2.1.3   Secondary Data on Evacuation Shelters

Secondary data on the distribution of official evacuation shelters were collected from the Jakarta Disaster Management Agency (BPBD Jakarta) and the Jakarta City Planning Department (DPK). The data were used to compare residents' preferences for shelter locations against the locations of official evacuation shelters. Table 18.1 lists all secondary data types and sources.

18.2.2   Data Analysis and Methods

18.2.2.1   Location of Twitter Users in or Near Evacuation Shelters

Content analysis was done in two steps: first, we checked the validity or relevance of the downloaded geo-located Twitter data; second, we performed content classification analysis of relevant tweets (Figure 18.4).

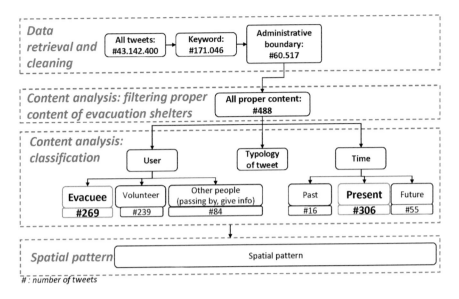

FIGURE 18.4 Flowchart of the analysis of Twitter data.

Atlas.ti™ software was used to check the validity – and thus relevance – of tweets to the topic of evacuation shelters. The purpose of this check was to filter for tweets that were contextually relevant to flood evacuation shelters in Jakarta. We used a deductive approach for content classification analysis, starting with predefined keywords regarded by an expert (in this case, the researcher in cooperation with Mrs Fitria Sudirman from Peta Jakarta) as being relevant. For instance, keywords like #banjir (flood) and #evakuasi (evacuation), and related words derived from the same roots (see also Section 18.2.1 above) (Holderness and Turpin, 2015), were used to determine relevant tweets.

Approximately 135,885 tweets were sent between December 2013 and March 2014, while 35,160 tweets were sent between December 2014 and March 2015. Subsequent data-frame clipping then yielded 60,517 tweets that were sent from within the administrative boundary of Jakarta Province (Figure 18.4 and Figure 18.5).

18.2.2.2 Spatial Pattern of Twitter Users in or Near Evacuation Shelters

People who send tweets with geo-location data indirectly disclose their location, as it can be collected from the geo-tagged tweet. However, there is a degree of inaccuracy in this. The Twitter location does not necessarily match the actual location of the person tweeting due to, for example, weak connectivity, the person tweeting being in motion or other forms of time lag between sending and reception. To deal with this inaccuracy, tweets can be grouped into spatial units in which the degree of accuracy required for data analysis is met.

TABLE 18.1

Types and Sources of Secondary Data Used

Data	Year	Source
Road	2014	Jakarta City Planning Department (DPK)
Provincial boundary	2014	Jakarta City Planning Department (DPK)
Flood area (flooded to a depth >10 cm)	2002, 2007, 2014/15	Jakarta Disaster Management Agency Jakarta City Planning Department (DPK)
Land use category	2014	Jakarta City Planning Department (DPK)
Locations of official evacuation shelters	2013/14; 2014/15	Jakarta City Planning Department (DPK)

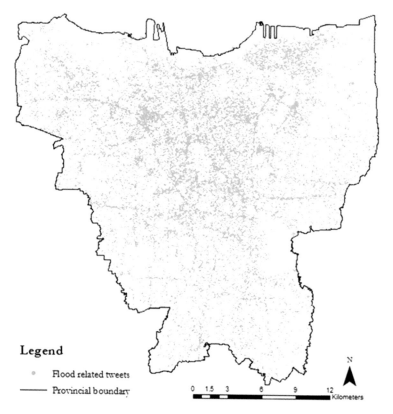

FIGURE 18.5 Source locations of 60,517 tweets related to the floods sent during the 2013/14 and 2014/15 floods in Jakarta Province.

The location of the Twitter data is obtained as a point feature, which can be ascribed to administrative boundaries or any other spatial unit (Poorthuis et al., 2014). The selection of the appropriate type of spatial unit is highly dependent on the purpose of the research, which in this case required using the smallest spatial unit available. Other scholars have used buildings or land-use categories when representing shelter sites (Chang and Liao, 2014; Gall, 2004; Kar and Hodgson, 2008).

It is necessary to consider the positional accuracy in order to select the most appropriate spatial unit. Many studies have investigated the accuracy of VGI (Goodchild and Li, 2012). Haklay (2010), for example, compared the data of Open Street Map with survey data, which showed an average deviation between the geo-location and actual location of 6 m. In our study, the accuracy was assessed using the content of tweets as control data (Comber et al., 2013). Using purposive sampling, i.e. tweets that clearly mentioned the location of the person in the text, we were able to test the distance between the geo-location and the actual location. The mean distance was used as a basis for choosing the spatial unit.

18.2.2.3 Shelter Preferences Among Twitter Users

Analysis of residents' preferences for locations of evacuation shelters was conducted using a qualitative, comparative method that considered the following data:

- land use categories;
- evacuation shelters officially classified as such by Jakarta's planning boards;
- results of the questionnaire (insufficient due to low response rate).

In our study, land use categories were used to categorise and characterise the location and preferences of evacuation shelters, because this is a categorization that is useful for urban planning and management. However, other studies might characterise evacuation shelters based on other features, such as height or age of the building, which may also be important characteristics. We determined the number of people tweeting from shelter sites located in each land-use category, and how many of them were at or near official evacuation shelters. In this way we were able to determine patterns of use of unofficial shelters and that of official evacuation shelters.

18.3 RESULTS

18.3.1 LOCATION OF TWITTER USERS IN OR NEAR EVACUATION SHELTERS

Our analysis shows that during 2013/2014 and 2014/2015 floods, 306 tweets could be recognised as coming from locations of evacuation shelters. By overlaying the tweet data on the flooding maps of 2013/2014 and 2014/2015, we were able analyse the spatial distribution of tweet locations.

The locations of tweets for 2013/2014 were clustered in the central area of Jakarta – the Kampung Pulo neighbourhood of Jatinegara District – which was most severely affected by the flood event (Figure 18.6a). Some areas in Kampung Pulo are located on the floodplain of Jakarta's biggest river, the Ciliwung River. For many years the floodplain has been the location of slums, inhabited by low-income residents. Like many other slums (Kit et al., 2011), Kampung Pulo occupies an area where the risk of flooding is high (Khomarudin et al., 2014). In response to the higher risks faced in Kampung Pulo as compared with other locations in Jakarta, voluntary organisations have provided aid to set up evacuation shelters.

In comparison with the 2013/2014 flood event, the area of flooding in 2014/2015 was smaller and the impact of the flood event lower, which is most likely also the reason why there were fewer tweets during the 2014/2015 event. According to BPBD Jakarta (2015), the time of inundation from December 2013 to March 2014 was on the 4th, 20th, 20th and 8th days of the consecutive months, respectively; whereas from December 2014 to March 2015 inundation began on the 5th, 2nd, 7th and 4th days of the consecutive months, respectively.

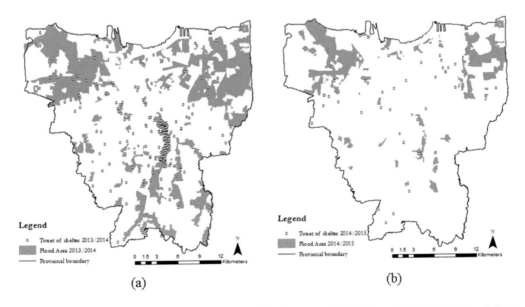

(a)   (b)

FIGURE 18.6 Tweets about evacuation shelters and flooded areas (a) 2013/14 and (b) 2014/15. (Data: Jakarta Disaster Management Agency (BPBD Jakarta); Jakarta Planning Board; Zook et al., 2016).

Only 48 tweets mentioned evacuation shelters during the 2014/2015 flooding, while there were 258 tweets in the 2013/2014 floods. The tweets were dispersed over several locations throughout the city, rather than being concentrated in the area of flooding, as was the case for the 2013/2014 flood event (Figure 18.6b).

18.3.2 Spatial Pattern of Twitter Users in or Near Evacuation Shelters

Of the 306 tweets related to evacuation shelters, 86 mentioned the detailed location of the shelter. These 86 tweets were used as a sample to calculate the mean distance between the geo-location and the actual location based on what was mentioned in the tweet. The average distance between the geo-location and the actual location was 188.28 m, ranging from 0 to 5,405 m.

Since the distances between the geo-location and the actual location were quite large, each tweet point was converted so that its location was recorded as being within one of the spatial units selected for further analysis. This procedure helped to group tweets from different geo-locations that mentioned the same location in the tweeted text. There are several types of spatial units that could accommodate the mean distance of 188.28 metres. Buffers, hexagons and land-use zones are some of them.

The first spatial unit considered was that of land-use zones, i.e. polygons ranging in size from 3 m² to 3 km². If, however, we directly converted the point of tweets into land-use zone locations using spatial joining without buffering, the result could be misleading when the evacuation shelter lies close to a border between different land-use categories. Tweets that were actually posted close to, but not from within, evacuation shelters (or have been posted from within but were recorded as being outside due to transmission speed) could already belong to a different land-use zone.

Hexagon tessellations are commonly used to simplify point data (Raposo, 2013). According to Birch, Oom and Beecham (2007), hexagon tessellations have several advantages, for example, over regular square grids. The nearest neighbourhood is more symmetrical in a hexagonal tessellation as compared to a rectangular grid since the length of each line of the hexagon contour is equal. The width of hexagons has conformity, meaning that using hexagons allows an area to be covered without overlapping neighbourhoods. Data can also be visualised more clearly.

In our study, the hexagon used had sides of an equal length of 200 m, based on the 188.28 m mean distance between geo-locations and actual locations. Following conversion, the 306 tweets from residents' evacuation shelters (points) were located in 215 hexagons representing residents' shelter sites.

18.3.3 Preferences for Evacuation Shelters Among Twitter Users

Twitter users' preferences for evacuation shelters were analysed using results from the questionnaire combined with the spatial pattern of the hexagons. Responses to the questionnaire came from people identified as evacuees – in contrast to volunteer workers, for example. In total, 269 relevant tweets sent from 184 Twitter accounts were identified as having been sent by evacuees. However, evacuees and volunteers could not always be differentiated. For example, when residents mentioned 'I am at an evacuation shelter' or only gave general information about the evacuation shelters, these have been classified as 'other people'.

In order to get information on the location preferences of evacuees, we sent the link to a questionnaire to the 184 Twitter accounts of evacuees. Several challenges were encountered in getting feedback from them. At first, people tended to ignore the questionnaire altogether, which led us to send several reminders. After six reminders over the course of six days, only three Twitter account users had provided feedback on their preferences. The low number of questionnaires returned lends support to one of our initial arguments: that obtaining information about preferences of shelter use is difficult to obtain through traditional methods of data collection, i.e. questionnaires. Another challenge was the limited number of characters (140) allowed in Twitter, which restricted the amount of information we could provide to introduce the questionnaire.

Subsequently, we used the distance of evacuation shelters from the flood area as one of the criteria determining residents' preferences in selecting an evacuation shelter (American Red Cross, 2002; FEMA, 2015; Kar and Hodgson, 2008). From the mean distance of the location of each resident's evacuation shelter to the nearest flood-prone area, we found that shelter sites in Jakarta were mostly located within flood-prone areas. About 60% of the sites of residents' evacuation shelters (hexagons) in Jakarta were within an area flooded in 2013/14 and/or 2014/15 (Figure 18.7). One explanation for this high number is that people tend to look for a safe location near their home. For example, some evacuees take refuge on the second floor of a neighbour's house.

Kongsomsaksakul, Yang and Chen (2005) mention that the ideal location for an evacuation shelter is outside the flooded area, but within 1 km of it. In the case of Jakarta, about 31% of residents' shelter sites (hexagons) meet this criterion – a fact that was also confirmed by answers in the questionnaire – indicating that evacuation shelters used by respondents were located very close to the flooded area (i.e. distances of between 200 m and 1 km).

Respondents mentioned that the main reason for their choice of evacuation shelters was accessibility, safety from flooding and proximity to their home. Accessibility is clearly an important factor for people considering where to go when they have to evacuate (CCCMCluster, 2014; Tai et al., 2010). One respondent added that the closeness of the evacuation shelter to their house allowed them to monitor conditions at their home at any time. The average distance from shelters to respondents' houses was 200–300 m. Another respondent, however, mentioned that his shelter was 2 km away from his house. He added that this shelter, provided by a religious organization, was the closest one he could reach.

All respondents stated that they travelled to evacuation shelters on foot – none used a car, motorbike or public transportation. This is consistent with research by Chang and Liao (2014), who found that people chose to walk rather than drive to evacuation shelters. In contrast, Kar and Hodgson (2008) assumed that people usually travelled by car to the shelters. The mode of transport used to reach evacuation shelters is also dependent on the type and impact of a flood. For example, one tweet used in this study stated that the person tweeting had to use a rescue boat to reach an

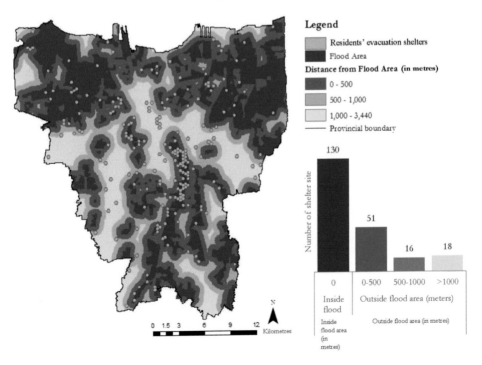

FIGURE 18.7　Distance between residents' evacuation shelters and flood areas. (Data: Jakarta Disaster Management Agency (BPBD Jakarta, 2013, 2015; Jakarta Planning Board, 2002, 2007; Zook et al., 2016).

evacuation shelter. According to our findings, however, the walking distance to the shelter should be the main factor considered when planning evacuation shelters in Jakarta.

Additionally, we identified the land-use types coinciding with the locations of residents' evacuation shelters on a land-use map. It was not always possible to locate the shelter precisely on a particular type of land use since the spatial unit for shelters are hexagons, and a hexagon can contain several types of land use. Thus, our analysis could only provide a general overview of types of land use preferred for the location of evacuation shelters.

However, results show that the evacuation shelters are mostly located in open/green spaces. This is also the land-use type most frequently used when siting official evacuation shelters provided by the government. Some of these official shelters were tents, rather than permanent buildings, e.g. the central evacuation shelter of Jakarta. This shelter was located in one of the largest open/green spaces in the city, where it also served as a logistics and coordination centre. Another tent shelter was located at the train station, erected in a park that is part of the buffer zone alongside the railway. The second most common land-use type for the location of unofficial shelters was residential land. Many people found shelter near their home, often provided by neighbours or family members. Several tweets sent by volunteers indicated that they had opened their houses as a temporary shelter for their neighbours. Offices were the third most common type of land use used for unofficial shelters, at sites mainly chosen by residents. Tweets indicate that several shelters were located in the basements of office buildings. Figure 18.8 shows the distribution of land-use types used for unofficial evacuation shelters.

18.3.4 COMPARISON OF SITES OF OFFICIAL AND UNOFFICIAL EVACUATION SHELTERS

By comparing the siting of unofficial evacuation shelters and those of official evacuation shelters, we were able to obtain an overview of people's use of official evacuation shelters. At the time of our study, there were 2,645 official evacuation shelters.

The spatial unit that shows the location of official evacuation shelters is based on land-use zones, which are different from the hexagons that identify the sites of unofficial evacuation shelters. Hence, it is not possible to determine with absolute certainty whether or not residents used the official evacuation shelters, even when the determined location of the tweeting resident was very close to an official shelter site. To deal with this issue, we analysed the spatial join between official and residents' shelter sites, assuming that the intersection of an official evacuation shelter with the hexagon of a residents' evacuation shelter site points to the use of an official evacuation shelter. Our analysis yields an intersection of 35.6%, indicating that about a third of the residents' shelter sites intersected with official evacuation shelters (Figure 18.9).

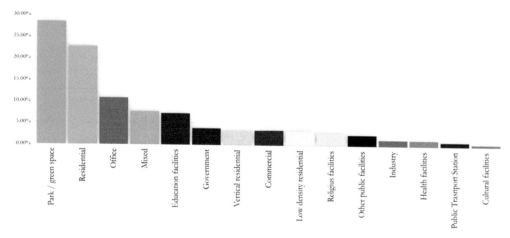

FIGURE 18.8 Type of land use at locations used for unofficial evacuation shelters. (Data: Jakarta Capital City Government, 2014).

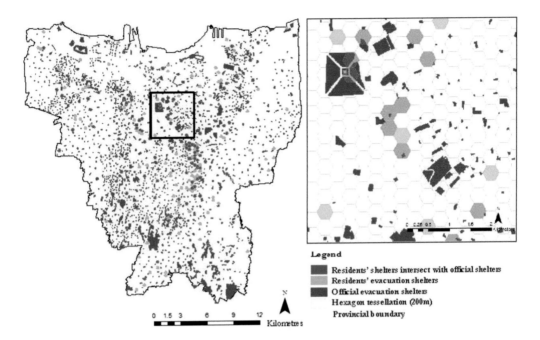

FIGURE 18.9 Intersection of residents' shelter sites with official evacuation shelters. The right inset shows a close-up as an example. Red polygons mark official evacuation shelters. Blue hexagons show residents' shelter sites that intersect with the location of official evacuation shelters. Green hexagons show residents' shelters that do not coincide with official shelters. (Data: Jakarta Capital City Government, 2014; Zook et al., 2016).

Our results show that 53.5% of the cases where official shelters and residents' shelter locations coincided were educational centres. Green space was the second most common type of land use upon which official and residents' shelters coincided – accounting for 29.6% of the total. The remaining cases concerned religious, health or sports centres.

In their daily lives, people use and are familiar with education centres, so they are aware of their locations. Moreover, education centres can be found in every neighbourhood of Jakarta: there are about 2,700 public schools and 4,100 private schools of all levels (DKI Jakarta Province Government, 2015).

Our results suggest that the sparse use of official evacuation shelters relates to people's lack of awareness. In response to the questionnaire, respondents mentioned that they never used the official evacuation shelters because they did not know they existed. One respondent mentioned that official evacuation shelters are only set up when flooding occurs. Moreover, if no evacuation shelters were provided by the government, respondents indicate that they would prefer to go to a neighbour's house or that of a family member that was safer than their house. Two of the respondents mentioned that they often shelter at sites known from their daily activities, one being a religious centre and the other the home of family members. Location preferences were strongly influenced by the familiarity that evacuees felt with the shelter site.

In this respect, our analysis shows that social networks in a community are also very important when disasters occur. Analysis of VGI data has provided important insights into the distribution and patterns of use of unofficial and official evacuation shelters during flood events in Jakarta.

18.4 DISCUSSION

Analysis shows that using VGI for disaster planning and management has more potential applications than just capturing real-time information (Erskine and Gregg, 2012). In this study, we have focused on information relating to evacuation shelters during flood events. As a result, the location of unofficial evacuation shelters used by evacuees during floods could be identified. We have shown that the general pattern of evacuation-shelter locations captured using VGI could provide important inputs for evacuation-shelter planning.

This research applies to a large area, some 662 km^2. Nevertheless, residents' preferences for evacuation-shelter locations could be determined in a relatively short time. Using just the VGI data, we were also able to map the distribution of evacuation shelters without conducting a field survey. Obtaining the same information using traditional data collection methods would have been time-consuming and costly (e.g. Mooney et al., 2011). Our study shows that VGI has potential as a cost-effective substitute for traditional data collection methods.

Another benefit of using VGI, and specifically Twitter datasets, is the ease it offers in selecting and accessing information. The analysis of Twitter content uncovered various types of information related to the evacuation shelters. Firstly, VGI analysis could identify different types of users based on the content of their tweets. In our research, Twitter users were classified as either volunteers, representatives of government agencies and NGOs or evacuees. Evacuees were then selected for a questionnaire on preferences regarding evacuation shelters. Thus, VGI was in this case also a plat-form for identifying potential survey participants as a sample of the population. Moreover, through content analysis – based on the use of verb tenses in tweeted sentences – we were also able to identify the time frame of tweets. We could, for example, identify who was in an evacuation shelter at the time they sent their tweet. Using this type of information, we could pinpoint the location of shelters more accurately.

Nevertheless, content analysis of tweets can be time-consuming and potentially could reduce the advantage of time-efficiency. Choosing proper keywords is therefore crucial, as are adjustments in keywords in the iterative process of content analysis. Several factors need to be considered. First, in filtering the content of the tweets, synonyms of each keyword should be considered. Some people may use another word with the same meaning as a more common alternative. Slang should also be included, especially if the users are young people. Moreover, adjectives, verbs and nouns formed from the same root-word all need to be included. Another factor to consider is the use of metaphors. Keywords are influenced by the characteristics of each language, and the same keyword may have many connotations. All these aspects were taken into account in the content analysis of tweets for our study. However, despite the care taken to guarantee consistent output in the content analysis, unorthodox use of language and particularly the use of metaphors could lead to the inclusion of tweets whose content is not relevant to the topic under consideration – a possibility that cannot be ruled out completely.

There are other drawbacks with using VGI. The problem of geographic (in)accuracy is currently a concern of many researchers. It was also an issue in this study. The accuracy assessment established considerable deviation between the geo-location supplied with the Twitter data and the actual location mentioned in the content of tweets. There are many possible explanations for this. One possible reason is that people tweet while moving; this would affect the geo-location, particularly if a time interval elapsed before the tweet was sent. It is also possible that people send tweets about their experiences in a shelter after moving away from that location. Poorthuis et al. (2014) argue that various technologies of geo-tagging have issues with accuracy. Many types of GPS and Wi-Fi also influence the accuracy of geo-tagging. Despite these shortcomings, our analysis shows that the advantages of using VGI data outweigh potential disadvantages.

Our study suggests that VGI can be used in the planning/siting of evacuation shelters using the feedback from targeted respondents. Evacuees were asked to answer a questionnaire sent to their Twitter account. Unfortunately, we received only three responses out of a total of 184 accounts to which the questionnaire was sent. When people were confronted with a long list of questions on an issue that had long passed, i.e. the flooding crisis, they did not respond. We conclude, therefore, that Twitter is not a useful platform for getting in-depth feedback from residents: careful content analysis of the tweets was the main source of information in this study. This result contradicts Brabham (2009), who maintains that VGI has the potential to elicit active public participation in urban planning projects. Our analysis shows that Twitter works best as a source of information provided voluntarily by residents. Requesting people to respond and participate more actively in urban planning initiatives was not successful in this case.

Overall, our analysis was able to capture residents' preferences regarding siting of evacuation shelters through VGI and location identification. The pattern analysis of these shelter locations was an important step in investigating why people prefer a certain shelter (both official and unofficial) or why not – a crucial piece of information for future planning of evacuation shelter sites. Some of the technical limitations encountered could be overcome, at least in part, by using VGI in combination with other approaches, such as participatory mapping. Goodchild and Li (2012) see the role of VGI mainly in the initial and hypothesis-generating step of research, this due to technological limitations (e.g. accuracy). Overall, however, our analysis provides important insights into how the planning, organization and notification of official residents' evacuation shelters can be improved.

18.5   CONCLUSION

This research described in this chapter was focused on using VGI in evacuation shelter planning/siting as one crucial part of emergency response.

Our results show that 35.6% of people who sent tweets from evacuation shelters may have been using an official evacuation shelter. Overall, the most frequently used land-use category for evacuation shelters was 'green/open spaces'. In the case of official shelters, this was followed by the land-use type 'schools and education centres'. We conclude that people only used the official shelters when they knew of the location from their daily activities. The issue of unfamiliarity goes a long way to explaining the failure by residents to use official shelters near their homes.

Overall, VGI is a useful approach for capturing residents' preferences for particular evacuation shelters, when used in conjunction with, for example, land-use data. VGI data provide a preliminary overview of any topic of interest in the form of data of a general nature that covers a broad area. Our research shows, however, that VGI should be combined with other approaches to fully understand residents' preferences for specific spatial planning problems.

REFERENCES

Akmalah, E., & Grigg, N. S. (2011). Jakarta flooding: Systems study of socio-technical forces. *Water International, 36*(6), 733–747. https://doi.org/10.1080/02508060.2011.610729

American Red Cross. (2002). Standards of Hurricane Evacuation Shelter Selection.

Birch, C. P. D., Oom, S. P., & Beecham, J. A. (2007). Rectangular and hexagonal grids used for observation, experiment and simulation in ecology. *Ecological Modelling, 206*(3–4), 347–359. https://doi.org/10.1016/j.ecolmodel.2007.03.041

BPBD Jakarta. (2015). Flood Map Jakarta (in Bahasa).

Brabham, D. C. (2009). Crowdsourcing the public participation process for planning projects. *Planning Theory, 8*(3), 242–262. https://doi.org/10.1177/1473095209104824

Carley, K. M., Malik, M., Landwehr, P. M., Pfeffer, J., & Kowalchuck, M. (2015). Crowd sourcing disaster management: The complex nature of Twitter usage in Padang Indonesia. *Safety Science, 90,* 48–61. https://doi.org/10.1016/j.ssci.2016.04.002

CCCMCluster. (2014). *The Mend Guide Comprehensive Guide for Planning Mass Evacuations in Natural Disasters.*

Chang, H.-S., & Liao, C.-H. (2014). Planning emergency shelter locations based on evacuation behavior. *Natural Hazards, 76*(3), 1551–1571. https://doi.org/10.1007/s11069-014-1557-x

Comber, A., See, L., Fritz, S., Van der Velde, M., Perger, C., & Foody, G. (2013). Using control data to determine the reliability of volunteered geographic information about land cover. *International Journal of Applied Earth Observation and Geoinformation, 23,* 37–48. https://doi.org/10.1016/j.jag.2012.11.002

DKI Jakarta Province Government. (2015). Number of Schools in Jakarta (in Bahasa).

Durahim, A. O., & Coşkun, M. (2015). #iamhappybecause: Gross National Happiness through Twitter analysis and big data. *Technological Forecasting and Social Change, 99,* 92–105. https://doi.org/10.1016/j.techfore.2015.06.035

Erskine, M. A., & Gregg, D. G. (2012). Utilizing volunteered geographic information to develop a real-time disaster mapping tool: A prototype and research framework. In *International Conference on Information Resources Management (Conf-IRM)* (p. Paper 27).

FEMA. (2015). *Safe Rooms for Tornadoes and Hurricanes: Guidance for Community and Residential Safe Rooms.* Washington D.C.

Firman, T., Surbakti, I. M., Idroes, I. C., & Simarmata, H. a. (2011). Potential climate-change related vulnerabilities in Jakarta: Challenges and current status. *Habitat International, 35*(2), 372–378. https://doi.org/10.1016/j.habitatint.2010.11.011

Gall, M. (2004). Where to go? Strategic modelling of access to emergency shelters in Mozambique. *Disasters, 28*(1), 82–97. https://doi.org/10.1111/j.0361-3666.2004.00244.x

Goodchild, M. F. (2007). Citizens as sensors: The world of volunteered geography. *GeoJournal, 69*(4), 211–221. https://doi.org/10.1007/s10708-007-9111-y

Goodchild, M. F., & Li, L. (2012). Assuring the quality of volunteered geographic information. *Spatial Statistics, 1,* 110–120. https://doi.org/10.1016/j.spasta.2012.03.002

Google. (2015). Google Maps.

Haklay, M. (2010). How good is volunteered geographical information? A comparative study of OpenStreetMap and ordnance survey datasets. *Environment and Planning B: Planning and Design, 37*(4), 682–703. https://doi.org/10.1068/b35097

Holderness, T., & Turpin, E. (2015). From Social Media to GeoSocial Intelligence: Crowdsourcing Civic Co-management for Flood Response in Jakarta, Indonesia. In S. Nepal, C. Paris, & D. Georgakopoulos (Eds.), *Social Media for Government Services* (pp. 115–133). Springer International Publishing. https://doi.org/10.1007/978-3-319-27237-5_6

Jakarta Capital City Government. (2014). Detailed Spatial Plan and Zoning Regulation (In Bahasa), Pub. L. No. 1/2014, 1.

Kar, B., & Hodgson, M. E. (2008). A GIS-based model to determine site suitability of emergency evacuation shelters. *Transactions in GIS, 12*(2), 227–248. https://doi.org/10.1111/j.1467-9671.2008.01097.x

Khomarudin, R., Suwarsono, Ambarwati, D. O., & Prabowo, G. (2014). Flood Event Evaluation in Kampung Pulo DKI Jakarta and Risk Reduction Analysis Using Unmanned Air Vehicle (UAV) and High Resolution Remote Sensing (English Abstract) (in Bahasa). In LAPAN (Ed.), *Seminar Nasional Penginderaan Jauh* (pp. 611–619). Jakarta: LAPAN.

Kit, O., Lüdeke, M., & Reckien, D. (2011). Assessment of Climate Change-Induced Vulnerability to Floods in Hyderabad, India, Using Remote Sensing Data. In K. Otto-Zimmermann (Ed.), *Resilient Cities: Cities and Adaptation to Climate 35 Change Proceedings of the Global Forum 2010* (pp. 35–44). Springer Netherlands. https://doi.org/10.1007/978-94-007-0785-6

Kongsomsaksakul, S., Yang, C., & Chen, A. (2005). Shelter location-allocation model for flood evacuation planning. *Journal of the Eastern Asia Society for Transportation Studies, 6,* 4237–4252.

Kreiner, K., & Neubauer, G. (2012). Social Media for Crisis Management: Problems and Challenges from an IT-Perspective. In *20th Conference of Interdisciplinary Information Management Talks.* Jindřichův Hradec.

Mooney, P., Sun, H., & Yan, L. (2011). VGI as a Dynamically Updating Data Source in Location-Based Services in Urban Environments. In *Proceedings of the 2nd International Workshop on Ubiquitous Crowdsourcing – UbiCrowd'11* (p. 13). New York, New York, USA: ACM Press. https://doi.org/10.1145/2030100.2030105

Perry, R. W. (1979). Incentives for evacuation in natural disaster research based community emergency planning. *Journal of the American Planning Association, 45*(4), 440–447. https://doi.org/10.1080/01944367908976988

Poorthuis, A., Zook, M., Shelton, T., Graham, M, and Stephens, M. 2016. Using Geotagged Digital Social Data in Geographic Research. In Key Methods in Geography. eds. Clifford, N., French, S., Cope, M., and Gillespie, T. London: Sage. 248–269.

Rahman, A., Mallick, F., Mondat, S., & Rahman, M. R. (2014). Flood Shelters in Bangladesh: Some Issues from the Users Perspective. In A. E. Collins & S. Akerkar (Eds.), *Hazards, Risks and Disasters in Society* (Illustrated, p. 425). Elsevier.

Raposo, P. (2013). Scale-specific automated line simplification by vertex clustering on a hexagonal tessellation. *Cartography and Geographic Information Science, 40*(5), 427–443. https://doi.org/10.1080/15230406.2013.803707

Schade, S., Díaz, L., Ostermann, F., Spinsanti, L., Luraschi, G., Cox, S., … De Longueville, B. (2011). Citizen-based sensing of crisis events: sensor web enablement for volunteered geographic information. *Applied Geomatics, 5*(1), 3–18. https://doi.org/10.1007/s12518-011-0056-y

Semiocast. (2012). Twitter reaches half a billion accounts – more than 140 million in the U.S.

Smart Insights. (2015). 2015 Social network popularity by country, *861*, 2014–2016.

Stein, R. M., Duenas-Osorio, L., & Subramanian, D. (2010). Who evacuates when hurricanes approach? The role of risk, information, and location. *Social Science Quarterly, 91*(3), 816–834. https://doi.org/10.1 111/j.1540-6237.2010.00721.x

Tai, C.-A., Lee, Y.-L., & Lin, C.-Y. (2010). Urban disaster prevention shelter location and evacuation behavior analysis. *Journal of Asian Architecture and Building Engineering, 9*(1), 215–220. https://doi. org/10.3130/jaabe.9.215

Takahashi, B., Tandoc, E. C., & Carmichael, C. (2015). Communicating on Twitter during a disaster: An analysis of tweets during Typhoon Haiyan in the Philippines. *Computers in Human Behavior, 50*, 392–398. https://doi.org/10.1016/j.chb.2015.04.020

Team Mirah Sakethi. (2010). Why flood in Jakarta: http://bpbd.jakarta.go.id/assets/attachment/study/buku_ mjb.pdf.

Wendling, C., Radisch, J., & Jacobzone, S. (2013). *The Use of Social Media in Risk and Crisis Communication. OECD Working Papers on Public Governance No. 24.* https://doi.org/10.1787/5k3v01fskp9s-en

World Population Review. (2018). Jakarta Population 2018. Retrieved from http://worldpopulationreview.com/ world-cities/jakarta-population/

Zook, M., Graham, M., Shelton, T., Stephens, M., & Poorthuis, A. (2016). floatingsheep: DOLLY.

19 Towards Equitable Urban Residential Resettlement in Kigali, Rwanda

Alice Nikuze, Richard Sliuzas, and Johannes Flacke

CONTENTS

19.1 Introduction 325
19.2 Methodology 326
    19.2.1 Study Area 326
    19.2.2 Data and Methods 327
        19.2.2.1 Data 329
        19.2.2.2 Methods and Analysis 330
19.3 Results 333
    19.3.1 Resettlement Requirements 333
        19.3.1.1 Economic Development 333
        19.3.1.2 Infrastructure and Access to Basic Services 334
        19.3.1.3 Protection of Social Networks 335
    19.3.2 Impoverishment Risks 335
        19.3.2.1 Landlessness and Joblessness 335
        19.3.2.2 Loss of Income from Houses for Renting 336
        19.3.2.3 Homelessness 337
        19.3.2.4 Marginalization 337
        19.3.2.5 Loss of Access to Common Facilities (Common Property Resources) 337
        19.3.2.6 Health Risks (Morbidity and Mortality, Food Insecurity) 337
        19.3.2.7 Risk of Social Disarticulation 338
        19.3.2.8 Loss of Mobility 338
        19.3.2.9 Uncertainty 339
    19.3.3 Strategies for Reversing Impoverishment Risk 339
    19.3.4 Potential Resettlement Sites 339
    19.3.5 Suitability of Potential Resettlement Sites 339
19.4 Discussion and Conclusions 341
References 342

19.1 INTRODUCTION

Rapid urbanization and frequent natural disasters have substantially increased the number of urban households – especially poor households – that have to be relocated away from their homes and communities (Satiroglu and Choi, 2015). Urbanization in developing countries is associated with many challenges, among them poverty and viral growth of informal settlements, many of which are often exposed to various natural hazards (Manirakiza, 2014). Resettlement is one option for reducing the risk of natural disasters that is being adopted both before and after the occurrence of a disaster (Correa et al., 2011). Governments and international agencies increasingly consider resettlement of vulnerable urban communities as a risk reduction strategy (Ibrahim et al., 2015). Disaster–risk mitigation is an indispensable instrument for protecting people's lives and assets, as well as expanding national and local economies. Poorly executed resettlement may, however, induce loss of livelihood and increase the vulnerability and impoverishment of the displaced population.

Population displacement and resettlement are recognised causes of various deprivation risks, raising issues of inequity and social injustice in the society (Cernea, 1997). According to Cernea (1997), resettlement often leads to impoverishment risks (referring to a worsening of people's welfare and livelihoods) that are manifested through eight interlinked phenomena: landlessness, homelessness, joblessness, loss of access to common property resources, marginalization, food insecurity, morbidity and mortality and social disarticulation. The most critical risks associated with urban displacement and resettlement are loss of employment, or sources of site-related income, and the uncertainty of finding new employment or income in the area of relocation. Recent work by Patel et al. (2015) clearly indicates that, along with other factors, the distance of the relocation site from a person's original place of residence and employment often becomes an obstacle to continuing prior employment – the original source of income of the displaced person. Such issues must be adequately addressed through policy and implementation.

According to ADB (1998), a suitable resettlement site provides opportunities for restoring the livelihoods of displaced households. For example, when Ibrahim et al. (2015) conducted a suitability analysis of resettlement sites for flood-disaster victims, they included environmental factors like climate, soil, slope and geology and also socio-economic factors such as accessibility to roads. Furthermore, they suggested that other socio-economic and infrastructural development issues should be considered in evaluating the suitability of sites for sustainable resettlement. They did not, however, mention which specific issues should be considered.

Given this background, our aim was to develop a methodology for resettlement-site selection that would help restore the livelihoods of affected urban communities and reduce the risks of impoverishment that are often associated with resettlement. We carried out our research in Kigali, the capital of Rwanda. There, the combination of steep terrain, poor drainage, large areas of wetlands, high-density informal settlements (with buildings constructed with non-durable materials) and heavy seasonal rainfall exacerbate the risk of disaster. Every year during the rainy seasons, water-related disasters cause loss of lives and damage to property. In line with its risk prevention and reduction policy, the Rwanda government has started to relocate the most vulnerable households from high-risk zones; a large number of households still need to be relocated.

19.2 METHODOLOGY

19.2.1 Study Area

BOX 19.1   Case Study Area

Kigali is the capital and the largest city of Rwanda. The majority of the country's urban population live in this city, which has an area of 730 km². The population of Kigali is growing fast: in 2017 the population growth rate was estimated to be on average 2.5%. The city is divided into three administrative districts – Gasabo, Kicukiro and Nyarugenge. Our study was carried out in *Gasabo* district, which covers an area of 429.3 km², approximately 60% of the total area of Kigali city. In 2012, the district's population was slightly more than 500,000 inhabitants. Most residential locations in Gasabo are in the immediate surroundings of Kigali CBD and other commercial areas, and they include large areas of unplanned high-density housing. Gasabo district was selected for our research because, according to the Rwandan Ministry of Disaster and Refugees, the most critical high-risk zones for disaster are to be found there. The informal settlement of Gatsata, for example, is located on steep slopes that are prone to landslides; it has been prioritised for relocation.

This study was conducted in Gasabo, the largest of the three districts in Kigali (Figure 19.1). Kigali city covers an area of 730 km^2 and comprises three administrative districts – Nyarugenge, Kicukiro and Gasabo – which for administrative purposes are further divided into sectors and villages. The landscape is characterised by a series of hills that are separated by wetlands (rivers, streams and marshes). Some hillslopes have inclines as much as 50%, while slopes in the valleys and wetlands are less than 2% (REMA, 2013). The scarcity of relatively flat, non-hazardous land has led to human settlement taking place on disaster-prone hills and in wetlands areas that are inappropriate for habitation. About 66% of inhabitants of Kigali live in informal settlements. Our study area, Gasabo, covers an area of 429.3 km^2, approximately 60% of the total area of Kigali city. In 2012, its population slightly exceeded 500,000 inhabitants (NISR, 2012). Most residential locations in Gasabo are in the immediate surroundings of Kigali CBD and other commercial areas, and they include large areas of unplanned high-density housing. Gasabo was selected because, according to the Rwandan Ministry of Disaster and Refugees, the most critical high-risk zones for disaster are to be found there. The informal settlement of Gatsata, for example, is concentrated on steep slopes that are prone to landslides. Gatsata has been prioritised for relocation, and as such we consider it in our study as a case of a community to be resettled.

19.2.2 Data and Methods

BOX 19.2   Methods Applied in the Chapter

This chapter illustrates the application of GIS methods, together with qualitative analysis and descriptive statistics, for analysing the suitability of potential resettlement sites in terms of minimizing risks of impoverishment of the resettled households. The GIS methods applied combine accessibility analysis (taking into account the effects of topographical features such as roads, slopes and rivers – done using ArcGIS™) and spatial multi criteria analysis performed in CommunityViz Scenario 360™.

Spatial data (administrative boundaries, road networks, public transport routes and bus stops, health facilities, schools, markets, strategic locations) and non-spatial data were used. The non-spatial data were gathered from a literature review, interviews with households and key informants and field observations. Using the spatial and non-spatial data, first we analysed resettlement requirements with respect to the socio-economic and cultural characteristics of the households to be displaced, as well as their potential risk of impoverishment. Next, we identified potential strategies for reversing risks, which we then used as criteria for analysing the suitability of potential resettlement sites. The results of suitability analysis were influenced by the criteria selected. For example, if non-spatial criteria such as affordability of services, e.g. the cost of public transport to the city centre (which was not considered in this study) is included, the suitability analysis would yield different results from those if it was not included. The methodology applied here in the case of a resettlement programme could be applied to any other planning process in which site suitability and accessibility are issues, for example, for analysing non-serviced and serviced areas of existing health facilities and identifying a suitable site for proposed new facilities.

In our study we followed three main steps to develop a methodology for equitable resettlement, i.e. an equitable resettlement process that is guided by concerns of fairness and justice by seeking to mitigate the various forms of impoverishment that are commonly associated with resettlement. The objective of such an equitable resettlement process is that resettled households should not experience increased, long-term economic and social deprivation as a result of resettlement. First, we identified potential resettlement requirements; second, we analysed the risks of various forms of

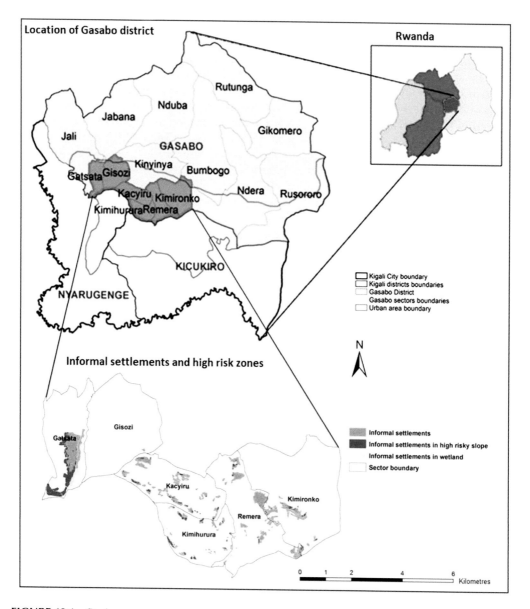

FIGURE 19.1   Study area.

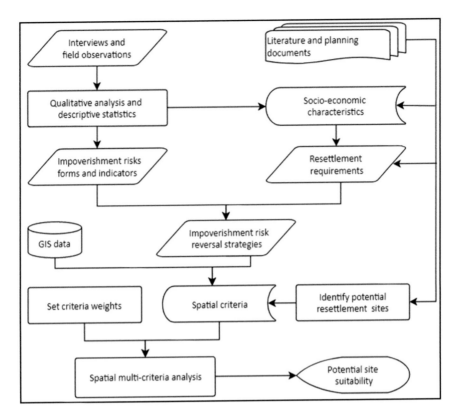

FIGURE 19.2   Methodological approach.

impoverishment and their indicators; and third, we developed a model for analysing the suitability of potential resettlement sites. To this end, we used a mixed-methods approach that combines qualitative and quantitative methods and techniques (Figure 19.2). In the following sections, the data, methods and analysis are described in detail.

19.2.2.1   Data

For the study, we used primary and secondary data collected from fieldwork conducted in the study area from 23 September until 12 October 2015. Primary sources included:

- interviews conducted with 90 households to be displaced;
- interviews conducted with 20 resettled households living on two resettlement sites: ten households in each;
- in-depth interviews with ten key informants;
- field observations in both the settlement to be displaced and the two resettlement sites chosen.

In this research, the pre-resettlement and post-resettlement characteristics of households' livelihood were fundamental in identifying crucial resettlement requirements that need to be met by future relocation programmes, as well as understanding the risk of impoverishment faced by the relocated households. For this reason, two questionnaires were used for the interviews: one for the households from Gatsata that were to be relocated, to capture their socio-economic and cultural characteristics, including their physical, social, financial, human and natural assets (Baud et al.,

2008), their perceptions about potential risks of impoverishment that would result from their relocation and their preferences regarding resettlement sites; and the other for the households already resettled on two existing resettlement sites, Bumbogo and Kinyinya, to collect data on their experiences with risk of impoverishment due to displacement. The latter interviews covered a wide range of topics, including employment, income, land location and housing, access to common property resources and social organization before and after resettlement. The households that had already undergone resettlement were selected to triangulate potential risks of impoverishment for those households from Gatsata that were to resettle. In doing this, we assumed that the characteristics of households to be resettled out of high-risk zones within the district were similar to those that had already been resettled.

Interviews with key informants involved in the planning and implementation of various resettlement programmes were set up to corroborate our findings from other research instruments in relation to resettlement requirements, impoverishment risks, their causes and possible recommendations for future relocation. Furthermore, at all sites we also intensively observed general social and economic conditions and the existing spatial location and conditions of physical elements that are important in the daily life of households, for example roads and transportation services and facilities. Secondary data used were mainly GIS data (Table 19.1) and data collected from a review of the literature and documents on national and international policy related to resettlement.

19.2.2.2 Methods and Analysis

First, we analysed resettlement requirements. With reference to Correa et al. (2011), we identified and summarised crucial resettlement requirements of Gatsata households that would have direct implications for the mitigation of risks of impoverishment: requirements in relation to economic development; access to infrastructure and basic services; and the protection of the social networks of affected households. To this end, descriptive statistics and qualitative data analyses of key elements of their socio-economic characteristics, social organization, occupational structure and access to public services – i.e. information defining which households need to be relocated, what they do for living and what they are likely to lose following displacement – were carried out. On the basis of these analyses and the general provisions of the Rwanda Human Settlement Policy, the

---

TABLE 19.1
Secondary Data Collected

Data	Source	Comment
Administrative (district, sector, village) boundaries of Kigali city	COK	
Transport infrastructure: road network, bus routes, bus stops	COK	Existing development (2013) and developments proposed in the master plan
DEM, 10 m resolution	COK	
Detailed land use and zoning, township boundaries, some important places in Kigali	COK	Existing developments (2013) and developments proposed in the master plan (2025)
Social infrastructure: trade centres, markets, schools, health centres	RNRA	Rwanda base map 2013
High risk slopes, wetlands, informal settlements	ITC; RHA	Informal settlements identification project 2013
Field study report on informal settlement, Kigali city	MIDIMAR	Informal settlements identification project 2013

COK: City of Kigali; RNRA: Rwanda National Resources Authority; RHA: Rwanda Housing Authority; ITC: Faculty of Geo-Information Science and Earth Observation, University of Twente; MIDIMAR: Ministry of Disaster Management and Refugees.

National Housing Policy, the Kigali City Master Plan and a literature review of international publications, the resettlement requirements for the relocation of the Gatsata community were identified.

Next, the potential impacts the resettled households would face and the likely risks of impoverishment of those to be displaced from Gatsata, such as loss of jobs, income, land and housing, access to common property resources and social disarticulation, were identified once again using descriptive statistics and methods of qualitative data analysis. An impoverishment risks model (Cernea, 1997) was used as a framework for summarizing the forms of risk identified and their indicators, although our study brought new forms to light.

Lastly, we identified mitigation strategies for impoverishment risks and used them to analyse the suitability of potential resettlement sites. As described by ADB (2012), the location of a resettlement site is a key element in minimizing the risks of impoverishment faced by relocated families. Using both ArcGIS™ and CommunityViz scenario 360™ (City Explained Inc., 2018), we developed a five-step GIS-based multi-criteria model for site suitability analysis:

Step 1 – Identify Suitability Criteria

These are criteria that we consider would be able to minimise impoverishment risks and therefore include the risk-reversal strategies obtained by reversing the impoverishment-risk indicators (see section 19.3.2) into positive actions to be taken (see Table 19.5). Further, we checked whether the strategies developed aligned with the identified resettlement requirements (see section 19.3.1). Although in our study we only proposed spatial strategies, we recognise that there are other, non-spatial factors that could also help to minimise the risk of impoverishment. For instance, loss of income from renting small houses (discussed in Section 19.3.2) should be compensated by other means.

Step 2 – Select Criteria for Spatial Multi-Criteria Model

Multi-criteria techniques are powerful tools for analysing the suitability of areas with respect to one or more activities (Kumar and Biswas, 2013). The main steps of suitability analysis include the establishment of criteria datasets, criteria weighting and aggregation of criteria (Ibrahim et al., 2015; Zucca et al., 2008). For our study, we selected suitability criteria (Table 19.2) from the potential risk-reversal strategies shown in Table 19.5.

Indicator maps, which were represented as distance maps in a GIS database, were prepared using an accessibility model (Zucca et al., 2008). To be able to consider the various topographical features present in the study area, e.g. rivers, slopes and road surfaces (paved/unpaved), and their effect on accessibility to services and facilities (Karou and Hull, 2014), an appropriate accessibility model (see Figure 19.3) was developed.

TABLE 19.2
Criteria for Site Suitability Analysis

Goal	Criteria	Indicator	Requirements	Impact
Improved livelihood; minimum impoverishment risks	Access to job opportunities	Distance to city centre Distance to commercial centres Distance to original settlement	Minimum distance, time and cost of transport	Reduce travel time and cost of transport
	Access to social facilities	Distance to health Distance to education Distance to market	Minimum distance, time and cost of transport	Reduce distances and cost of transport
	Access to public transport	Distance to bus route Distance to bus stop	Minimum distance, time and cost of transport	Reduce distances and cost of transport

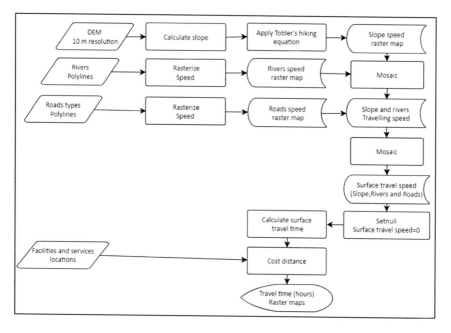

FIGURE 19.3   Accessibility model.

In this accessibility analysis, we compared two modes of transport – walking and public transport – both of which are frequently used by low-income populations. Table 19.3 summarises the speeds that were used in both modes of transport for each type of surface.

Speed in relation to terrain was calculated using the hiking equation (Tobler, 1993). For both walking and bus transport modes, rivers overrule slope, while roads overrule the combination of slope and rivers. In public transport mode, an average speed of 21 km/h (Niyonsenga, 2012) was used as an extra factor for the 2012 bus-route network, and this overrules the three other factors considered in the walking mode.

Step 3 – Criteria Weighting

The third step is criteria weighting, using a stepwise method of ranking preferences (Voogd, 1982). We used the preferences named by the households because allowing people affected by policies to participate in the design is important to them, as well as being a strategy to minimise risk of impoverishment. Once the ranking of the preferences was completed, ranked sum weights were calculated using the following formula:

$$W_k = \frac{n - p_k + 1}{\displaystyle\sum_{j=1}^{n} \left(n - p_j + 1\right)}$$

where:

W_k is the *k*th criterion weight

n is the number of criteria under consideration ($k = 1,2,3,\ldots,n$)

p_k is the rank position of the criterion.

Table 19.4 presents the relative importance (weighting) of the selected criteria.

Step 4 – Identify Potential Resettlement Sites

Potential resettlement sites were identified from the land-use zoning proposed in the Kigali master plan for Year 2025. Land for potential resettlement sites had to meet three criteria, which were:

TABLE 19.3
Selected Speeds According to the Type of Surface and Mode of Transport

	Speed (km/h)	
Surface Type	Walking	Public Transport
Slope	Tobler's hiking equation	
Rivers	0	0
Paved roads	5	5
Unpaved roads	3	3
Public transport network	Not applicable	21

- the land had to be zoned for residential use;
- the land had to be proposed for high-density development; and
- that land had to be affordable for low-income groups.

19.3 RESULTS

This section presents results related to: resettlement requirements (subsection 19.3.1) that enable economic development, access to infrastructure and basic services, and those for providing protection of social networks of affected households. In the following subsections, the identified impoverishment risks (subsection 19.3.2); strategies for reducing these risks (subsection 19.3.3); potential resettlement sites (subsection 19.3.4); and the results of suitability of potential resettlement sites (subsection 19.3.5) are also presented.

19.3.1 RESETTLEMENT REQUIREMENTS

19.3.1.1 Economic Development
The main objective of economic development is to re-establish economic activities and the incomes of affected households. The socio-economic characteristics of affected households in Gatsata's

TABLE 19.4
Suitability Criteria Weightings

Rank	Weight	Subset	Indicator	Rank	Partial Weightings	Overall Weightings
1	0.667	Site location preferences	Distance to city centre	1	0.5	0.33
			Distance to commercial centres	2	0.33	0.22
			Distance to original settlement (Gatsata area)	3	0.167	0.11
2	0.33	Social and public services	Distance to markets	1	0.33	0.11
			Distance to schools	2	0.267	0.089
			Primary schools	*1*	*0.133*	*0.045*
			Secondary schools	*2*	*0.09*	*0.029*
			Universities	*3*	*0.044*	*0.015*
			Distance to health facilities	3	0.2	0.067
			Distance to bus route	4	0.133	0.044
			Distance to bus stop	5	0.067	0.022

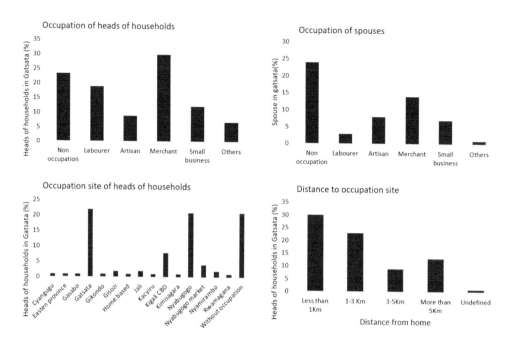

FIGURE 19.4 Selected characteristics related to economic development and income generation for the affected households from Gatsata.

high-risk zone highlight the link between the income-generating activities of the households and their surrounding environment. As seen in Figure 19.4, most heads of Gatsata households have jobs in the same neighbourhood or in the nearby commercial area of Nyabugogo. They mostly travel a distance of 1–3 km to Nyabugogo, where they or their spouses are self-employed as merchants or owners of small businesses. The educational level among the workers revealed that most do not possess any special skills beyond those that allow them to find work in the informal economy. Therefore, any resettlement programme should ensure that low-skilled jobs are available in new resettlement areas. Specifically, for merchants, the resettlement program should ensure that the new site also has a need for additional market services. If insufficient demand is available in or near the new settlement, impoverishment risk will be high. This requirement is also recognised in the Rwanda National Housing Policy, which encourages access to housing for the low-income urban population by supporting their housing within or near economically viable neighbourhoods.

19.3.1.2   Infrastructure and Access to Basic Services

Access to markets, schools and health and sanitation services are universal basic human rights. Since the households that need to be relocated live in a high-risk, informal settlement, there is a need to improve their general living environment, especially in relation to sanitation. Nevertheless, because of the location of their settlement, these households benefit from opportunities that are available in their surrounding environments. Our analysis reveals that most of the school-age children are enrolled in a public primary school to which they walk since they generally live nearby. The study also revealed that households in Gatsata also have relatively good access to healthcare and other basic services such as public transport. In general, because Gatsata is located in an inner-city area, its residents are able to take maximum advantage of the public facilities available; there are more facilities in the inner city than in other parts of the district. National statistics confirm that accessibility to basic services in Gasabo is higher in urban areas than in rural ones (NISR, 2011). Thus, meeting the principle of improving or at least maintaining current levels of service can only be achieved if well-serviced relocation sites are available. There is a need to ensure that availability in terms of the quality and quantity of services such as schools, health facilities and markets present in the resettlement area can meet demand, such that children will continue their education and continue to use the walking mode within acceptable distance. Likewise, basic healthcare services should be no more than one hour's travel away, as proposed by the Ministry of Health (2009).

19.3.1.3 Protection of Social Networks

Social networks can influence the choice of a household about where to live. People often prefer to settle close to their relatives and close, trusted friends, from whom they can get assistance such as watching over children when parents are not around. Many people may also rely on such networks for financial assistance. The time spent by a household in a place is a factor in forming such social networks. Figure 19.5 shows that almost half of the interviewed households have resided in their settlement for more than ten years. In that time, they were able to construct both social and economic networks. The study also revealed strong social and economic networks based on community financial organization (mutual money lending), as well as community trading (merchants–customers). Those networks and ties are key ingredients for a community's socio-economic development and well-being, and they can be easily damaged by poorly planned and implemented resettlement.

19.3.2 Impoverishment Risks

19.3.2.1 Landlessness and Joblessness

According to Cernea (1997), 'land is the foundation upon which people's productive activities, commercial activities, social networks and livelihoods are constructed and its expropriation is a form of impoverishment'. Patel et al. (2015) showed that the importance of land in urban contexts is given by its location with respect to opportunities for livelihoods and common facilities such as education, health services and markets. Figure 19.6 shows the impact of resettlement on the occupations of heads of households resettled in Kinyinya and Bumbogo.

The number of heads of households with no occupation increased after resettlement, while the number of labourers and artisans before relocation decreased slightly after resettlement. All those who had been merchants and small business owners before relocation completely stopped with those activities after resettlement. These changes in occupations are attributed to the increased distance to a place of job opportunities, lack of job and business opportunities in the new area,

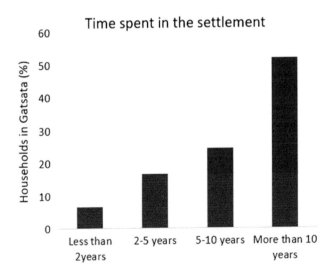

FIGURE 19.5 Time living in high-risk zones of Gatsata.

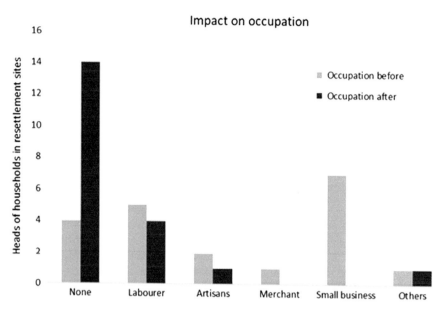

FIGURE 19.6   Impact of resettlement on occupations of head of households relocated on sites in Kinyinya and Bumbogo.

increased distance to market places, increased distances from customers, unfamiliar communities, cost of transport to areas of job opportunities and loss of capital through reconstruction.

Other scholars have reported similar findings of landlessness and risk of joblessness after relocation of slum dwellers. Hunter and Posel (2012) emphasised that most households displaced from informal urban settlements face the risk of job loss and a weakening of their livelihoods. Many informal settlements also accommodate immigrants who come to seek jobs and other income opportunities. Because of their low and irregular income, they often prefer to live closer to their job location and income-generating activities to avoid the cost of transportation to and from their place of work. Therefore, relocation far from their original settlements and urban areas offering job opportunities is likely to lead to loss of income and increased risk of impoverishment.

The households to be relocated are living in locations that provide them strategic access to opportunities for income generation and to basic services. These households are therefore likely to face risks of becoming landless as a result of their relocation away from the Gatsata area. Further, our study revealed a remarkable link between this area and the income-generating activities of its residents. Gatsata settlement is located near Nyabugogo, an economically vigorous area upon which the majority of households to be resettled depend. If, therefore, those households are relocated far from Nyabugogo, they are likely to lose accessibility to income opportunities and basic services. As a result, their landlessness may contribute to their joblessness. Households resettled in Kinyinya and Bumbogo sites reported significant impoverishment as a result of these issues.

19.3.2.2   Loss of Income from Houses for Renting

Many households from Gatsata are likely to lose significant income from renting houses. More than 75% of the households sampled have on average one to three small houses that they rent out. For some households, those rental houses also secure the capital they need for running their small businesses or merchant activities. Loss of their rental houses would bring serious financial complications, including loss of income, loss of a substantial amount of invested capital and, as a consequence, a drop in their credit-worthiness.

19.3.2.3 Homelessness

Each of the households that were resettled in Kinyinya and Bumbogo received a house as a form of compensation, yet some of these beneficiaries closed their houses and returned to the city to look for jobs. Because they could not afford the cost of transport, they returned to the city and rented houses in areas offering job opportunities, leaving their own, new houses unused. Thus, with relocation far from job opportunities, the risk of homelessness might become a grim reality – for households to be resettled from Gatsata as well. Clearly, the risks of homelessness that relocated households face shows that compensation in terms of land, cash or houses only cannot adequately deal with the losses caused by displacement – a conclusion that has been demonstrated by other scholars (Cernea, 2003).

19.3.2.4 Marginalization

Households to be displaced from Gatsata are likely to become marginalised if they are not able to use their previous skills, maintain their income-generating activities and quickly regain their full economic strength. Experience of the households that have already been resettled in Bumbogo and Kinyinya shows that 18 out of 20 heads of households interviewed were not able to use their previous skills after resettlement. Their loss of income increased the vulnerability of these households. The majority of interviewed households in Kinyinya replied that they did not find any job opportunities in their resettlement area except for running a small business or becoming a merchant. However, starting a small business was also not possible for them because their income decreased after relocation and their loss of capital and income was a barrier for establishing such activities. In the Bumbogo resettlement site, which is more of a rural area than Kinyinya, respondents mentioned that working as a daily agricultural labourer was the only available job opportunity in the area. Such work is, however, poorly paid and seasonal. Moreover, labouring opportunities do not fit their skills, which were more attuned to jobs in an urban economy.

19.3.2.5 Loss of Access to Common Facilities (Common Property Resources)

Access to common property was interpreted in the context of Gasabo as access to facilities such as public primary schools, public health centres and public hospitals. We found that 18 of the 20 resettled households sampled in Kinyinya and Bumbogo sites had had children studying at local primary and secondary schools. Some of these households were obliged to transfer their children to new primary schools closer to their resettlement sites. Unlike in Kinyinya, most of the households resettled in Bumbogo complain that the distance to primary schools increased significantly after relocation. For the residents of both Kinyinya and Bumbogo, having children attending secondary schools and university means high transport costs because most secondary schools and universities are concentrated in the inner city.

Concerning accessibility to health facilities, the Bumbogo resettlement site is located adjacent to a health centre, while most respondents at the Kinyinya site reported that there was no change in accessibility when compared with their previous settlements. However, most households to be resettled from Gatsata perceive that they might experience loss of accessibility to health facilities after relocation in view of Gatsata's strategic location with – potentially – access to numerous basic services in the vicinity.

19.3.2.6 Health Risks (Morbidity and Mortality, Food Insecurity)

Health risks as a consequence of malnutrition go in hand with food insecurity. We found that drop in income, increased distance to main markets and fairs and change in monthly expenditure are indicators for health risks that may be induced by poorly executed relocation. Nyabugogo is one of the main wholesale markets for food crops in Kigali where people can find fresh fruit and vegetables at low prices. When accessibility to fairs and markets is reduced and at the same time income decreases, there is a risk of undernourishment. The evidence from households already resettled reveals that monthly expenditures in nine of the 20 resettled households increased, while ten households reported a decrease in monthly expenditures. The explanation for this has two linked causes. On the one hand, households replied that their expenditures decreased because they no longer had the money to spend. On the other hand, an increase in monthly expenditures was associated with a higher cost of living resulting from the higher price of commodities found in local shops: there were no markets near the resettlement sites where inhabitants could find cheap fresh vegetables, and the cost of locally supplied food was high due to increases in transport costs. In both cases, because of loss of income and higher food prices due to a lack of markets, the resettled households were deprived of some types of foods they used to eat before relocation; in some cases, these households only ate once a day. The increase in expenditures for them meant a change in living conditions that had negatively impacted their health. In the context of Gasabo's resettlement programme, these results show how health risks are linked to risks of landlessness and joblessness. Terminski (2015) has also argued that health risks might be a consequence of landlessness or reduced access to common resources.

19.3.2.7 Risk of Social Disarticulation

Households that need to be resettled from Gatsata are likely to face the risk of destruction of their social and economic structure. In the time they have spent in the Gatsata area, they have been able to build social and financial community structures. And regular customers of shop owners also come from the same neighbourhood. Looking at what was experienced in Bumbogo and Kinyinya, 14 households interviewed indicated that they could not maintain their social and economic networks simply because of the increased distance from their original settlements. The distance brought higher transportation costs while their income had decreased. Only six respondents out of the 20 households interviewed in Bumbogo and Kinyinya replied that they had been able to maintain contact with their relatives and friends solely by telephone. Our study also reveals that many resettled households experienced risks of loss of livelihood because of being resettled far from their customers. This shows how the risk of social disarticulation is associated with the risk of becoming jobless, and vice versa. Similar findings have been reported by Patel et al. (2015), who found that shop owners, artisans and small businessmen resettled in Ahmedabad (India) experienced stronger effects of displacement through loss of customers.

19.3.2.8 Loss of Mobility

Reduced personal mobility, as a result of an inefficient transport system, is a new risk of impoverishment to be associated with the Gasabo resettlement programme. Resettled households in both Kinyinya and Bumbogo need to reserve additional funds to travel to their workplaces, the city centre, markets, secondary schools and other places that form part of their daily lives. Moreover, residents of Kinyinya have to walk 15–20 minutes to reach the nearest bus stop, where they must queue for a bus for an indefinite period, inducing stress among commuters. Eighty percent of respondents from Kinyinya reported that the distance to the bus stop had increased compared to their previous residence. Mobility in Bumbogo is worse still: there is no bus stop nearby and residents reported that they must use a motorbike to travel about 10 km over unpaved roads to the nearest bus stop, adding substantially to their mobility costs. Reduced mobility for Bumbogo residents exacerbates the effects of other impoverishment risks, as important trips may be postponed or even cancelled, further contributing to impoverishment.

Our research has revealed that the reduced mobility of resettled households has reduced their household budget and weakened their social networks. Some households in Bumbogo reported that their children stay with relatives or friends in the city to be close to secondary schools and to reduce transportation costs. Similar findings of limited access to schools in resettlement sites, leading to separation of children and parents, were discussed by Takesada et al., (2008). Reduced mobility poses a significant threat to the well-being of households who may be resettled from Gatsata.

19.3.2.9 Uncertainty

Uncertainty is another impoverishment risk that was identified in the Gasabo resettlement programme. According to Patel et al. (2015) in their study of Ahmedabad (India), this form of impoverishment occurs when resettlement distracts a community from pursuing income-generating activities and a strong perception of negative impacts and stress arises due to a lack of community participation and transparency in a resettlement programme. We found such a situation in Gatsata. A lack of information about the resettlement programme, such as where households were to be resettled, what compensation would be offered and an absence of policy guidelines, magnified the negative perceptions respondents had about their relocation. Uncertainty was high, especially among those households that had small houses for renting. They perceived resettlement as a direct threat to a major source of household income.

19.3.3 Strategies for Reversing Impoverishment Risk

Spatial strategies for reversing potential impoverishment risks were identified. Most of the resettlement risks identified had a clear spatial component. Issues concerning access and mobility are, therefore, of paramount importance when selecting potential resettlement sites; see Table 19.5. In addition, the identification and analysis of potential sites should be a collaborative process that involves the City of Kigali and targeted households so that concerns about uncertainty can be addressed through a more transparent decision-making process. Level of accessibility is a function of modes of transport, so access to bus services should be considered in addition to walking when evaluating resettlement sites.

19.3.4 Potential Resettlement Sites

Figure 19.7 shows potential resettlement sites in Gasabo. Potential residential areas in Gasabo were identified on the City of Kigali master plan (City of Kigali, 2013). Three residential types, one of which was not available in Gasabo, were found to be the most appropriate for resettlement of low-income households:

- low rise residential district (R2);
- low rise residential district (R2A);
- high rise residential district (R4) (not available).

19.3.5 Suitability of Potential Resettlement Sites

Figure 19.8 clearly shows that the suitability of potential sites is very high near to Kigali's CBD and decreases toward the periphery. Due to the increased mobility provided by public transport, the

TABLE 19.5

Spatial Strategies for Reversal of Impoverishment Risk

Risks	Reversal Strategies
Landlessness, joblessness and homelessness	Stabilise income generation by prioritizing relocation sites with access to the CBD and other commercial centres that provide opportunities for small businesses and trading
	Minimise transport costs, prioritise locations close to previous locations of residence
Health and education	Prioritise locations with access to established cheap fresh-food markets, public health services and primary and secondary schools
Social disarticulation	Facilitate the maintenance of social networks by prioritizing locations with ready access to households' previous residential locations
Mobility	Prioritise locations that have easy connections with, or can be easily connected to, bus services

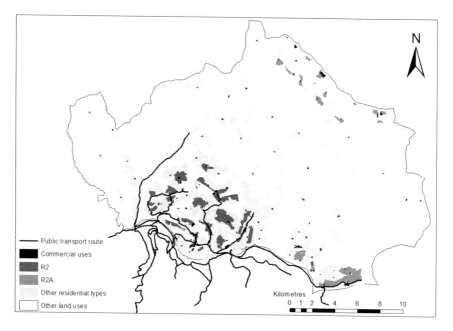

FIGURE 19.7   Identified potential resettlement sites in Gasabo.

use of this mode generates more 'highly suitable' and 'suitable sites'. Many potential sites are only moderately suitable for walking because they are a bit too far away from the city centre, Nyabugogo or the original site of Gatsata, but these sites become much more accessible when travelling by bus. Nevertheless, even when using the public transport, the northern periphery of Gasabo is far from suitable because the level of public transportation services is low (Niyonsenga, 2012) and, as a peri-urban area, it has few basic services due to its low population density and more rural lifestyle. These suitability results show that improving the quality of transportation services could increase the options for potential resettlement sites. If both transport modes are taken into account – walking

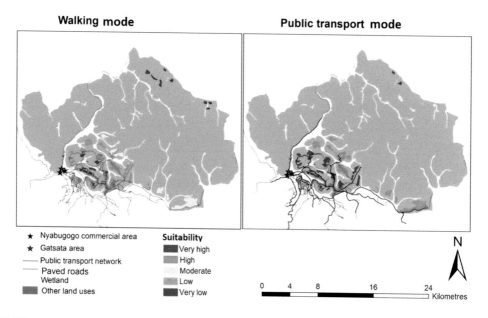

FIGURE 19.8   Potential resettlement sites' suitability in both walking and public transport modes.

and bus transport – the southern region of Gatsata would be the most suitable area for resettlement of Gatsata's households. It is part of the existing urban area of Kigali and directly adjacent to the city's core, and this part of the district offers many locational advantages due to the concentration of basic facilities there, as well as enjoying the possible benefits from basic facilities in the neighbouring districts of Nyarugenge and Kicukiro.

19.4 DISCUSSION AND CONCLUSIONS

Our research links the socio-economic characteristics of affected households with their resettlement requirements and the risks of impoverishment they are likely to face, and it shows how these risks might be minimised by systematic analysis of the suitability and the selection of potential resettlement sites. The findings show that households' original choice to settle in Gatsata was mainly linked to their workplaces and accessibility to basic services. The productive activities, jobs, social networks and general livelihood of these households are linked to the location of their place of residence and its surrounding environment. Location matters: it is a critical factor in determining the impact of resettlement on resettled households. Our study shows that all eight forms of impoverishment proposed by Cernea (1997) have emerged in the case of those already resettled in Gasabo and could also emerge for the households to be relocated away from the Gatsata high-risk area. Among all the risks, landlessness and joblessness are the two most critical relocation risks to be faced. Relative location plays an important role in most of the impoverishment risks identified: distances, time and cost aspects associated with mobility are critical, particularly for low-income households. The argument that the socio-economic problems faced by resettled communities are linked to the spatial location of the resettlement sites is therefore confirmed. Finding an appropriate site that provides good access to all amenities and basic services is crucial for successful resettlement. Uncertainty among the households and reduced mobility emerge from our study as additional forms of risk in the Gasabo district resettlement program.

The systematic analysis of socio-economic characteristics and impoverishment risks of households to be relocated allows spatially based risk-reversal strategies to be developed. The identification of potential resettlement sites should explicitly consider their location, taking into account the two modes of transport most used by low-income groups: walking and buses. Such an approach emphasises that relocation take place close to the original settlement, the CBD or other vigorous commercial areas to reduce the range of potential resettlement risks. Our study shows that the impoverishment risks are interdependent and that risk-reversal strategies can also be built upon the same interdependencies. The stated preferences for selected strategies and their spatial indicators show that households are very much concerned with the location of the resettlement site which offers access to job opportunities before the basic infrastructure to be provided. Households emphasised three criteria: being near to the city centre (CBD); being near their original settlement; and being near to other commercial centres could contribute in some way to the suitability of the resettlement location.

Suitability analysis revealed a large amount of 'highly suitable' and 'suitable' land in the southern part of the district (around the core of Kigali city) for both the walking and public transportation modes. Improving the efficiency of public transport would further increase levels of suitability of potential resettlement sites, but investments in mobility should be synchronised with resettlement, or even precede it, in order to avoid impoverishment of the resettled population.

Notwithstanding the foregoing considerations, we acknowledge that some other factors also need to be considered. For instance, in the case of Gasabo, the limited availability of cheap land close to the original settlement and the CBD will need to be taken into account as the cost of land is an important component in total project cost. Nevertheless, choosing cheaper, more peripheral locations does not reduce total cost, but rather transforms and transfers it to the relocated households, most of whom can ill afford the additional burden. Also, resettlement policy could be tied to residential zones identified in the Kigali city's new master plan and be used as a means for implementing the plan. Suitability analysis of resettlement sites involves a number of criteria, including compliance with land-use plans, safety, location with respect to workplace opportunities, access to public services and land values – as recommended by Correa et al., (2011). Our study did not include land value and land tenure issues due to a lack of relevant data. Land value and tenure are also essential criteria that influence the selection of the suitable sites. Current owners of suitable land need to be involved as stakeholders in the process of final site selection in order to minimise possible resistance. Further, the 'highly suitable' areas might be located in areas that are already built up (urban areas). Therefore, there is a trade-off to be made between choosing for redevelopment of existing housing areas (as per Kigali city's master plan) or acquiring available vacant land. In all the cases, areas close to the original site could be the potential candidates since they can reduce the indirect effects of resettlement, such as increased travel costs to employment opportunities.

Consideration of standards or stakeholder agreement regarding acceptable travel time is essential because this will influence the outcomes of our multi-criteria model for site suitability analysis. For instance, in Rwanda the standard of geographic accessibility states that no one should have to travel more than one hour for basic healthcare (Ministry of Health, 2009). Therefore, areas within a one-hour travel time to healthcare facilities score higher than other areas. The use of such threshold will affect the number of initially suitable areas. Because of that, the oversimplified practice in which planners consider only "the closer, the better" as sufficient for suitability analysis should be avoided. Furthermore, the amount of spatially available opportunities constitutes an uncertainty that would affect the outcomes of the site suitability analysis. If sites do not possess the services and infrastructure required, then these would need to be provided before relocation, to ensure site suitability.

Only spatial indicators were used to develop a spatial multi-criteria model of site suitability. However, we recognise that strategies to overcome risks of impoverishment following a resettlement process need to address more than just spatial factors. Other, non-spatial dimensions of 'access' (Obrist et al., 2007), such as affordability of, for instance, food on available markets and adequacy (synchronizing the organization of supplied resources with the need and expectations of the relocated households) of, for instance, public transport schedules and market opening times, should be considered when devising strategies for reversal of risks of impoverishment. In addition, the loss of income coming from renting out small houses, rising levels of uncertainty and the availability of public transport all depend on non-spatial factors and related interventions.

Suitability analysis here assumes that people will always travel to the nearest facilities available for particular services. In practice, however, they might be inclined to use more distant facilities and services thought to provide better quality services or perhaps offer cheaper prices in the case of markets. Despite the above limitations, the results from the overall methodology and the model of site suitability we developed are useful and have the potential to support more equitable resettlement since they provide a basis for the evaluation of resettlement sites. The analysis shows that efficient public transport can significantly increase suitability levels of potential resettlement sites. A coordinated group of policies for resettlement planning that include public transportation services will, therefore, influence the suitability of potential resettlement sites more than any single action.

REFERENCES

ADB. (1998). *Handbook on Resettlement: A Guide to Good Practice.* Manila: Asian Development Bank.

ADB. (2012). *Involuntary Resettlement Safeguards: A Planning and Implementation Good Practice Sourcebook-Draft working document.* Malina: Asian Development Bank. Retrieved from http://www. adb.org/documents/involuntary-resettlement-safeguards-planning-and-implementation-good-practice-sourcebook-d

Baud, I., Sridharan, N., & Pfeffer, K. (2008). Mapping urban poverty for local governance in an Indian mega-city: the case of Delhi. *Urban Studies*, 45(7), 1385–1412. https://doi.org/10.1177/0042098008090679

Cernea, M. (1997). The risks and reconstruction model for resettling displaced populations. *World Development*, 25(10), 1569–1587. https://doi.org/10.1016/S0305-750X(97)00054-5

Cernea, M. M. (2003). For a new economics of resettlement: a sociological critique of the compensation principle. *International Social Science Journal*, 55(175), 37–45. https://doi.org/10.1111/1468-2451.5501019_3

City Explained Inc. (2018). CommunityViz Scenario 360. Retrieved January 8, 2018, from http://communit yviz.city-explained.com/communityviz/scenario360.html

City of Kigali. (2013). *Gasabo District zoning plan*. Kigali: City of Kigali. Retrieved from http://www.mast erplan2013.kigalicity.gov.rw/downloads/Docs/RWF1101_11_Gasabo_Zoning Report_04062013-s.pdf

Correa, E., Ramírez, F., & Sanahuja, H. (2011). *Populations at Risk of Disaster A Resettlement Guide*. Washington, DC: The World Bank and Global Facility for Disaster Reduction and Recovery (GFDRR).

Hunter, M., & Posel, D. (2012). Here to work: the socioeconomic characteristics of informal dwellers in post-apartheid South Africa. *Environment and Urbanization*, 24(1), 285–304.

Ibrahim, I., Muibi, K. H., Alaga, A. T., Babatimehin, O., Ige-Olumide, O., Mustapha, O.-O. O., & Hafeez, S. A. (2015). Suitability analysis of resettlement sites for flood disaster victims in Lokoja and environs. *World Environment*, 5(3), 101–111. https://doi.org/DOI: 10.5923/j.env.20150503.02

Karou, S., & Hull, A. (2014). Accessibility modelling: predicting the impact of planned transport infrastructure on accessibility patterns in Edinburgh, UK. *Journal of Transport Geography*, 35, 1–11. https://doi.org/10.1016/j.jtrangeo.2014.01.002

Kumar, M., & Biswas, V. (2013). Identification of potential sites for urban development using GIS based multi criteria evaluation technique. A case study of Shimla Municipal Area, Shimla District, Himachal Pradesh, India. *Journal of Settlement and Spatial Planning*, 4(1), 45–51. Retrieved from http://geografie.ubbcluj.ro/ccau/jssp/arhiva_1_2013/05JSSP012013.pdf

Manirakiza, V. (2014). Promoting inclusive approaches to address urbanisation challenges in Kigali. *African Review of Economics and Finance*, 6(1), 161–180. Retrieved from https://www.ajol.info/index.php/aref/article/view/106931/96838

Ministry of Health. (2009). *Health Sector Strategic Plan: July 2009–June 2012*. Kigali: Rwandan Ministry of Health. Retrieved from http://www.internationalhealthpartnership.net/fileadmin/uploads/ihp/Documents/Country_Pages/Rwanda/Rwanda Health Sector Strategic Plan 2009-2012[1].pdf

NISR. (2011). *EICV3 District Profile: Gasabo*. Kigali: National Institute of Statistics of Rwanda (NISR). Retrieved from statistics.gov.rw/system/files/user_uploads/files/books/Gasabo.pdf

NISR. (2012). *Rwanda Fourth Population and Housing Census. District Profile: Gasabo*. Kigali: National Institute of Statistics of Rwanda (NISR) for Ministry of Finance and Economic Planning of Rwanda (MINECOFIN).

Niyonsenga, D. (2012). *Assessing Public Transport Supply for Kigali, Rwanda* (MSc Thesis). Enschede: University of Twente Faculty of Geo-Information and Earth Observation (ITC). Retrieved from http://www.itc.nl/library/papers_2012/msc/upm/niyonsenga.pdf

Obrist, B., Iteba, N., Lengeler, C., Makemba, A., Mshana, C., Nathan, R., … Mshinda, H. (2007). Access to health care in contexts of livelihood insecurity: a framework for analysis and action. *PLoS Medicine*, 4(10), e308. https://doi.org/10.1371/journal.pmed.0040308

Patel, S., Sliuzas, R., & Mathur, N. (2015). The risk of impoverishment in urban development-induced displacement and resettlement in Ahmedabad. *Environment and Urbanization*, 27(1), 231–256. https://doi.org/10.1177/0956247815569128

REMA. (2013). *Kigali State of Environment and Outlook*. Kigali: Rwanda Environment Management Authority (REMA). Retrieved from http://na.unep.net/siouxfalls/publications/Kigali_SOE.pdf

Satiroglu, I., & Choi, N. (2015). *Development-Induced Displacement and Resettlement: New Perspectives on Persisting Problems*. Hoboken: Taylor and Francis.

Takesada, N., Manatunge, J., & Herath, I. L. (2008). Resettler choices and long-term consequences of involuntary resettlement caused by construction of Kotmale Dam in Sri Lanka. *Lakes and Reservoirs: Research and Management*, 13(3), 245–254. https://doi.org/10.1111/j.1440-1770.2008.00374.x

Terminski, B. (2015). *Development-Induced Displacement and Resettlement: Causes, Consequences, and Socio-legal Context*. Stuttgart: ibidem-Verlag.

Tobler, W. (1993). *Three Presentations on Geographical Analysis and Modeling: Non-Isotropic Geographic Modeling; Speculations on the Geometry of Geography; Global Spatial Analysis*. Santa Barbara: National Center for Geographic Information and Analysis. Retrieved from http://www.geodyssey.com/papers/tobler93.html

Voogd, H. (1982). *Multicriteria Evaluation for Urban and Regional Planning* (PhD Thesis). Delft: Delftsche Uitgevers Maatschappij.

Zucca, A., Sharifi, A. M., & Fabbri, A. G. (2008). Application of spatial multi-criteria analysis to site selection for a local park: a case study in the Bergamo Province, Italy. *Journal of Environmental Management*, 88(4), 752–69. https://doi.org/10.1016/j.jenvman.2007.04.026

Index

A

Accessibility analysis model, 15
Accessibility-centred approach, 10
Activity-based indicators, 150
Activity–Structure–Intensity–Fuel (ASIF) framework, 16
Actual TOD index, 268–269, 271
    computing indicators and, 272, 274
    criteria, 273, 275–276
    inferences and recommendations, 278–279
    train stations in the City Region Arnhem–Nijmegen
        highest-scoring, 279
        values, 277
Agenda for Sustainable Development (2030), 2
Agent-based models (ABMs), 206
Ahmedabad Municipal Corporation (AMC), 246–247,
    252–253, 257, 261
Ahmedabad (India), resettling slum dwellers, 246
    analysis of changes in IRR, 249–261
        dynamics of impoverishment, 261
        education and health facilities, 255
        health risks, 255, 257
        housing and basic services, 252–253
        joblessness, 253–254
        landlessness, 250–251
        marginalization, 257–258
        school dropouts, 255, 257
        social disarticulation, 260–261
    BSUP. *See* Basic Services for Urban Poor (BSUP) sites
    displacement and resettlement in, 246–247
    method, 247–249
    resettled households, 248, 262
Artificial neural networks (ANNs), 206
Assessment systems, 226
Avoid–Shift–Improve (ASI) strategy, 10

B

Basic Services for Urban Poor (BSUP) sites, 246–249
    households
        with access to banking, community saving and
            insurance services, 260
        with basic services, 253
        distance matrix, 258
        distances to markets, 253
        distances to school and travel expenditures by,
            252, 257
        with positive perception, 260–261
        with sanitation problems, 259
        surveyed at, 252
    housing policy, 262
    locations, 250
    programme aim, 261
    residence status on, 251
    solid waste management, 259
    status of community facilities at, 256
    transportation, 254
    workers experiencing unemployment, 254

Below Poverty Line (BPL) card, 255
    changes in access to, 259
Bicycle level of service (BLOS) assessment, 226
Birmingham (UK), built-environment attributes
    case study area, 65–66
    data analysis, 68–71
        GIS measures of macro built-environment
            attributes, 68–70
        Pearson correlations, 70–71
        pedestrian-route network, 71
        statistical correlations, 70
    electoral wards, 66
    geographic location, 65
    macro built-environment attributes. *See* Macro built-
        environment attributes
    micro built-environment attributes, 64–65, 74–75
    neighbourhood built-environment attributes. *See*
        Neighbourhood built-environment attributes
    outdoor walking levels of older adults in, 63–64, 78
    results, 72–76
    study/studies
        limitations, 78–79
        previous, 77
    survey, 66–67
Borda Count method, 274
Bus Rapid Transit (BRT) system, 270–271

C

Carbon emissions module, 16
Cellular automata (CA) model, 206, 300, 302
Chawls (slum-like housing), 47, 59–60
Child-Friendly Cities, 165
Childhood
    beyond dichotomous framework, 164
    and city, 164–165
Choice phase process, 31, 33
Christchurch (New Zealand), cycle programme, 226–227
    cycle network, 227
    improvements for future, 241–242
    MCA. *See* Multiple criteria analyses (MCAs), cycle
        route design
    regional cycling projects, 227
    stakeholders. *See* Stakeholders, cycle route design
    study area, 226–228
City development plan (CDP), 47–48
The City Region Arnhem–Nijmegen (the Netherlands),
    TOD. *See also* Transit-oriented
        development (TOD)
    case study area, 268–269
    inferences and recommendations, 278–279
    location, 270
    methodology, 269–277
        computing indicators and TOD indices,
            272–274, 277
        identification of indicators, 271–272
        transit system, 270–271
    results, 277–278

Climate change, 9
    transport systems in, 10
Collaborative decision-making processes
    housing development and, 30–32
    implementation, 34–37, 35
Community integration, 186
CommunityViz® PSS* extension, 37
    dynamic assessment model implemented in, 37
    SMCE implementation in, 36
Compact city, 2, 30
Compensatory design criteria, 229
Competitive city, 2
Computer-Assisted Qualitative Data Analysis (CAQDAS)
        software, 171, 190
Connectivity index, 146
Constraint design criteria, 229
The Convention on the Right of the Child, 164–166
'Corredor Aurora Cañas' project, 33

D

Dar es Salaam (Tanzania). See Street as public space
        (Msasani Bonde la Mpunga, Dar es Salaam)
Data collection methods, 169–171
Decision Support Systems (DSSs), 31–32
Decision-tree modelling, 206
Design phase process, 31, 33, 34–36
Detroit Area Study (DAS) model, 127
Differential Global Positioning System (DGPS) survey, 54
Digital elevation model (DEM), 50, 54
    with LSTs and building footprints, 57
Digital On-line Life and You (DOLLY) archive, 311
Disaster–risk mitigation, 325–326
    resettlement sites for flood-disaster victims, 326
Discrete choice experiments (DCEs), 285
Drinking-water quality, 285
Dual city, 184

E

Earth observation (EO) methods, 50
Economic input variables, 37
Electricity-based transport system, 11
English Indices of Deprivation 2010, 46
Enschede (The Netherlands), children's perception, 164
    childhood and city, 164–165
    methods, 167–171
        data collection, 170–171
        data preparation and analysis, 171
        mixed-method approach, 169–170
        qGIS approach, 169–170, 177–179
    participatory planning and children's insights, 165–166
    qualities of local living environment, 166–169
    results, 172–177
        general perceptions, 172–173
        perception, place and meaning, 175, 177
        registered perceptions, 174
        spatial perceptions, 173–175
Environmental health issues. See Health problems
        associated with environment
Environmental quality of KD. See Kalyan–Dombivli
        (KD), India, environmental quality
Environment-friendly modes, 17
'Ermou 1900' festival (Nicosia), 118

Ethiopia, 124–125
Evacuation shelters, Jakarta (Indonesia), 311–315
    data analysis and methods, 312–315
        location of Twitter users, 312–313
        preferences of Twitter users, 314–315
        spatial pattern of Twitter users, 313–314
    data retrieval and processing, 311–312
        preferences on evacuation shelters, 312
        twitter data, 311–312
    distance between flood areas and, 317
    non-official, 308, 318–319
    official, 308, 318–319
Expansion, urban growth, 205, 211
    drivers of, 215–216
    model evaluation for, 220
    model parameters for, 217
    model validation, 221

F

Fractal-based models, 206
Functional cluster of activities, 143
Functional indicators, 150
Functional integration, 186
Future-oriented approaches, 297

G

Gated communities, 185, 197–198
    perceptions of unplanned neighbourhood by, 199
Gender in urban environments, 142
    transport and planning for safety, 144
Geo-computational approach, 295
Geographic information system (GIS), 2–3, 35, 64
    -based land use and transportation data, 64, 66
    children's perceptions of living environment, 163,
        166–169
    environmental health issues, 84
    limitations, 72
    measures of macro built-environment attributes,
        68–70, 72–73
    software applications, 30
Geo-information data technologies, 30
Geo-Picto Narratives, 172, 176
Germany, 86
GIS-based planning support tool, 10
    for KMC to access LCT plans, 13–20, 25
        accessibility analysis model, 15
        carbon emissions module, 16
        operational framework, 15
        scenarios, 16–20
Global integration index, 146
Ground Control Points (GCPs), 54
Groundwater in Yogyakarta (Indonesia), 283–284,
        286, 292
Growing Up In Cities (GUIC) project, 166
Guatemala City (Guatemala), sustainable
        development, 29–30
    case study and data, 33–34
    collaborative planning
        housing development and, 30–32
        implementation, 34–37
    land suitability and sites for development, 40
    MEHUD housing projects, 30, 33–34, 39

participatory planning for housing projects, 32–33
results, 38–40, 42
site ranks on sensitivity analysis, 42

H

Health problems associated with environment, 84
    context, 85–86
    Interactive-CuBA approach, 92–96
    interactive GIS-based support systems, 87, 96–98
    ISUSS approach, 88–92
        conceptual framework, 88–89
        for Dortmund workshop, 89–92
        indicator maps, 89
    methodology, 86–88
    stakeholders, 84
    wicked problem, 84
    works, 85
Hillier's approach, 146
Housing input variables, 36–37
Hydroelectricity, 11

I

Impoverishment risk and reconstruction (IRR)
        model, 247
    analysis of changes in, 249–261
        dynamics of impoverishment, 261
        education and health facilities, 255
        health risks, 255, 257
        housing and basic services, 252–253
        joblessness, 253–254
        landlessness, 250–251
        marginalization, 257–258
        school dropouts, 255, 257
        social disarticulation, 260–261
    parameters of, 249, 263
    residential resettlement, 330–331, 335–336
        accessibility model, 332
        criteria weighting, 332–333
        health risks, 337–338
        homelessness, 337
        landlessness and joblessness, 335–336
        loss of access to common facilities, 337
        loss of income, 336
        loss of mobility, 338
        marginalization, 337
        reversing strategies, 339
        risk of social disarticulation, 338
        sites, 332–333
        site suitability criteria, 331
        spatial multi-criteria model, 331–332
        uncertainty, 339
Inclusive city, 2
Indoor environment, 46
Infilling urban growth, 205, 211–212
    drivers of, 213–215
    model evaluation for, 218
    model parameters of, 216
    model results for, 220
    model validation, 221
Information system, 2. *See also* Geographic information
        system (GIS)
Institutional fragmentation, 184

Integrated Planning- and Decision-Support System
        (IPDSS), 32
Intelligence phase process, 31–32, 34–36
Intended outcome phase of ISUSS approach, 89
Interactive Cumulative Burdens Assessment (Interactive-
        CuBA), 85, 87–88, 92–96
Interactive GIS-based support systems, 87
    knowledge co-production and social learning, 96–98
Interactive Spatial Understanding Support System
        (ISUSS), 84, 87–92
    conceptual framework, 88–89
    for Dortmund workshop, 89–92
    indicator maps, 89
Intermodal network (integrated transportation system), 13
Istanbul (Turkey), city morphology, 147–148
    actors, 142
    driving forces, 143
    fieldwork and data compilation, 149
    foundation, 143
    linkages, 144
    methodology, 149–153
        attractiveness index and SMCE, 152, 156, 159
        urban morphology indicators, 149–150, 153
    results, 153–159
        indicator, 155–157, 159
        survey results and analysis, 153–154
        tracing women's movements, 154
    sketch maps, 150
    study area, 147, 148
    and women's perceptions of city. *See* Travel behaviour
        of women in Istanbul

J

The Jakarta Disaster Management Agency, 310
Jakarta (Indonesia), flood evacuation shelter
    case study area, 308–310
    data and methods. *See* Evacuation shelters, Jakarta
        (Indonesia)
    map, 309
    results (Twitter users for evacuation shelters), 315–319
        location, 315–316
        preferences, 316–318
        spatial pattern, 316
Jawaharlal Nehru National Urban Renewal Mission
        (JNNURM), 47–48
Jufo-Salus research project, 86, 88

K

Kalyan–Dombivli (KD), India, environmental quality, 46
    for aggregation levels, 50
    case study, 47
    components, 49
    data sets and methodology, 49–55
        building density, 53
        data aggregation, 55
        distribution of greenery, 52–53
        sources, 51
        street geometry, 54
        urban thermal environment, 51–52
    location, 48
    poor, 46
    results, 55–59

building density, 56
    distribution of greenery, 55–56
    index of environmental quality, 58
    street geometry, 56–58
    sub-ward level analysis, 58–59
    urban thermal environment, 55–56
  selection of environmental aspects for, 47–49
Kampala (Uganda), flood models, 295
  background, 296–299
  case study area, 296
  metropolitan area, 297–298
  problem conceptualization, 299–300
  projecting drivers of urban growth, 300–302
  scenario planning, 297–298, 301
  uncertainties, 296
  urban growth scenarios, 302–304
    flood impacts in Upper Lubigi, 304
    future urban development, 302–303
    land cover projections for 2025, 303
    metropolitan urban growth, 303–304
Kampala Physical Development Plan (KPDP), 295, 301
Kathmandu Sustainable Urban Transport (KSUT)
    Project, 12, 19
Kathmandu valley/Kathmandu Metropolitan City
    (KMC), urban growth, 10, 206
  case study, 11–12
  data collection, 12–13
  data preparation for accessibility model, 13
  materials and methods, 206–211
    data sources, 207–209
    identifying patterns, 208, 210
    input variables for SLR, 211
    SLR model, 210–211
    study area, 206–208
  model evaluation, 216–220
    for expansion, 220
    for infill growth, 218
    observed and predicted growth, 218–219
    for overall urban growth, 216–218
  planning support tool, 13–20
    accessibility analysis model, 15
    carbon emissions module, 16
    scenarios, 16–20
  results, 211–221
    driving factors. *See* Urban growth, driving
        factors of
    spatial pattern of infilling and expansion,
        211–212, 214
    spatio-temporal pattern, 211
  validation of prediction model, 220–221
Kathmandu Valley Town Development Committee
    (KVTDC), 208
Kigali (Rwanda), residential resettlement
  case study area, 326–328
  data and methods, 327–333
    data, 329–330
    interviews for households, 329–330
    methods and analysis, 330–333
  Gasabo, 340
  Gatsata, 335–336, 341
  impoverishment risks, 335–339. *See also*
        Impoverishment risk and reconstruction (IRR)
        model, residential resettlement
    health risks, 337–338

    homelessness, 337
    landlessness and joblessness, 335–336
    loss of access to common facilities, 337
    loss of income, 336
    loss of mobility, 338
    marginalization, 337
    reversing strategies, 339
    risk of social disarticulation, 338
    uncertainty, 339
  methodological approach, 329
  potential resettlement sites, 339–341
  resettlement requirements, 333–335
    economic development, 333–334
    infrastructure and access to basic services,
        334–335
    protection of social network, 335
  socio-economic characteristics, 341
Kirkos sub-city (Addis Ababa). *See* Quality of life
    (QoL) in Kirkos

L

Land-cover classes of Kathmandu valley, 208, 210,
    214–215
Land surface temperatures (LSTs), 49–50, 59
    index of environmental quality, 58
    for October 1999, 55
Land use input variables, 36
Land-use planning systems, 295, 297
    scenarios, 297, 301
Latin America, 29
Linear maximum standardization, 230
Linkages, transport, 143
Living environment, 46
Local climate zones (LCZ), 46
Low-carbon development (LCD), 9
Low-carbon transport (LCT), 10, 12
    planning support tool for KMC, 13–20
        accessibility analysis model, 15
        carbon emissions module, 16
        scenarios, 16–20
Low-density suburbanization, 2
Low-rise/high-density clustered housing, 30–31

M

Macro built-environment attributes, 64–65
    correlations among GIS measured, 73–75
    data for measuring, 67–68
    GIS measures of, 68–70, 72–73
Main criteria weights (MCW), 237
Micro built-environment attributes, 64–65
    correlations, 74–75
Mixed-method approach, 169–170
Movement economy concept, 145–146, 157
Msasani Bonde la Mpunga, Dar es Salaam. *See* Street
        as public space (*Msasani Bonde la Mpunga*,
        Dar es Salaam)
Multi-Criteria Evaluation (MCE), 37–38
Multi-criteria techniques, 331
Multinomial logit (MNL) model, 291
Multiple criteria analyses (MCAs), cycle route design,
        226–227
    application for cycle route design, 233–241

criteria hierarchy and performance measures, 234
data sets for study area, 236
ranks converted to MCW, 237–238
ranks converted to sub-criteria weights,
238–241
route segment and junction scores map, 235–236
modifying for cycle route design, 230–233
preference ranking of stakeholders, 232
sensitivity analysis, 233
target populations as stakeholders, 230–232
TOD, 268, 274
traditional methods, 228–230
compensatory criteria, 229
constraint criteria, 229
criteria standardization, 230
Multiple regression analysis, QoL, 132–133
Mumbai Metropolitan Region (MMR), 47
The Municipal Enterprise of Housing and Urban
Development (MEHUD), 30, 33–34, 39

N

Natural disasters and urbanization, 325
Natural environment, 46, 49
Natural movement concept, 145–146, 157
Neighbourhood built-environment attributes, 64
correlations, 72, 74
data on, 67–68
home-based neighbourhood, 69
non-residential land use, 70
and older adults outdoor walking levels. *See*
Birmingham (UK), built-environment
attributes
perception of interaction between, 196
planned, 185
social interaction with, 195
unplanned, 196
Nicosia (Cyprus), divided city, 103
buffer zone/green line, 104, 107
case study, 106–110
the *Ayios Dhometios* (Metehan) crossing, 107
Bandabulya market, 108
Buyuk Han market, 108
Faneromeni Square, 107
the *Ledra Street* crossing, 107, 112, 117–118
Lokmaci Street, 108
Markou Drakou Street, 110, 118
the *Municipal Gardens*, 110, 112, 117
dead zone, 103
defined, 104
'Ermou 1900' festival, 118
methodology, 106–112
public spaces, 103–104
activities/events dimension of, 116
diversity in, 116
location of crossings and, 107
map of selected, 115
occupant density of, 115
social interaction. *See* Social interaction,
public spaces
public spaces in. *See* Public spaces in divided cities
results, 112
Non-official evacuation shelters, 308, 320
sites comparison, 318–319

Normalised Difference Vegetation Index (NDVI), 49,
52–53, 59
distribution of greenery, 55
index of environmental quality, 58

O

Official evacuation shelters, 308
sites comparison, 318–319
Outdoor environment, 46
Outlying expansion, urban growth, 205

P

Participatory mapping technique, 154
Participatory planning, children, 165–166
PDAM *(Perusahaan Daerah Air Minum)*/Piped Water
Supply Company, 283–286, 291, 293
Pearson correlations, 70
classification, 71
QoL, 132
Peta Jakarta system, 310, 312
Physical indicators, 150, 158
Piped water systems in Yogyakarta (Indonesia), 283, 285,
291–292
Place of contact, public spaces, 106, 118
Places of transit, public spaces, 105
'Plan Guatemala 2020,' 29–30
Planned neighbourhoods, 185, 197
Planning Support Systems (PSSs), 31–32, 85
Policy instruments (transport system), 17, 19
Post-classification refinements, 208
Potential TOD index, 268–271
computing indicators and, 274, 277
criteria and indicators for, 273, 275–276
inferences and recommendations, 279–280
scores and poor access to transit, 280
values throughout the City Region
Arnhem–Nijmegen, 278
Preparation phase of ISUSS approach, 89
PRESTO Programme (Europe), 225
Process phase of ISUSS approach, 89
Public spaces in divided cities, 103–104
activities/events dimension of, 116
diversity in, 116
location of crossings and, 107
map of selected, 115
occupant density of, 115
social interaction. *See* Social interaction, public spaces
Public transport (PT) system, 11, 13, 17
scenarios in 2020
accessibility, 20–22
emissions, 22–23
sensitivity analysis, 23–24

Q

Qualitative (qGIS) approach, 169–170
reflection and results, 177–179
Quality of life (QoL) in Kirkos, 121–122
domains and attributes, 122–124, 135
methodology, 124–128
case study, 124–126
data analysis, 127–128

data set and sampling procedure, 126
results, 128–137
    context and policy implications, 137
    correlation matrix, 133
    factor analysis, 134, 136
    factor loading matrix, 136
    family income, 134
    Kebele administrations, 137
    multiple regression analysis, 132–133
    relation between QoL and satisfaction, 132, 134
    statistics and domain satisfaction, 128–129
    variation of satisfaction, 130–132

R

Rail-based transit services, 270
Rainfall–runoff approach, 299
Rank–sum method, 274
Real-time disaster response and management, 308
Regulatory instruments (transport system), 17
Remote sensing (Resourcesat IRS P6), 52
Resettlement of slum dwellers in Ahmedabad (India), 246
    analysis of changes in IRR, 249–261
        dynamics of impoverishment, 261
        education and health facilities, 255
        health risks, 255, 257
        housing and basic services, 252–253
        joblessness, 253–254
        landlessness, 250–251
        marginalization, 257–258
        school dropouts, 255, 257
        social disarticulation, 260–261
    BSUP. *See* Basic Services for Urban Poor (BSUP) sites
    displacement and resettlement in, 246–247
    method, 247–249
    resettled households, 248, 262
Residential resettlement of Kigali. *See* Kigali (Rwanda), residential resettlement
Resident welfare associations (RWAs), 260–262
Resilient city, 2
Rwanda Human Settlement Policy, 330

S

Social connectedness, QoL, 130
    defined, 124
    variability in, 130
    variation and satisfaction for, 132
Social fragmentation, 184
Social interaction, public spaces, 104–106, 114
    index of, 111–115, 117
Social media
    applications, 308
    during disasters, 307
    Peta Jakarta system, 310, 312
    Twitter. *See* Twitter (for evacuation shelters)
Social security card (Adhar card), 255
    changes in access to, 259
Socio-economic amenities, 250
Socio-economic status (SES), 83
Space syntax approach, 146
Space–time accessibility, 26
Spatial Decision Support Systems (SDSS), 32, 85
Spatial fragmentation, 184

Spatial logistic regression (SLR) model, 206, 210–211
    1999 independent variables, 212
    input variables for, 211
    variables and descriptions for, 213
Spatial MCA (SMCA) tool, 268
Spatial multi-criteria evaluation (SMCE), 35–36, 145, 150, 331–332
    attractiveness index and, 152
    implementation in CommunityViz®, 36
    of site suitability, 342
Stakeholders, 34–35, 39
    agreement, 342
    cycle route design
        Christchurch cycling focus group, 240
        criteria, 231–232
        linear maximum standardization, 230
        preference ranking, 232
        target populations as, 230–232, 241
    environmental health issues, 84
    ISUSS approach, 89, 91, 97
    weighting scheme, 39, 41
Stated choice experiments, 285, 289–290
    choice set, 290
    on water provision choice, 289–290, 292–293
State-of-the-art remote sensing techniques, 46
Statistical urban growth models, 206
Street as public space (*Msasani Bonde la Mpunga*, Dar es Salaam), 184–186
    case study, 187
    community integration, 197
    data analysis method, 189–190
    functional integration, 197–198
    land use in 2014, 187, 191
    methodology
        interviews, 189
        measuring street qualities, 186, 188
        selection of street and indicators, 188, 189
    primary and secondary data, 188–189
    results, 190–197
        land use characteristics, 190
        residents' perception of streets, 193–197
        selected street profile, 190–192
        street connectivity level, 191, 193–194
    symbolic integration, 198–199
Suitability analysis, 342
Sustainable city, 2
Sustainable housing development, 30–32, 38
Symbolic integration, 186
System Development Life Cycle (SDLC), 87

T

Target populations, 225–226, 240
    as stakeholders, 230–232, 241
Transit-oriented development (TOD)
    defined, 267
    index/indices, 267–268. *See also* Actual TOD index, Potential TOD index
    objectives, 268
    plans, 267
    transit-related indicators, 274
Transport system of cities, 142
    driving forces, 143
    electricity-based, 11

in Istanbul, 148
in low-carbon cities, 10
public, 11, 13, 17
sustainable, 11
urban planning/design
    disciplines of transport and, 144–145
    transport planning, 144
Travel behaviour of women in Istanbul, 142
areas of fear, 156, 159
gender-related issues in safety, 144
mobility, 143
perception of travel, 144, 146
survey, 149, 153–154
transport planning, 144
Trolley bus system, 17
penetration with PT improvement plans, 20
renewal of, 19
Twitter (for evacuation shelters), 310
data, 311–313
impressions during 2014/15 flooding period, 312
as source of information, 320–321
tweets about evacuation shelters, 315
users in/near evacuation shelters, 320
    location, 312–314
    shelter preferences among, 312, 314–315
    spatial pattern, 313–314, 316
Twitter Application Program Interface (API), 311

U

Unplanned neighbourhoods, 185
perceptions of gated community, 198
Upper Lubigi sub-catchment area (Kampala),
    298–299, 304
Urban Agenda of the United Nations, 2
Urban complexity, 142
Urban Cycleway Program, 227
Urban development projects, 245–247
Urban fragmentation, 184
concepts of, 184–185
manifestation in African cities, 185
sense of belonging, 194–195
Urban growth
driving factors of, 213–216
    drivers of expansion, 215–216
    drivers of infill growth, 213–215
    drivers of overall urban growth, 213, 216
Kampala, 295
    cellular automata model, 300
    model, 302
    projecting drivers of, 300–302

scenarios. *See* Kampala (Uganda), flood models,
        urban growth scenarios
in Kathmandu valley. *See* Kathmandu valley/
        Kathmandu Metropolitan City (KMC),
        urban growth
Moran's index, 211, 213, 215
types, 205, 215
Urbanization, 1–3, 205
challenges, 2
information, 2
KMC, 11
and natural disaster, 325
Urban landscape, elements of, 143
Urban morphology, 143, 145
attractive index, 159
and mobility indicators, 149–153, 159
Urban planning/design technique
approaches, 145–146
    individual movements, 146, 150
    space syntax, 145
disciplines of transport and, 144–145
transport planning, 144
Urban spatial arrangements, 143

V

Volunteered geographic information (VGI), 308, 311
analysis, 319–320
drawbacks, 320
planning/siting of evacuation, 320
role, 321

W

Water supply in Yogyakarta (Indonesia), 285–286
case study area, 284
drinking-water quality, 285
findings and recommendations, 292–293
groundwater, 283–284, 286, 292
literature review, 284–285
    perception and behaviour of water users, 284–285
    stated choice experiments. *See* Stated choice
        experiments
methodology, 286–289
    cluster sampling, 287–288
    households sample, 288
    location of water users, 289
    Tegalrejo District, 288
PDAM, 283–286, 291, 293
piped water systems, 283–284, 292
results, 290–292